Copyright © 2023 by Trient Press

All rights reserved. No part of this publication may be reproduced, distributed, or transmitted in any form or by any means, including photocopying, recording, or other electronic or mechanical methods, without the prior written permission of the publisher, except in the case of brief quotations embodied in critical reviews and certain other noncommercial uses permitted by copyright law. For permission requests, write to the publisher, addressed "Attention: Permissions Coordinator," at the address below.

Criminal copyright infringement, including infringement without monetary gain, is investigated by the FBI and is punishable by up to five years in federal prison and a fine of $250,000.

Except for the original story material written by the author, all songs, song titles, and lyrics mentioned in the novel From Data to Disruption: How AI is Changing Business Forever are the exclusive property of the respective artists, songwriters, and copyright holder.

Trient Press
3375 S Rainbow Blvd
#81710, SMB 13135
Las Vegas, NV 89180

Ordering Information:
Quantity sales. Special discounts are available on quantity purchases by corporations, associations, and others. For details, contact the publisher at the address above.
Orders by U.S. trade bookstores and wholesalers. Please contact Trient Press: Tel: (775) 996-3844; or visit www.trientpress.com.

Printed in the United States of America

Publisher's Cataloging-in-Publication data
Ruscsak, M.L.
A title of a book : From Data to Disruption: How AI is Changing Business Forever

ISBN
Hard Cover       979-8-88990-132-7
Paper Back       979-8-88990-133-4
Ebook            979-8-88990-134-1

Part 1: Introduction: From Data to Disruption

Chapter 1: AI and Its Fundamental Concepts
Chapter 2: The Significance of AI in the Modern World
Chapter 3: Unleashing the Potential: How AI Transforms Industries
Chapter 4: The Disruptive Force of AI
Chapter 5: Ethical Considerations and Challenges

Part 2: Understanding AI: A Primer for Readers

Chapter 6: Overview of AI Technologies.
Chapter 7: Unveiling the Essence of AI
Chapter 8: The Building Blocks of AI
Chapter 9: How AI Algorithms Work in Simple Terms
Chapter 10: The Development and Rapid Growth of AI
Chapter 11: The Role of Data in AI

# Part 1: Introduction: From Data to Disruption

In the ever-evolving landscape of technological advancements, one phenomenon stands out as the vanguard of innovation and transformation: Artificial Intelligence (AI). A creation born from the convergence of human ingenuity and computational power, AI has emerged as a powerful force with the potential to reshape the fabric of our societies and industries. In this comprehensive exploration, we embark on a journey through the realm of AI, delving into its fundamental concepts, significance in the modern world, and its profound impact on industries worldwide.

Chapter 1: AI and Its Fundamental Concepts

In this foundational chapter, we lay the groundwork for our understanding of AI by exploring its core concepts and principles. Delving into the origins and history of AI, we trace its development from early theoretical musings to the present day, where it has become an integral part of our daily lives. From the rudimentary beginnings of AI to the more sophisticated machine learning and neural networks, we demystify the essence of artificial intelligence.

Chapter 2: The Significance of AI in the Modern World

As AI continues its ascendancy, its significance in the modern world cannot be overstated. In this chapter, we examine the pervasive influence of AI across diverse domains, from finance to healthcare, from e-commerce to transportation. We discuss how AI-driven innovations have redefined business models, enhanced customer experiences, and unlocked new opportunities for growth and efficiency. The rise of AI has ushered us into an era of unprecedented potential and disruption.

Chapter 3: Unleashing the Potential: How AI Transforms Industries

AI, as a formidable agent of change, has penetrated industries with unbridled vigor. In this chapter, we delve into how AI transforms various sectors. We explore the e-commerce revolution, where AI-powered recommendation systems and personalized marketing drive customer engagement to new heights. In healthcare, AI is revolutionizing medical imaging, diagnostics, and drug discovery, empowering healthcare providers with unprecedented precision and efficiency. Banking and finance embrace AI for fraud detection, risk assessment, and algorithmic trading, altering the landscape of the financial sector. We also witness AI's influence in supply chain management, transportation, agriculture, and energy, unlocking untold possibilities in these domains.

Chapter 4: The Disruptive Force of AI

The impact of AI is not confined to mere enhancement; it is a disruptive force reshaping the status quo. In this chapter, we explore AI's ability to challenge traditional business models and foster a culture of innovation and adaptation. We examine how AI's data-driven insights and decision-making capabilities have enabled businesses to achieve unparalleled efficiency and competitiveness. Through compelling examples, we reveal how AI has set the stage for revolutionary change, beckoning us into an age of extraordinary possibilities.

Chapter 5: Ethical Considerations and Challenges

As we venture further into the realm of AI, we cannot overlook the ethical considerations and challenges it presents. This chapter delves into the ethical dilemmas surrounding AI's development and deployment, from issues of bias in algorithms to concerns over data privacy and security. We confront the complex interplay between AI and human values, recognizing the need for responsible AI development that prioritizes fairness, transparency, and societal benefit.

Throughout this journey, we strive to elucidate complex concepts in a clear and accessible manner. Drawing on examples and case studies from E-commerce, Healthcare, Banking, and Finance, we provide a well-rounded understanding of AI's profound impact across industries. As we embrace diverse perspectives, we present counterarguments and dissenting opinions in a balanced and objective manner, inviting our readers to embark on a profound exploration of the transformative power of AI.

In each subsequent chapter, we delve deeper into the intricacies of AI, exploring its technical underpinnings, real-world applications, and the interplay between AI and blockchain technology. Our aim is to engage readers in thought-provoking discussions, inspiring them to push the boundaries of human innovation and envision a future where AI fosters a world of greater knowledge, creativity, and progress.

# Chapter 1: AI and Its Fundamental Concepts

In the ever-expanding universe of technological innovation, one phenomenon shines as a beacon of intellectual prowess and paradigmatic transformation: Artificial Intelligence (AI). As we embark on a profound expedition into the intricacies of AI, we find ourselves standing at the crossroads of computer science, machine learning, and cognitive neuroscience. This chapter serves as our portal to the realm of AI, where we shall unveil its fundamental concepts, trace its evolutionary trajectory, and acquaint ourselves with the terminology that adorns its domain.

Section 1: Introducing Artificial Intelligence (AI)

In this inaugural section, we tread upon the path of enlightenment, unveiling the essence of AI in all its intellectual grandeur. Within the luminous corridors of academia, AI stands as a formidable pursuit - an ambitious endeavor to bestow machines with cognitive prowess akin to human intelligence. We shall embark on an odyssey to decipher AI's philosophical origins, stretching back to antiquity's ancient tales of artificial life. From those mystical narratives, we traverse the temporal tapestry, arriving at the dawning of formal AI in the mid-20th century. We expound upon the canonical pioneers of AI, championed by illustrious minds such as Alan Turing and John McCarthy.

Section 2: The Evolution of AI

As stargazers of AI's evolutionary celestial, we bear witness to its undulating trajectory through time and circumstance. From the heady days of its inception, AI sauntered into the labyrinth of an "AI Winter," where funding and faith waned amidst unmet promises. But as the seasons of technology waxed anew, AI reemerged, emboldened by newfound prowess, nurtured by computational renaissance. We behold its contemporary prowess, honed by a tempest of data and computational might. Indeed, the epochs have witnessed the resurrection and rejuvenation of AI, raising it from dormancy to the grand stage of technological ascendancy.

Section 3: Key AI Terminology

Within the hallowed halls of AI, a lexicon of captivating terminology flourishes. In this enlightening section, we unfurl the tapestry of essential AI concepts that underscore its domain. Machine learning, the fulcrum upon which AI stands, grants machines the power to discern patterns, infer insights, and learn without explicit instruction. Neural networks, mirroring the intricacies of the human brain, engender remarkable feats of pattern recognition and decision-making. Natural Language

Processing (NLP), a literary wizardry, empowers AI to comprehend and converse in the tongues of humanity. And within the realm of sight, computer vision confers upon AI the gift of visual perception and understanding.

Section 4: Types of AI and AI's Future Prospects

As we traverse AI's expansive dominion, we come upon the forked paths of its types: Weak AI and Strong AI. The former, a realm of narrow focus, excels in distinct tasks, yet lacks the expansive intellect of humanity. The latter, a prospect of mythical allure, dreams of general AI - a creation that rivals the human mind in all its kaleidoscopic complexity. In this mélange of AI possibilities, we envision the forthcoming epoch of AI's destiny. A future infused with the marvels of robotics, the wonders of autonomous vehicles, and the bounties of creativity. The horizons beckon, and the intrigue of AI's future prospects dances before us.

Conclusion: Embracing the AI Journey

With minds primed for intellectual odyssey, we reflect upon our sojourn through AI's fundamental concepts. From its genesis in ancient mythos to its technological ascendancy, AI reveals itself as a multidisciplinary pursuit, invoking elements from computer science, machine learning, and cognitive psychology. As we embrace the lexicon of AI, its neural networks and natural language prowess awaken the imagination to the countless possibilities that lie ahead. With fervent ardor and resolute inquisitiveness, we stride forth to unravel the mysteries that AI unfolds, seeking the elusive spark that shall illuminate humanity's path to greater understanding and innovation. The chapters that lie ahead shall usher us deeper into the labyrinth of AI's profound impact, as we explore its influence across E-commerce, Healthcare, Banking, and Finance. Brace yourself, dear reader, for this voyage of thought shall transport you to the frontiers of AI's transformative potential.

# Section 1: Introducing Artificial Intelligence (AI)

Amidst the labyrinthine pathways of human curiosity and technological prowess, one enigmatic creation looms large: Artificial Intelligence (AI). In the annals of academic inquiry and engineering innovation, AI stands as an extraordinary pursuit - a quest to bestow machines with cognitive faculties akin to human intelligence. Embarking upon a scholarly odyssey, we traverse the historical and conceptual origins of AI, navigating through epochs of mythical wonder, foundational theories, and contemporary advancements that propel this captivating field.

The genesis of AI finds roots in the timeless tales of artificial life, tales where craftsmen and magicians breathed sentience into inanimate creations. These myths of antiquity foreshadowed humanity's enduring fascination with imbuing machines with semblances of consciousness. However, it was in the intellectual renaissance of the 20th century that AI found formal footing. Distinguished visionaries, among them the iconic Alan Turing and John McCarthy, laid the cornerstones of AI as a discipline of scientific inquiry.

Venturing beyond the thresholds of antiquity, we traverse the tempestuous epochs of AI's historical trajectory. Cast into the shadow of an "AI Winter," where funding dwindled and ambitions waned, AI faced challenges that tested its mettle. Yet, it was from the crucible of adversity that AI emerged, resilient and rejuvenated, surging forward with newfound vigor. The advancements in computational power, the abundance of data, and the trailblazing breakthroughs in machine learning reignited AI's flame, propelling it into the forefront of modern technological ascendancy.

Within the realm of AI, a lexicon of captivating terminology unfurls. Herein lies the fulcrum of AI's magic: Machine Learning. This transformative concept grants machines the capacity to discern patterns, infer insights, and refine their performance through iterative learning, devoid of explicit programming. With neural networks, AI mimics the intricate workings of the human brain, unveiling remarkable feats of pattern recognition and decision-making. While the sorcery of AI extends further with Natural Language Processing (NLP), empowering machines to comprehend and converse fluently in the languages of humanity. And behold, Computer Vision, where AI's sight beholds the nuances of visual perception, unlocking a new dimension of understanding.

As we tread deeper into the labyrinth of AI, a forked path emerges, diverging between Weak AI and Strong AI. Within the confines of Weak AI, AI systems excel in specialized tasks, showcasing prowess in chess, language translation, and autonomous vehicles. Yet, the elusive allure of Strong AI beckons - a prospect of profound significance - wherein AI aspires to emulate the rich tapestry of human cognition. The contemplation of Strong AI evokes a vision of machines that possess the boundless creativity and adaptability inherent in human intelligence.

This section is an invitation to embark on a journey of intellectual discovery, where the boundaries of human ingenuity meld with the realm of AI's transformative potential. The chapters that lie ahead shall unveil the profound impact of AI across E-commerce, Healthcare, Banking, and Finance, where the symphony of AI's capabilities converges with diverse domains. As we venture forth, let our minds resonate with curiosity, as the wonders of AI beckon us toward the frontiers of innovation and understanding.

**Defining AI: Unraveling the essence of AI and its mission to replicate human intelligence and problem-solving capabilities through machines.**

In the grand tapestry of technological progress, one thread stands out as an audacious endeavor - Artificial Intelligence (AI). A cerebral creation fusing human intelligence with computational prowess, AI embodies the pursuit of endowing machines with cognitive faculties akin to the human mind. In this chapter, we embark on an intellectual voyage to grasp the quintessence of AI - a voyage that unearths its conceptual underpinnings, examines its defining characteristics, and contemplates its transformative potential across the domains of E-commerce, Healthcare, Banking, and Finance.

The Quest for AI

At the vanguard of human endeavor lies a momentous quest that transcends the boundaries of science and philosophy - the quest to create Artificial Intelligence (AI). This captivating pursuit resonates with humanity's deepest yearnings, as it endeavors to materialize its most profound aspirations: the realization of machines that possess the ability to think, learn, and solve problems autonomously, mirroring the very essence of human intelligence.

The genesis of this quest finds echoes in the annals of ancient myths and legends, where tales of mythical creatures and golems embodied humanity's yearning for artificial life. From the mechanical automatons of ancient Greece to the homunculi of alchemy, ancient cultures indulged in tales of beings endowed with animated existence. These age-old narratives foretell humanity's enduring fascination with the notion of breathing life into inanimate matter, an archetype that would later manifest in the quest for AI.

The voyage of AI, as a formal and scientific pursuit, commenced during the mid-20th century, amidst the intellectual renaissance of the computer age. Visionaries like Alan Turing and John McCarthy stepped forth as trailblazers, seeking to encode intelligence into machines and engender an era of unprecedented technological progress. Alan Turing's seminal work in the 1930s laid the theoretical foundations for computing and the concept of "Turing machines" - a theoretical model capable of performing any calculation that could be described by an algorithm. This pivotal work set the stage for the formal exploration of AI in the subsequent decades.

John McCarthy, who coined the term "Artificial Intelligence" in 1956, orchestrated the seminal Dartmouth Conference, marking the birth of AI as a distinct academic discipline. At this historic gathering, the visionaries envisioned a future where machines would emulate human intelligence, exhibiting the ability to reason, learn, and adapt. The quest for AI began in earnest, guided by a convergence of brilliant minds, where a rich tapestry of disciplines, including mathematics, logic, computer science, and cognitive psychology, wove together into a unified pursuit.

As the quest for AI unfolded, it encountered challenges and triumphs, ebbs, and flows. The 1970s and 1980s saw periods of skepticism, often referred to as "AI Winters," where progress stalled due to high expectations and unmet promises. However, resilience and determination spurred the field forward, and the AI renaissance of the 21st century saw a resurgence of interest and advancement. Factors such as exponential growth in computational power, the availability of vast data sets, and groundbreaking developments in machine learning algorithms reignited the spark of AI's potential.

The philosophical underpinnings of AI's quest remain profound, touching upon questions of consciousness, intelligence, and the nature of cognition itself. As AI ventures into the realm of general intelligence, known as "Strong AI," it faces philosophical debates about the possibility of machines possessing subjective experiences and self-awareness. These questions, deeply rooted in the philosophy of mind and artificial consciousness, continue to challenge our understanding of intelligence and machine capabilities.

The quest for AI remains a dynamic and evolving journey, fueled by unyielding curiosity and the relentless pursuit of knowledge. It stands as a testament to the indomitable spirit of human exploration, where humanity's profound desire to understand its own intelligence finds expression in the creation of machines that may one day share in the marvels of human cognition. As we traverse the depths of AI's profound impact across diverse fields, we find ourselves on the cusp of an extraordinary era, where the quest for AI converges with real-world applications, shaping the course of human progress and innovation.

AI's Cognitive Canvas

Within the tapestry of Artificial Intelligence (AI), its cognitive canvas unfolds - a realm where machines endeavor to mirror the profound faculties of human intelligence and problem-solving. Drawing inspiration from the intricate workings of the human brain, AI embarks on a multi-faceted journey into cognitive domains, embracing a diverse array of methodologies that unlock its cognitive potential. This chapter delves into the essence of AI's cognitive prowess, exploring the transformative techniques of Machine Learning and the neural wonders of Neural Networks.

✧ Machine Learning - The Engine of Cognitive Empowerment

At the core of AI's cognitive prowess lies the transformative technique of Machine Learning. Unlike traditional programming paradigms, which rely on explicit instructions, Machine Learning empowers AI systems with the ability to learn and improve from experience - an autonomous evolution of intelligence. Through this method, machines become adept at identifying patterns and correlations in vast data sets, extracting invaluable insights without human intervention.

Machine Learning finds manifold applications across diverse fields. In the realm of E-commerce, it fuels the engines of recommendation systems, deftly curating personalized shopping experiences for customers. Healthcare witnesses the prowess of Machine Learning in medical image analysis, where AI aids in diagnosing ailments, detecting anomalies, and expediting patient care. In the financial realm, Machine Learning algorithms navigate the complexities of risk assessment, fraud detection, and algorithmic trading, elevating the precision of decision-making.

✧ Neural Networks - Unveiling the Miracles of Cognition

Gazing through the lens of Neural Networks, AI peers into the architectural brilliance of the human brain. Emulating the intricate web of neurons and synapses, Neural Networks unravel the secrets of pattern recognition, image classification, and natural language understanding. Convolutional Neural Networks (CNNs) excel in image recognition, as seen in self-driving cars perceiving their surroundings, and healthcare systems interpreting medical scans with unprecedented accuracy.

In the domain of natural language, Recurrent Neural Networks (RNNs) breathe life into chatbots, enabling human-like conversations and language translation. Long Short-Term Memory (LSTM) networks, a variant of RNNs, wield temporal context, allowing AI to grasp nuanced meaning from sequential data.

The synergy of Machine Learning and Neural Networks is a cornerstone of AI's cognitive canvas. By marrying the ability to learn from data with the complex architecture of Neural Networks, AI unravels the mysteries of human cognition in ways previously deemed unimaginable.

### ❖ Embarking into Uncharted Territory

As AI traverses the cognitive canvas, it encounters both challenges and triumphs. Amidst the marvels of cognitive replication, questions arise about AI's decision-making transparency, bias, and accountability. The "black box" nature of certain AI models raises concerns about explainability, where understanding the rationale behind AI's decisions becomes crucial for adoption in critical domains like healthcare and finance.

AI's ascent into Strong AI - the realm of machines rivaling human cognition - brings ethical quandaries about the boundaries of machine intelligence, raising questions about the implications of potentially sentient machines.

### ❖ Conclusion: Illuminating the Path of AI's Cognitive Odyssey

The cognitive canvas of AI reveals a captivating panorama of human endeavor and technological innovation. Machine Learning, with its autonomous learning capabilities, shapes the landscape of data-driven insights and informed decision-making. Neural Networks, drawing inspiration from the brain's intricacies, orchestrate feats of perception and language understanding.

As AI embarks into uncharted territory, pondering the implications of Strong AI and addressing the ethical dimensions of its transformative power becomes an imperative. As we traverse AI's cognitive journey, we uncover the symphony of human ingenuity and AI's computational prowess, enacting a masterpiece where human and machine intelligence entwine in the quest for progress and understanding. The chapters that lie ahead shall illuminate AI's transformative impact across E-commerce, Healthcare, Banking, and Finance, where its cognitive prowess manifests into real-world applications, resonating with the marvels of human cognition.

The Linguistic Marvels of NLP

Enter the Realm of Natural Language Processing (NLP): Unveiling the Art of Language Comprehension

In the vast expanse of AI's cognitive pursuits, a realm of exquisite artistry emerges - Natural Language Processing (NLP). At the heart of this domain lies the audacious endeavor to empower machines with the ability to comprehend, interpret, and generate human language - a quintessential conduit of human expression that has eluded machines for eons. NLP stands as a testament to AI's unwavering commitment to bridge the chasm between human communication and the realm of machines.

### ✦ The Complexity of Language and the Quest for NLP

Language, a symphony of words and meaning, has been an integral aspect of human identity and culture since time immemorial. However, deciphering the nuances and subtleties of human language has posed a formidable challenge for machines. Human expressions are rich with ambiguity, context, figurative language, and cultural references, making language processing a profound and intricate pursuit.

AI's quest for NLP arose from the realization that unlocking the intricacies of language would unlock unprecedented potential in human-machine interactions. Pioneering researchers recognized that for AI to truly understand and communicate with humans, it must traverse the labyrinth of language comprehension and expression.

### ✦ Dismantling Barriers with NLP

NLP stands as a testament to AI's prowess in dismantling the barriers that have separated humans and machines in the realm of language. From the inception of simple chatbots that could answer pre-programmed responses, NLP has evolved into an art that allows machines to engage in dynamic and meaningful conversations with humans.

At the forefront of NLP's transformative power are the developments in Natural Language Understanding (NLU). NLU algorithms dissect the intricate structures of human sentences, discerning syntactic patterns, semantic meaning, and intent. With this newfound ability, chatbots now engage users in intelligent conversations, addressing queries, and providing personalized responses, transcending the realm of scripted interactions.

### ✦ Crossing Linguistic Boundaries with Translation Engines

The enchantment of NLP extends beyond linguistic barriers. Translation engines, empowered by NLP, now act as linguistic bridges, allowing communication across diverse languages. AI-driven translation models, often employing deep learning techniques, have reached astonishing levels of accuracy, enabling seamless and real-time language translations across written and spoken languages.

In a world where global communication is vital for commerce, diplomacy, and cultural exchange, NLP-driven translation technology plays an indispensable role in fostering mutual understanding and cooperation.

✧ Gauging Emotions with Sentiment Analysis

Beyond the mere understanding of language, NLP extends its reach into the realm of emotions with sentiment analysis. By analyzing textual input, NLP algorithms gauge sentiments, emotions, and attitudes expressed by individuals or groups. This ability is of immense value in various domains, including market research, customer feedback analysis, and social media monitoring.

With sentiment analysis, businesses can assess customer satisfaction, public perception of products or services, and emerging trends, enabling them to adapt and respond proactively to the sentiments of their audience.

✧ Conclusion: The Language Nexus of NLP

As AI delves into the intricacies of Natural Language Processing, it embarks on a journey of understanding and expression, where the barriers of human-machine communication fade. NLP has unlocked the potential for chatbots to converse engagingly, translation engines to unite global communities, and sentiment analysis to decipher human emotions.

The chapters ahead shall illuminate the profound impact of NLP across diverse fields, revealing how AI's comprehension of human language has transformed E-commerce customer service, enabled advanced Healthcare diagnostics, revolutionized Banking and Finance interactions, and fostered cross-cultural connections. The pursuit of NLP continues to break new ground, propelling AI ever closer to the heart of human expression and understanding, where the intricate art of language converges with the computational prowess of machines.

The Vision of Computer Vision

Amidst the Artistic Tapestry: Unveiling the Marvels of Computer Vision in AI

Within the grand tapestry of AI's creative endeavors, a captivating spectacle unfolds - Computer Vision. This wondrous domain grants machines the gift of sight, enabling them to perceive and interpret the complexities of images and videos. As AI's vision comes to life, machines gain a profound understanding of their visual surroundings, unlocking a realm of possibilities across diverse fields. From recognizing faces to detecting anomalies in medical imaging, and from empowering autonomous vehicles to interpreting visual data in E-commerce analytics, Computer Vision stands as an awe-inspiring testament to AI's ability to comprehend the visual world.

### ❖ The Quest for Vision

The human faculty of sight, a marvel of natural evolution, has long been regarded as an elusive realm for machines to grasp. Computer Vision emerges as a domain of AI that endeavors to bridge this perceptual gap, imbuing machines with the ability to see, interpret, and make sense of visual data. The journey of Computer Vision traces back to the early years of AI research when the ambition to endow machines with the "eyes" to understand the world gained traction.

As technology advanced, so did the promise of Computer Vision, bolstered by the exponential growth in computational power, vast amounts of visual data, and groundbreaking algorithms. From simple image recognition tasks to the understanding of complex visual scenes, AI's vision journeyed towards ever-greater sophistication.

### ❖ The Vast Canvas of Applications

Computer Vision, like a master artist, paints a vast canvas of applications across various domains, each stroke a transformative breakthrough. In the realm of E-commerce, Computer Vision enables augmented reality experiences, allowing customers to virtually try on clothing and accessories before making a purchase. Additionally, AI-powered visual search empowers shoppers to discover products from images, revolutionizing the way consumers engage with E-commerce platforms.

In the Healthcare sector, Computer Vision assumes a vital role in medical imaging analysis. AI-driven tools can detect and diagnose diseases from X-rays, MRIs, and CT scans with unprecedented accuracy, aiding healthcare professionals in timely and accurate diagnoses. Computer Vision also plays a pivotal role in pathology and histology, revolutionizing disease detection and prognosis.

### ❖ Vision Unleashed: Autonomous Vehicles

One of the most transformative applications of Computer Vision emerges in the domain of autonomous vehicles. By equipping vehicles with visual perception capabilities, AI enables cars to navigate complex environments, detect obstacles, and make real-time decisions, ensuring safety and efficiency on the roads.

The fusion of Computer Vision with other AI techniques, such as LiDAR and radar, allows autonomous vehicles to perceive the world with remarkable precision. From pedestrian detection to lane tracking and object recognition, AI's vision lays the foundation for the future of transportation.

### ✧   Unlocking Insights in E-commerce Analytics

In the digital realm of E-commerce, Computer Vision finds yet another compelling application - the realm of visual data analysis. By interpreting product images, AI can extract valuable insights for retailers and businesses. Visual analytics tools can analyze consumer behavior, identify popular products, and assess the effectiveness of marketing strategies.

Additionally, Computer Vision enhances quality control processes in manufacturing industries by automatically detecting defects in products during production, reducing errors and ensuring consistency.

### ✧   Conclusion: The Visionary Revelation

Computer Vision, a breathtaking revelation in AI's artistic tapestry, elevates the capabilities of machines to interpret the visual world. From recognizing faces to diagnosing diseases, from empowering autonomous vehicles to revolutionizing E-commerce analytics, Computer Vision showcases the transformative potential of AI's perception.

As we venture forth into the following chapters, the profound impact of Computer Vision will unfold across E-commerce, Healthcare, Banking, and Finance, where AI's visual understanding finds expression in real-world applications. The synergy between human ingenuity and AI's computational prowess weaves an extraordinary tableau, where the vision of machines converges with human potential, reshaping industries and illuminating new pathways to progress.

The Chameleon of AI: Weak vs. Strong AI

As We Traverse the Terrain of AI: Unraveling the Enigma of Weak AI and the Quest for Strong AI

Within the captivating landscape of AI's intellectual exploration, a fundamental dichotomy emerges - that of Weak AI and Strong AI. These two divergent realms define the capabilities of AI systems and the scope of their cognitive prowess. As we delve into this dichotomy, we encounter machines that excel at specific tasks with superhuman precision, yet the pursuit of general intelligence beckons as an enigma yet to be fully unraveled, signifying the ultimate frontier in AI's cognitive odyssey.

- **The Power of Weak AI**

Within the bounds of Weak AI, we witness the remarkable achievements of machines that specialize in performing specific tasks with exceptional accuracy and efficiency. Weak AI systems exhibit cognitive abilities in narrow domains, far surpassing human capabilities in these specific tasks. These specialized systems thrive on vast datasets and sophisticated algorithms, processing information at speeds beyond human capacity.

Examples of Weak AI abound across various industries. In the realm of E-commerce, recommendation engines excel at predicting user preferences and suggesting personalized products, boosting customer satisfaction and sales. In Healthcare, AI-driven diagnostic tools analyze medical data to identify diseases with higher accuracy than human doctors alone. In the financial world, trading algorithms execute transactions at lightning speeds, optimizing investment strategies and outperforming human traders.

- **The Quest for Strong AI**

Strong AI, on the other hand, represents the holy grail of AI's cognitive odyssey - a quest to create machines with general intelligence that rivals or surpasses human cognitive abilities across diverse domains. Strong AI aspires to understand, learn, and adapt to an extensive range of tasks and situations, effectively mirroring human cognition in all its complexity.

The journey towards Strong AI remains an enigma, as it involves unraveling the mysteries of human consciousness, intuition, and emotions. The grand challenge lies in imbuing machines with qualities that go beyond raw computational power - the essence of human understanding, empathy, and creativity. Philosophical debates emerge about whether Strong AI is even achievable, raising questions about the nature of consciousness and the potential ethical implications of creating machines with self-awareness.

- **The Blurred Boundaries**

The boundaries between Weak AI and Strong AI often appear blurred, as AI systems with narrow capabilities can sometimes appear to exhibit human-like intelligence. Chatbots that hold engaging conversations or virtual assistants that comprehend natural language can create the illusion of broader intelligence. Yet, these systems remain constrained by their programmed algorithms and lack genuine comprehension or consciousness.

The challenge of Strong AI involves the development of AI systems that possess not just functional intelligence but a deeper understanding of context, creativity, and the ability to adapt to unforeseen situations.

✧ Conclusion: Navigating the Cognitive Frontier

As AI navigates the terrain of Weak AI and contemplates the quest for Strong AI, it unravels the depths of human potential and the possibilities of machine intelligence. Weak AI has already transformed industries and unlocked new realms of efficiency and precision. However, the path towards Strong AI remains elusive, shrouded in complexity and mystery.

The chapters ahead will illuminate the profound impact of AI's cognitive capabilities across E-commerce, Healthcare, Banking, and Finance, revealing how AI's transformational power shapes the future of human-machine interactions. The enigma of Strong AI beckons, inviting humanity to embark on an intellectual odyssey, where the pursuit of general intelligence converges with the potential to redefine human existence in ways that have yet to be fully imagined.

Conclusion: Pondering the Embodied Brilliance of AI

Our expedition into the defining essence of AI leads us to the precipice of human ingenuity. The aspiration to replicate human intelligence within machines, once confined to ancient myths, now stands as a tangible endeavor of the 21st century. From the methodologies of Machine Learning to the linguistic marvels of NLP and the perceptual vision of Computer Vision, AI embodies the ingenuity of humanity's quest for knowledge. The chapters ahead unfurl the profound impact of AI across E-commerce, Healthcare, Banking, and Finance, where its transformative potential materializes into real-world applications. As we venture forth, our minds aglow with curiosity, we find ourselves on the cusp of a new era, where the marriage of human intellect and AI's computational prowess forges an unprecedented symphony of innovation and progress.

**The Origins of AI: Tracing the conceptual origins of AI, from ancient myths to the formalization of the field in the 1950s.**

In the pursuit of understanding the genesis of Artificial Intelligence (AI), we embark on a captivating journey through time, tracing the conceptual origins of this extraordinary field. From the enchanting myths of ancient civilizations to the formalization of AI as a distinct academic discipline in the 1950s, we unravel the rich tapestry of ideas and inspirations that laid the groundwork for the transformative power of AI that we witness today.

# Mythical Origins - Golems and Automatons

## The Roots of AI: Unraveling the Mythical Inspirations

The seeds of Artificial Intelligence were sown in the fertile soil of human imagination, finding roots in the myths and folklore of ancient civilizations. In these tales, the yearning to create artificial life and bestow intelligence upon non-living matter came alive through mesmerizing stories of golems and automatons. These ancient narratives provide a profound insight into humanity's enduring fascination with the idea of creating beings with artificial intelligence.

- ✧ The Enchantment of the Golem

In the heart of ancient Hebrew folklore, the legend of the golem casts a spell of awe and wonder. The golem was a creature crafted from lifeless matter, brought to life through mystical incantations and divine knowledge. This animated being, shaped from the dust of the earth, moved with a semblance of life and followed the commands of its creator.

The tale of the golem symbolizes humanity's eternal desire to transcend the boundaries of the natural world and create beings that mimic life itself. It echoes the very essence of AI's conceptual origins - the aspiration to fashion intelligent beings that can understand, learn, and act with purpose.

- ✧ The Marvels of Hephaestus' Automatons

In the realm of ancient Greek myths, the divine blacksmith Hephaestus stood as a master craftsman who forged wondrous automatons. These mechanical beings, fashioned from bronze and endowed with remarkable abilities, served and assisted the gods on Mount Olympus. Among Hephaestus' creations was Talos, a giant bronze automaton that guarded the island of Crete.

The tales of Hephaestus' automatons foreshadowed the idea of intelligent machines serving human needs, akin to the vision of AI-powered assistants and autonomous robots in modern times. The myths highlight humanity's yearning to create beings that not only resemble humans in form but also possess cognitive faculties and capabilities.

- ✧ The Uniting Threads

The common thread that weaves through these myths lies in the very essence of humanity's quest for AI. Across cultures and civilizations, the desire to create artificial life and imbue inanimate matter with intelligence resonates profoundly. The myths evoke a sense of wonder and curiosity about the possibility of endowing machines with the ability to think, reason, and perceive the world around them.

The ancient narratives serve as a testament to human creativity and ingenuity, inspiring the pioneers of AI in their pursuit of creating intelligent machines. They embodied the vision that would eventually give rise to the formal discipline of AI, where the scientific exploration of intelligence in machines would become a reality.

Conclusion: The Enduring Quest

As we gaze upon the mythical inspirations that fueled the roots of AI, we glimpse the timeless yearning of humanity to bring forth intelligent beings from the realm of the imagination into the realm of reality. The myths of golems and automatons reflect the ever-present desire to transcend the boundaries of what is known and venture into the realm of the extraordinary.

The chapters that lie ahead will illuminate the wondrous achievements of AI across E-commerce, Healthcare, Banking, and Finance, where the seeds of ancient myths have blossomed into transformative technologies. The journey of AI continues, a tribute to the ancient dreams that forged the very quest of AI - to breathe life and intelligence into machines, forever captivating the human spirit with the magic of possibilities.

**The Seeds of Modern AI - From Turing to Cybernetics**

The mid-20th century heralded a momentous period of transformative innovation in computing and cybernetics, giving rise to the formal seeds of Artificial Intelligence (AI) as a scientific discipline. At the forefront of this intellectual revolution stood Alan Turing, an extraordinary figure often regarded as the father of computer science. Turing's seminal contributions to the conceptual foundations of AI and his groundbreaking work in cryptography during World War II left an indelible mark on the trajectory of AI's evolution.

- Turing's Turing Machine - A Blueprint for Computation

In 1936, Alan Turing introduced a groundbreaking concept that would pave the way for the future exploration of intelligence in machines - the "Turing machine." This theoretical device laid the foundation for understanding the fundamental principles of computation. The Turing machine served as a blueprint for any computation that could be described by an algorithm, signifying the universality of computation and its potential implications for mimicking human thought processes.

Turing's visionary notion of a universal machine capable of carrying out any computation marked a pivotal moment in the history of computing. It not only laid the groundwork for the development of modern computers but also provided a theoretical framework for the formalization of AI as a distinct discipline.

- ✧ The Enigma of Cryptanalysis - Turing's War Efforts

During World War II, Turing's brilliance shone forth in his critical contributions to cryptanalysis, the art of breaking codes and ciphers. Leading a team at Bletchley Park, Turing played a pivotal role in cracking the Enigma code used by the Germans, a feat considered crucial to the Allied victory. The success of this effort demonstrated the transformative power of machine computation in tackling complex and seemingly insurmountable problems.

The wartime application of machine computation to decrypt enemy messages underscored the potential of computers to assist human intelligence in solving complex tasks. This profound insight would significantly influence the trajectory of AI research, kindling the idea of creating machines that could augment and extend human cognitive capabilities.

- ✧ The Turing Test - Probing Machine Intelligence

In 1950, Alan Turing introduced the concept of the "Turing test," a seminal milestone in AI's journey towards assessing machine intelligence. In a thought-provoking paper titled "Computing Machinery and Intelligence," Turing proposed a criterion to evaluate whether a machine could exhibit intelligent behavior indistinguishable from that of a human.

The Turing test presented a provocative challenge, questioning the nature of intelligence and the possibility of creating machines that could mimic human conversation and thought processes. While the Turing test remains a subject of ongoing debate and criticism, it became a benchmark for measuring the progress of AI research and ignited discussions on the nature of cognition and consciousness.

Conclusion: The Turing Legacy - Forging the Path of AI

Alan Turing's remarkable legacy reverberates throughout the annals of AI history. His pioneering work on the Turing machine laid the groundwork for the computational nature of intelligence, while his contributions to cryptanalysis during World War II showcased the transformative potential of machine computation.

Turing's vision of a universal machine capable of emulating any computation and his probing question of machine intelligence through the Turing test continue to inspire AI researchers and enthusiasts alike. As we journey further into the chapters ahead, the profound impact of Turing's intellectual contributions will become apparent across E-commerce, Healthcare, Banking, and Finance, where the seeds of AI's birth blossom into an extraordinary tapestry of innovation and progress. Turing's visionary brilliance endures, forever guiding the path of AI's exploration, and illuminating the ever-evolving frontiers of human-machine intelligence.

## The Dartmouth Conference and the Birth of AI

In the annals of AI history, a transformative event stands as a defining moment - the legendary Dartmouth Conference of the summer of 1956. This pivotal gathering of brilliant minds, including John McCarthy, Marvin Minsky, Allen Newell, and Herbert A. Simon, among others, coalesced the quest for Artificial Intelligence (AI) into a formal and distinct academic discipline.

- ✧ The Gathering of Visionaries

In the idyllic setting of Dartmouth College, a select group of visionaries assembled to explore an audacious idea - the creation of machines that could exhibit human-like intelligence. John McCarthy, an eminent computer scientist, played a central role in convening this historic gathering. He envisioned a future where machines transcended mere calculators and algorithms to simulate the essence of human cognition.

Among the participants were Marvin Minsky, a prodigious thinker and pioneer of AI, and Allen Newell and Herbert A. Simon, renowned for their groundbreaking work on the logic theorist, the first AI program. Together, they formed a constellation of intellects, each contributing unique perspectives to the unfolding vision of AI.

- ✧ Coining the Term "Artificial Intelligence"

At the Dartmouth Conference, the conceptual framework of AI crystallized, and a momentous milestone was reached. John McCarthy, during a late-night brainstorming session, proposed the term "Artificial Intelligence" to encapsulate the quest to imbue machines with intelligence akin to human cognition. This pivotal moment marked the formalization of AI as a distinct academic discipline.

The term "Artificial Intelligence" was pregnant with significance, encompassing the very essence of the quest at hand - the creation of intelligence that was not organic but carefully crafted by human ingenuity. This marked a turning point in the history of AI, signaling the beginning of a scientific exploration that would transform the landscape of technology and human existence.

### ✧ The Vision of Human-like Cognition

The visionaries at the Dartmouth Conference envisioned a bold future where machines would not merely perform pre-programmed tasks but would learn from experience, adapt to novel situations, and solve complex problems autonomously. Their ambition extended beyond narrow domains of expertise, embracing the dream of "General AI" - machines possessing the breadth of cognitive capabilities exhibited by humans.

The idea of a machine that could think, reason, and learn with human-like capabilities ignited the imaginations of the attendees. They foresaw AI as a partner to human intelligence, augmenting human abilities and extending the frontiers of knowledge and understanding.

Conclusion: The Enduring Legacy of Dartmouth

The Dartmouth Conference cast a monumental and lasting legacy on the field of AI. In a few short weeks of intense collaboration and visionary thinking, the seeds of AI were sown, bearing fruit in the decades to come. The formalization of AI as an academic discipline, coupled with the audacious vision of human-like cognition, set the stage for the transformative advancements that AI would soon achieve.

The chapters that lie ahead will illuminate the remarkable achievements of AI across E-commerce, Healthcare, Banking, and Finance, where the seeds planted at Dartmouth blossomed into groundbreaking technologies. The vision of the Dartmouth Conference endures, inspiring AI researchers to continue the pursuit of machines that emulate human intelligence and paving the way for a future where the boundaries between humans and machines blur, forever changing the fabric of our existence.

## The AI Winters and Resurgence

The journey of Artificial Intelligence (AI) has been a tale of resilience, marked by alternating waves of exuberance and skepticism. In the decades that followed its formalization, AI experienced periods of soaring optimism during the 1970s and 1980s. Ambitious expectations for AI's potential to revolutionize industries and solve complex problems permeated the scientific community and captivated the public. However, these grand promises were met with technological limitations, leading to disillusionment and periods known as "AI Winters," where AI research stalled in some circles.

### ✧ The Era of Optimism - AI's Ambitious Promises

In the 1970s and 1980s, AI enthusiasts were buoyed by the promise of machines emulating human-like intelligence. Researchers and media alike heralded AI as the harbinger of a new era, where machines would perform tasks with unparalleled precision and complexity. The potential applications of AI seemed boundless, from automating mundane tasks to transforming healthcare diagnostics and driving economic growth.

AI researchers pursued ambitious projects, hoping to build systems capable of reasoning, understanding natural language, and mimicking human problem-solving. However, as the technological capabilities of the time were put to the test, it became apparent that the promised AI revolution was yet to materialize.

### ✧ The Chill of AI Winters - Reality Meets Expectations

Amidst the enthusiasm, the initial AI projects faced challenges that dampened the fervor. AI researchers encountered obstacles in building intelligent systems that could generalize beyond specific tasks and adapt to changing environments. As a result, some projects failed to deliver on their lofty promises, leading to skepticism and reduced funding.

During the AI Winters, the public's perception of AI swung from soaring optimism to doubt. Some critics questioned whether AI could ever achieve human-like intelligence or if it was merely a mirage, perpetually out of reach. The limitations of existing technologies, combined with the inflated expectations, brought AI research to a standstill in certain circles.

### ✧ The 21st Century Renaissance - AI Rises Anew

The spirit of AI endured, and the 21st century brought about a remarkable resurgence. The advent of the internet and exponential growth in computational power unlocked a new era of possibilities for AI. The emergence of Big Data, fueled by the proliferation of digital information, provided abundant resources for AI systems to learn and improve.

Machine learning, particularly deep learning, revolutionized AI research by enabling systems to learn from vast datasets and extract patterns previously beyond human comprehension. AI-powered applications began permeating various industries, from virtual assistants in E-commerce to medical diagnosis and fraud detection in Healthcare and Banking.

> ✧ AI's Modern Renaissance - A Testament to Human Ingenuity

The 21st-century AI renaissance stands as a testament to human ingenuity and the pursuit of knowledge. The setbacks of the AI Winters did not deter the determination of AI researchers and enthusiasts. Instead, they spurred the development of more sophisticated algorithms and approaches, enabling AI to achieve unprecedented feats.

As AI continues to evolve, it has transformed how we conduct business in E-commerce, optimize healthcare services, and manage financial systems in Banking and Finance. AI's impact reverberates across diverse fields, revolutionizing industries and enhancing human capabilities.

Conclusion: The Journey Unfolds

The ebb and flow of AI's progress paint a vivid portrait of technological evolution. From the AI Winters of skepticism to the 21st-century resurgence, AI's journey is one of perseverance, resilience, and remarkable achievement. As we venture deeper into the chapters ahead, we will witness AI's profound impact on various industries, illuminating a future where human intelligence collaborates with the brilliance of machines to conquer new frontiers of possibility. The journey of AI continues, beckoning us to explore its ever-expanding horizon and discover the boundless potential of human-machine collaboration.

**Conclusion: An Ever-Evolving Odyssey**

The journey to understand the origins of AI traverses the realms of ancient myth, the brilliance of human thought, and the convergence of computation and intelligence. From the golems and automatons of antiquity to the formalization of AI as a scientific discipline, we witness an ever-evolving odyssey that has culminated in the remarkable AI systems we encounter today.

As we journey deeper into the chapters that lie ahead, the conceptual origins of AI will resonate with the transformative impact of AI across E-commerce, Healthcare, Banking, and Finance. The quest to replicate human intelligence through machines endures, fueling an ever-evolving odyssey of human ingenuity and technological progress, where AI's cognitive tapestry continues to weave profound change across diverse fields.

# The Pillars of AI: Identifying the core pillars of AI research, including symbolic AI, machine learning, and natural language processing.

The realm of Artificial Intelligence (AI) stands upon a robust edifice, supported by pillars of profound research and innovation. As we traverse the landscape of AI, we encounter three fundamental pillars that have shaped its evolution: Symbolic AI, Machine Learning, and Natural Language Processing (NLP). Each pillar represents a distinct approach to understanding and replicating human intelligence in machines, together forming the bedrock upon which AI's transformative power rests.

## Symbolic AI - The Power of Logic and Reasoning

The world of artificial intelligence is a vast and variegated landscape, teeming with diverse models and methodologies, each with its unique capabilities and peculiarities. Among these numerous approaches, Symbolic AI, colloquially known as Good Old-Fashioned AI (GOFAI), holds a special place. Despite the advent of newer and arguably more advanced methods, Symbolic AI continues to play a critical role in a variety of fields, from e-commerce and healthcare to banking and finance.

The distinctive feature of Symbolic AI lies in its aspiration to encode human intelligence into machines, employing symbolic representations and logic. This quest is steeped in the traditions of classical logic and formal reasoning, aiming to encode not just information, but wisdom – the ability to infer, deduce, and conclude from encoded knowledge.

- ✧ Symbolic AI and E-commerce

In the realm of e-commerce, Symbolic AI breathes life into expert systems that leverage rules and logic to furnish personalised recommendations for customers, mirroring the personalised advice given by a knowledgeable salesperson in a brick-and-mortar store. Consider an online bookstore. Instead of merely suggesting books based on browsing history, a system underpinned by Symbolic AI could assess the user's reading habits, preferred genres, past purchases, and even review sentiments to recommend books that align with their reading preferences. Here, symbolic reasoning enables the system to infer customer preferences in a nuanced manner, rather than merely reacting to explicit input.

- ✧ Symbolic AI and Healthcare

In healthcare, symbolic AI facilitates decision-making processes by aiding in medical diagnosis. This approach reflects the syllogistic reasoning often used by medical professionals, whereby symptoms are logically linked to potential diseases.

For instance, if fever and cough are symptoms associated with a common cold, and a patient exhibits these symptoms, the system can conclude that the patient might have a common cold.

However, medicine is a complex field with multitudes of overlapping symptoms and conditions, necessitating more sophisticated symbolic reasoning. In such cases, these systems may utilise probabilistic logic, effectively handling the uncertainty associated with medical diagnosis. For example, the system may determine that a patient with certain symptoms has a 70% chance of having disease X, a 20% chance of having disease Y, and a 10% chance of having disease Z.

- Symbolic AI and Banking & Finance

In banking and finance, Symbolic AI can support fraud detection by establishing rules based on historical fraud patterns. If a transaction meets these rules, the system could flag it as potential fraud, aiding in proactive fraud prevention. For example, rapid, high-value transactions from a previously low-activity account could be flagged as suspicious.

Moreover, Symbolic AI assists in risk assessment and credit scoring. By encoding rules and knowledge about financial behaviour, these systems can assess an individual's or a company's creditworthiness and assign a risk score.

Despite the prowess of Symbolic AI, it is not without its limitations. These systems lack the capability to learn and adapt from new information unless the rules are manually updated, which limits their flexibility and adaptability. Moreover, they struggle with tasks requiring perception and pattern recognition, areas where sub-symbolic approaches, such as neural networks, excel.

However, these limitations do not diminish the value of Symbolic AI. Hybrid systems, combining symbolic and sub-symbolic AI, could leverage the strengths of both methodologies. In these hybrid models, symbolic AI could be used for tasks requiring logical reasoning and rule-based decisions, while sub-symbolic AI could handle tasks related to perception and pattern recognition.

In conclusion, although Symbolic AI may seem like an antiquated approach in the era of deep learning and neural networks, it remains a valuable asset in the AI arsenal. With its roots deeply grounded in the principles of logic and formal reasoning, Symbolic AI continues to serve diverse fields with its distinctive capabilities, offering insight into the fundamental understanding of intelligence in both humans and machines.

# Machine Learning - The Art of Learning from Data

Artificial Intelligence (AI) has traversed a dynamic path to reach its current state of sophistication and ubiquity. One of its primary pillars, responsible for much of this transformative journey, is Machine Learning (ML). Machine Learning is the linchpin that allows machines to glean insights from colossal volumes of data, decipher patterns, and make predictions without the necessity for explicit programming. This chapter aims to expound upon the multifarious nature of Machine Learning, focusing on its principal categories: supervised learning, unsupervised learning, and reinforcement learning.

### ✧ The Taxonomy of Machine Learning

Supervised Learning stands as the most prevalent form of Machine Learning, where models are trained on a labelled dataset. Labelled data implies that each instance in the training set contains the input features and the corresponding output or target. Using this mapping of input-output pairs, the model discerns the underlying function that best predicts the output from the input.

Unsupervised Learning, in stark contrast, deals with unlabelled data. Here, the primary goal is to discover inherent structures and patterns in the dataset. Unsupervised learning techniques include clustering, where data is grouped based on similarities, and dimensionality reduction, where high-dimensional data is simplified without significant loss of information.

Reinforcement Learning, meanwhile, is an approach where an agent learns to make decisions by interacting with an environment. The agent performs actions and receives feedback in the form of rewards or penalties, which guide the learning process. Over time, the agent learns the optimal strategy, called a policy, to obtain the maximum cumulative reward.

### ✧ Machine Learning in E-commerce

In the E-commerce industry, Machine Learning plays a vital role in personalising the user experience, primarily through recommendation systems. Such systems utilise techniques like collaborative filtering and content-based filtering to suggest products to a user. For instance, collaborative filtering takes into account the preferences of similar users to recommend products, while content-based filtering suggests items similar to the ones that a user has interacted with or purchased in the past. These recommendation systems bolster customer satisfaction and can significantly enhance sales and revenue.

- Machine Learning in Banking and Finance

Within the realm of banking and finance, Machine Learning exhibits enormous utility in fraud detection systems. Such systems use unsupervised learning algorithms to identify anomalous patterns indicative of fraudulent activity within vast pools of transaction data. Techniques such as outlier detection or anomaly detection can flag suspicious activities that deviate from the norm, thereby bolstering the security of financial transactions.

Further, Machine Learning aids in credit scoring by predicting the risk associated with lending to a particular individual or business. Supervised learning models trained on historical data, containing various features of borrowers and the information whether they defaulted or not, can predict the likelihood of a borrower defaulting in the future.

However, it is imperative to note that while Machine Learning possesses substantial potential, it is not without challenges. Issues such as overfitting, where a model learns the training data too well and performs poorly on unseen data, and the black-box nature of some algorithms, where the decision-making process is not transparent, are notable areas of concern.

Despite these challenges, Machine Learning's transformative potential is unquestionable. As it continues to evolve and mature, the influence of Machine Learning is bound to permeate even more profoundly into various fields, driving unprecedented levels of efficiency, personalisation, and intelligent decision-making.

## Natural Language Processing (NLP) - Unraveling the Tapestry of Human Language

Natural Language Processing (NLP), a fascinating subfield of Artificial Intelligence (AI), is primarily concerned with the interaction between machines and human language. As its name implies, Natural Language Processing imparts machines with the capabilities to understand, interpret, generate, and even to some degree, appreciate human language in all its complexity and nuance.

Language, being a cornerstone of human intelligence and interaction, is laden with subtleties that embody cultural intricacies, semantic associations, syntactic rules, and emotive connotations. Empowering machines to comprehend these nuances paves the way for more sophisticated human-machine interactions and facilitates numerous language-based tasks.

- Natural Language Processing and E-commerce

Within the context of E-commerce, Natural Language Processing has been instrumental in enhancing customer engagement and support through the use of chatbots and virtual assistants. These intelligent entities, powered by NLP, can interpret customer queries, respond in a human-like manner, and provide relevant solutions or recommendations. By handling routine customer inquiries and support tasks, these systems can significantly enhance the efficiency of customer service operations while providing a personalized and engaging customer experience.

For instance, a customer might inquire, "What is the status of my order?" The NLP-driven system needs to understand the intent behind the query, access the relevant information, and provide a coherent response. Furthermore, it can handle more complex tasks, like processing return requests or providing product recommendations based on the customer's past purchases and preferences.

- Natural Language Processing and Healthcare

In the healthcare domain, Natural Language Processing has a compelling role in aiding research by parsing through enormous volumes of scientific literature. Medical research is continually evolving, and keeping abreast of the latest advancements is an insurmountable task for any individual. However, NLP can automate this process by scanning through thousands of research papers, extracting relevant information, and summarizing findings.

For instance, in the wake of a pandemic, researchers around the globe produce a multitude of studies exploring various facets of the disease. An NLP system can sift through these articles to extract valuable insights, such as symptoms, transmission rates, effective treatment strategies, and so on. This not only accelerates the pace of medical discoveries but also ensures that healthcare professionals are always equipped with the latest knowledge.

Natural Language Processing, though powerful, faces numerous challenges, chiefly because human language is highly ambiguous and context-dependent. Interpretations can significantly vary based on cultural backgrounds, personal experiences, or the context in which a statement is made. Nevertheless, advancements in deep learning and neural networks, coupled with increasing computational power and data availability, have catalyzed progress in this field, bringing us closer to the goal of creating machines that can truly understand and engage in human language.

In conclusion, Natural Language Processing stands at the intersection of linguistics, computer science, and AI, endeavoring to unravel the intricacies of human language. With its myriad applications across diverse fields, NLP is redefining human-machine interactions and unlocking new avenues for research and development in the quest for more intelligent and intuitive artificial systems.

## The Synergy of Pillars - Building AI's Future

Artificial Intelligence, in its multifaceted existence, encapsulates a triad of distinct yet complementary pillars - Symbolic AI, Machine Learning, and Natural Language Processing. These pillars, while individually contributing to AI's progress, create a dynamic synergy when they converge. This chapter discusses the integration of these foundational components, painting a coherent picture of AI's future landscape.

- ✧ Convergence for Sophistication

AI's quest to emulate human intelligence has led to the development of diverse approaches, each unravelling different facets of intelligence. Symbolic AI focuses on the formal logic and rule-based reasoning aspect, embodying the element of deductive intelligence. Machine Learning, on the other hand, centres on learning from data and experience, mirroring human abilities of learning and adaptation. Natural Language Processing, finally, seeks to navigate the intricacies of human language, reflecting our unique capacity for linguistic communication.

However, it is when these elements intertwine that we truly witness AI's potential. For instance, consider a virtual assistant in an e-commerce setting, such as a chatbot. The chatbot would use Natural Language Processing to understand customer queries, Machine Learning to personalize its responses based on the customer's history and preferences, and Symbolic AI to follow logical procedures, such as processing a return request.

- ✧ Towards Strong AI: The Grand Aspiration

The seamless integration of these pillars fuels the pursuit of Strong AI or Artificial General Intelligence (AGI) - the conceptual endpoint where machines exhibit cognitive abilities on par with, or even surpassing, human intelligence. In AGI, machines would not only execute tasks that typically require human intelligence but also understand, learn, adapt, and apply knowledge across diverse domains, similar to how humans transfer knowledge from one context to another.

While we stand at a considerable distance from realizing AGI, the convergence of Symbolic AI, Machine Learning, and Natural Language Processing brings us a step closer. It enhances the sophistication of AI systems, enabling them to handle tasks of increased complexity and exhibit more 'human-like' intelligence.

- Ethical and Societal Considerations

As we journey towards this horizon of hyper-intelligent machines, it becomes incumbent upon us to remain vigilant of the ethical and societal implications that accompany AI's advancements. Concerns about privacy, security, employment, and fairness surface with AI's increasing pervasiveness. Ensuring transparency, accountability, and equitable access to AI's benefits are crucial to mitigating these concerns.

Further, as AI systems become more intelligent and autonomous, questions about machine ethics and rights could arise. As such, it is vital to establish robust ethical and legal frameworks to govern the development and deployment of AI technologies.

In conclusion, the future of AI rests on the synergy of its foundational pillars - Symbolic AI, Machine Learning, and Natural Language Processing. While the aspiration of achieving Strong AI continues to motivate progress, maintaining a conscientious focus on the ethical and societal implications is equally crucial. This delicate balancing act between technological advancement and ethical awareness will be the key to ensuring that AI emerges as a force for good, a true catalyst for human progress.

## Conclusion: A Journey of Discovery

The pillars of AI represent a voyage of discovery, an ongoing odyssey to unlock the mysteries of human intelligence and propel machines into realms of unparalleled cognition. As we explore AI's applications in E-commerce, Healthcare, Banking, and Finance, we witness the potency of these pillars in transforming industries, empowering decision-making, and redefining the human-machine relationship.

The chapters that follow will showcase how these pillars unite to elevate AI from theoretical abstraction to practical reality, illuminating the path towards an AI-driven future, brimming with innovation and boundless possibilities. Embrace the journey as we delve deeper into the captivating world of AI, where the convergence of technology and human ingenuity marks the dawn of a new era in human history.

# Section 2: The Evolution of AI

As we venture into the second section of our exploration into the enigmatic realm of Artificial Intelligence, we shift our focus from the foundational pillars of AI to its temporal journey. This section illuminates the profound evolution of AI, tracking its metamorphosis from a nascent concept in the minds of visionary thinkers to a transformative force that permeates every facet of modern existence.

- From Inception to Reality: A Historical Perspective

The journey of AI traces back to the mid-20th century, when the idea of creating an artifact capable of emulating human intelligence was first propounded. It is necessary to delve into this historical narrative, to appreciate the leaps and bounds AI has made since its conception. This exploration also allows us to acknowledge the visionary pioneers who, despite the technological constraints of their time, dared to dream of a world where machines could think. It is their imagination and foresight that provided the initial impetus for the development of AI.

- Epochs of AI: Winter and Renaissance

The narrative of AI is characterized by alternating periods of intense interest and disillusionment, commonly referred to as AI's 'summers' and 'winters'. Each 'winter' was succeeded by a 'summer', a resurgence of interest, spurred by breakthroughs that challenged previous limitations and expanded the realm of the possible. Understanding these epochs elucidates the cyclical nature of scientific discovery, and how each 'winter', rather than signalling failure, set the stage for a future 'summer' by highlighting challenges to be overcome.

- Modern AI: A Technological Revolution

As we approach the present day, the narrative shifts from historical analysis to the examination of contemporary AI. This phase bears witness to an explosion of innovation, as AI pervades every industry, from e-commerce and healthcare to banking and finance, catalysing unprecedented transformation. The examination of contemporary applications and advancements in AI allows us to appreciate the extent to which AI has infiltrated our lives and altered our societal fabric.

Tomorrow's AI: Speculating the Future

Finally, we cast our gaze into the future, speculating on the potential developments in AI. While this is inherently speculative, given the unpredictable nature of technological advancement, it is nonetheless a necessary exploration. Predicting future trends allows us to prepare for the opportunities and challenges they might present, ensuring that we harness AI's potential responsibly and ethically.

In conclusion, this section invites the reader on a temporal voyage, tracing the evolution of AI from its conception to its current state and beyond. It presents an enlightening narrative, filled with triumphs and setbacks, breakthroughs and challenges, and dreams realized and those yet to be fulfilled. It illustrates how AI, once a figment of science fiction, has become an inextricable part of our reality, and holds the potential to shape our future. It is a testament to human ingenuity, ambition, and the relentless quest for knowledge.

**The Early Pioneers: Exploring the foundational work of early AI visionaries, such as Alan Turing and John McCarthy.**

Artificial Intelligence, despite its contemporary pervasiveness, has its roots in the visionary work of pioneering thinkers who envisaged a future where machines would be capable of emulating human thought. Two figures that stand out in the annals of AI's early history are Alan Turing and John McCarthy. Their groundbreaking work laid the groundwork for the expansive field of AI as we know it today.

- Alan Turing: The Forefather of Modern Computing

Alan Turing, a mathematician, cryptanalyst, and theoretical biologist, is widely recognized as the forefather of modern computing. His contribution to AI, although indirect, is monumental. During the Second World War, Turing designed the Bombe, a machine that deciphered encrypted messages sent by the German Enigma machines, demonstrating that machines could perform tasks traditionally requiring human intelligence and ingenuity.

Turing's conceptual contribution to AI was encapsulated in his seminal 1950 paper, "Computing Machinery and Intelligence", where he proposed the 'Turing Test'. This test, aimed at assessing a machine's ability to exhibit intelligent behaviour indistinguishable from that of a human, was a profound step in challenging the fundamental question: 'Can machines think?' Turing's work effectively set the stage for the birth of AI, influencing subsequent generations of researchers.

- John McCarthy: The Father of AI

While Turing sowed the seeds of AI, it was John McCarthy, a computer scientist, who brought these seeds to fruition. McCarthy coined the term 'Artificial Intelligence' in 1956 and organized the Dartmouth Conference, the first formal gathering of AI researchers. This conference marked the birth of AI as an independent field of study.

McCarthy's contributions to AI extend beyond these organizational achievements. He developed the LISP programming language, a pivotal tool in the AI field for decades due to its flexibility and expressive power. Furthermore, McCarthy's work on knowledge representation, logic programming, and the 'Advice Taker' system were instrumental in shaping what we now know as Symbolic AI.

- Legacy of the Pioneers

The contributions of Turing and McCarthy continue to influence modern AI. Turing's philosophical inquiries into machine intelligence set the tone for the discourse around the ethical and societal implications of AI. His thoughts on the potential of machines to mimic human intelligence continue to steer conversations on Strong AI and consciousness in machines.

McCarthy's technical contributions, particularly regarding Symbolic AI, still resonate in modern AI applications. His advocacy for the use of mathematical logic in AI has influenced the development of expert systems and rule-based AI that we see in fields like e-commerce and healthcare.

In conclusion, the early visionaries of AI set the foundation upon which the edifice of modern AI stands. Their audacious imaginations, rigorous research, and pioneering developments catalysed the journey of AI from theoretical concept to practical reality. By appreciating their groundbreaking work, we can better understand AI's historical trajectory and appreciate the visionary thinking that continues to propel the field forward.

## AI's Journey Through the "AI Winter": Investigating the periods of stagnation and reduced funding that shaped the perception of AI in the 1970s and 1980s.

Despite the profound advancements that have characterized the field of Artificial Intelligence, its trajectory has not always been an ascendant one. In the annals of AI's evolution, there have been periods of exuberant optimism, as well as periods of disillusionment and skepticism. One such era that stands out for its influence on the perception of AI was the "AI Winter", a term coined to describe periods of stagnation and reduced funding in AI research during the 1970s and 1980s.

## ✧ The Genesis of the AI Winter

AI's genesis was steeped in an atmosphere of buoyant optimism. Alan Turing's work had already opened the door to the prospect of machines mimicking human intelligence. John McCarthy, who is credited with coining the term "Artificial Intelligence," further stoked the flames of this optimism with his pioneering work. The Dartmouth Workshop of 1956, where McCarthy, Marvin Minsky, Claude Shannon, and Nathaniel Rochester gathered to explore the potential of AI, spurred an ambitious vision of the future. The consensus was that machines could, within a reasonable timeframe, carry out tasks that required human-like intelligence.

In the early stages, this optimism was justified as computers started to execute complex mathematical equations, demonstrating an ability to solve problems previously thought to be the sole purview of human intellect. These accomplishments inspired confidence among researchers and investors alike, leading to generous funding for AI research, predominantly from governmental sources. This led to an era of ambitious AI projects aiming for the realization of general AI or machines possessing cognitive abilities on par with humans.

However, as the late 1960s approached, it became increasingly evident that the initial optimism had been somewhat premature. The challenges involved in scaling up AI projects became clear. Language translation projects, which were among the most promising AI initiatives of the time, fell far short of expectations. The landmark report "ALPAC" in 1966, which critiqued the inefficiencies and overstated promises of machine translation research, catalyzed the onset of skepticism.

The intricacies of natural language, the nuances of human communication, and the complexity of real-world problem-solving proved to be far more challenging to encode into machines than initially envisioned. The promised breakthroughs, like machines understanding natural language or autonomously improving their knowledge, did not materialize. The gap between expectations and tangible results started to widen.

This led to an environment of disenchantment among funding bodies and the scientific community, primarily those outside the AI research field. Skepticism grew about the feasibility of AI's grand vision and the justification for the substantial investment it required. The AI community's failure to fulfill its ambitious promises led to doubts about AI's potential, culminating in the onset of the first AI Winter in the 1970s. Consequently, funding was significantly reduced, slowing the pace of AI development and marking the beginning of a challenging era for AI research.

- Impact of the AI Winter

The AI Winter, encompassing two primary periods of slowdown in the 1970s and the 1980s, marked a turning point in the trajectory of AI research and development. The optimism and abundant funding that had hitherto characterized the AI landscape gave way to a climate of skepticism and financial austerity. This transformation had tangible ramifications for both academia and the burgeoning AI industry.

Primary funders, such as the Defense Advanced Research Projects Agency (DARPA) in the United States, which had previously furnished generous financial support, turned circumspect. The conspicuous disparity between the promises of AI researchers and the actual deliverables sparked a reassessment of AI's potential and the subsequent contraction in funding. Universities saw a significant reduction in their research grants, leading to a noticeable decrease in the number of AI research projects. Concurrently, the industrial sphere, which had been nurturing early AI enterprises in anticipation of a technologically advanced future, also witnessed a contraction. Many nascent AI companies, unable to sustain their operations in the face of diminished investment and market skepticism, were compelled to close their doors.

However, it is essential to note that while the AI Winter was a period of retrenchment, it was not an era of complete stagnation. Rather, it heralded a shift in the focus of AI research. The grand aspiration of developing a general intelligence - an AI system capable of comprehending and learning any intellectual task that a human being can - receded into the background. Instead, researchers turned their attention to more specific, narrowly defined problems that seemed more within reach and offered practical, tangible benefits.

This period witnessed the blossoming of various AI subfields, such as expert systems and rule-based AI. Expert systems, also known as knowledge-based systems, simulate the decision-making ability of a human expert. They make use of a knowledge base, acquired from human experts, and an inference engine that applies logical rules to the knowledge base to answer queries. Their capability to provide reasoned explanations for their suggestions, resembling the thought process of human experts, found practical utility in a variety of fields.

For instance, in e-commerce, expert systems became instrumental in personalized recommendation engines. By analyzing user behavior and preferences based on a set of pre-programmed rules, they provided tailored suggestions, enhancing the user experience and augmenting sales. In healthcare, rule-based AI systems proved beneficial in diagnostic processes, where they helped deduce diseases based on a set of symptoms following a pre-established rule set, thereby aiding healthcare professionals.

Thus, while the AI Winter was a period of funding drought and diminished enthusiasm, it led to a recalibration of research focus, fostering advancements in narrower AI domains that continue to offer value across various industries.

◇ Lessons from the AI Winter

The AI Winter, a stark chapter in the annals of AI development, bequeaths a trove of lessons that continue to hold relevance in the present narrative of AI research and advancement. It stands as a poignant testament to the oscillations between unbridled optimism and sobering realism that can characterize the development of transformative technologies.

The first, and perhaps the most salient, lesson is the imperative for a pragmatic and grounded approach to AI research and development. The initial phases of AI research were marked by a fervor that bordered on hubris, with researchers confidently forecasting the advent of machines that could mimic, and potentially outperform, human intelligence within a few decades. However, these projections did not take into account the daunting complexity of the human intellect and the corresponding challenges in replicating it in silicon. When these ambitious promises were not fulfilled, the ensuing disillusionment precipitated a severe contraction in interest and funding, leading to the AI Winter.

This underscores the importance of setting realistic expectations and making incremental progress in AI development. While audacious goals can stimulate innovation, they should be tempered by an understanding of the intricate technical challenges and the time and effort needed to surmount them. A measured approach, based on careful research, meticulous planning, and a recognition of the inherent complexities, is more likely to yield sustainable progress and enduring success.

The AI Winter also illuminates the dynamic interplay between technological development and societal perception. It demonstrates how the trajectory of a scientific field can be significantly influenced by factors beyond the purely technical. Public sentiment, shaped by the gap between expectations and reality, exerted a significant impact on the course of AI development. This underscores the importance of honest and transparent communication with the public and stakeholders about the potentials and limitations of AI. Managing expectations, while keeping the discourse factual and avoiding hyperbole, can help foster public understanding and support, crucial for the steady evolution of the field.

Moreover, the relationship between technological advancement and funding availability is starkly illustrated by the AI Winter. The retrenchment of funding bodies, in response to the perceived failures of AI, precipitated a slowdown in research and innovation. This highlights the importance of diverse and stable funding sources to ensure the continuity and resilience of research endeavors. It also underlines the role of government and institutional policies in fostering a supportive ecosystem for technological advancement.

In conclusion, the AI Winter serves as a valuable lesson in the journey of AI. By reminding us of the pitfalls of overpromising and underdelivering, the importance of public perception, and the role of funding, it provides a historical lens to guide current and future AI initiatives, ensuring they are rooted in pragmatism, transparency, and resilience.

❖ The Thawing of the AI Winter and the Dawn of a New Era

The late 1980s and early 1990s marked a turning point in the narrative of AI, as the frost of the AI Winter began to recede. This phase of renewal was engendered by a confluence of factors that, together, rekindled interest in AI and spurred a revival that has continued into the present era of burgeoning AI applications.

One of the primary catalysts of this thaw was the advent and maturation of machine learning techniques. The limitations of rule-based systems and expert systems, which had been the focus during the AI Winter, had become increasingly evident. These systems struggled with tasks that required dealing with uncertainty, learning from new data, or scaling to large and complex problem domains. The emergence of new learning-based paradigms, including neural networks and Bayesian models, promised to address these limitations. These methodologies offered a fresh perspective: instead of attempting to encode human knowledge into machines, why not let machines learn from data, mirroring the learning process in humans? This shift represented a fundamental rethinking of the approach to AI and opened up exciting new possibilities for the field.

Simultaneously, the increasing availability of digital data offered fertile ground for these new machine learning techniques to flourish. As more aspects of human life became digitized, from commerce and communication to entertainment and education, vast troves of data were being generated. This data, harnessed effectively, could be used to train machine learning models, enabling them to discern patterns and make predictions. The advent of the internet, in particular, accelerated this trend, as it spawned a multitude of online activities and interactions that generated an unprecedented volume and variety of data.

In addition, the rise of more powerful computing hardware and the decreasing cost of storage made it feasible to store and process these large datasets. Advances in parallel computing and the advent of graphics processing units (GPUs), which were well-suited for the matrix operations common in many machine learning algorithms, facilitated the training of more complex models on larger datasets.

Moreover, the revival of AI was fostered by a renewed inflow of funding. As the potential of the internet began to be realized and the digital economy started to take shape, both private and public investors began to perceive the commercial and strategic value of AI. This led to a resurgence of funding for AI research, both in academic institutions and in the burgeoning tech industry, catalyzing a new wave of innovation and development in the field.

Taken together, these factors ushered in the end of the AI Winter and set the stage for the current era of AI abundance. The confluence of new machine learning techniques, the proliferation of digital data, advances in computing power, and a resurgence of funding led to a renaissance in AI research and applications. This has propelled us into an age where AI systems, once the stuff of science fiction, are now becoming an integral part of our everyday lives, transforming industries and reshaping the way we live, work, and play.

In summary, the AI Winter represents a crucial phase in the history of AI. It offers valuable insights into the challenges faced by early AI researchers and the subsequent course correction that led to a more pragmatic and focused approach to AI research. Reflecting upon this period of AI's evolution imparts vital lessons that continue to shape our approach to AI's development, ensuring we remain cognizant of the delicate balance between ambition and realism in this dynamic field.

**AI Renaissance: Analyzing the revival of AI in the 21st century, driven by advancements in computing power, data availability, and breakthroughs in machine learning.**

The 21st century has borne witness to a remarkable resurgence in the field of Artificial Intelligence (AI), commonly referred to as the AI Renaissance. Markedly different from the nascent stages of AI development and the subsequent AI Winter, this period has been characterized by the profound expansion of AI technologies, enabled by significant advancements in computing power, data availability, and breakthroughs in machine learning. This renaissance has positioned AI as a transformative force across diverse sectors, from healthcare and banking to e-commerce and beyond.

- Revolutionizing Computing Power

One of the key catalysts for the AI Renaissance has been the spectacular progress in computing power. The emergence of advanced computing hardware, such as graphics processing units (GPUs) and tensor processing units (TPUs), has revolutionized the capacity to process and analyze vast amounts of data. These specialized chips, designed to handle the high throughput of data required by AI applications, have facilitated the execution of complex machine learning algorithms. For instance, the advent of GPUs has been particularly significant for the training of deep learning models, which require parallel processing capabilities to manage their matrix-based computations efficiently.

Moreover, the development of cloud computing infrastructure has provided AI researchers and practitioners with scalable, on-demand computing resources. This has democratized access to high-performance computing, enabling even small startups and individual researchers to train sophisticated AI models, a feat once reserved for well-funded organizations.

- The Data Deluge

The explosion in digital data availability has been another key driver of the AI Renaissance. The advent of the internet and subsequent digitalization of many aspects of human life have led to the generation of massive datasets. From social media interactions and online transactions in e-commerce to electronic health records in healthcare, these digital traces of human activity offer a rich source of data for training AI models.

Simultaneously, advances in data storage and processing technologies, such as Hadoop and Apache Spark, have made it feasible to store and handle these large datasets. Data, in its abundance, has become the new oil, feeding the machine learning algorithms that underpin AI and enabling them to identify patterns, make predictions, and glean insights.

- Machine Learning Breakthroughs

Lastly, breakthroughs in machine learning methodologies have provided the intellectual fuel for the AI Renaissance. The development of deep learning, a subset of machine learning inspired by the neural networks of the human brain, has been particularly transformative. Deep learning models, such as convolutional neural networks (CNNs) and recurrent neural networks (RNNs), have excelled in tasks such as image and speech recognition, outperforming traditional machine learning models and even human benchmarks in some cases.

Additionally, the rise of reinforcement learning, where AI systems learn to make decisions by interacting with their environment and receiving feedback, has opened up new possibilities for AI, from mastering complex games like Go and Poker to developing self-driving cars.

The synergistic convergence of these factors - advanced computing power, abundant data, and machine learning breakthroughs - has propelled us into the AI Renaissance, a period marked by rapid advancements and expansive applications of AI. As we navigate this era, it becomes imperative to acknowledge and address the ethical and societal implications posed by AI, to ensure that this powerful technology is harnessed responsibly for the betterment of humanity.

Yet, even as we celebrate the remarkable feats achieved during the AI Renaissance, we must be cognizant of the challenges that lie ahead. The quest for Strong AI, or Artificial General Intelligence, remains an ambitious and unresolved pursuit. As AI continues its onward march, the journey promises to be as intriguing and illuminating as the destination itself. The AI Renaissance, thus, represents not an end, but a new and exciting chapter in the ongoing saga of human ingenuity and invention.

# Conclusion: Embracing the AI Journey

As we stand on the precipice of this enlightening exploration of Artificial Intelligence (AI), we can reflect on the journey thus far, traversing the core concepts, the fascinating history, and the transformative potential of this cutting-edge field. The labyrinthine world of AI, with its complex terminologies and paradigms, its exhilarating heights of innovation, and its sobering periods of stagnation, reveals itself as an ever-evolving narrative of human ingenuity and aspiration.

We have delved into the foundational pillars of AI, scrutinized the ingenious work of its early visionaries, navigated the chilling corridors of the AI Winter, and reveled in the radiant dawn of the AI Renaissance. We have striven to comprehend the intricate tapestry of Machine Learning, Neural Networks, Natural Language Processing, and Computer Vision. In doing so, we have opened our minds to the infinite possibilities that these technologies represent.

AI is not merely a collection of algorithms and data. It is a reflection of our endeavor to decode the essence of intelligence, a testament to our ambition to mold a future that is as awe-inspiring as it is fraught with challenges. Its applications span across industries and disciplines, from E-commerce and Healthcare to Banking and

Finance, illuminating the profound potential of AI to revolutionize the fabric of our society.

But our journey does not end here. The world of AI is dynamic, with novel developments and discoveries continually reshaping our understanding of what is possible. As we progress through subsequent chapters, we will delve deeper into the intricate maze of AI, exploring its profound impact, grappling with its ethical implications, and envisioning its future possibilities.

The AI journey is a mirror reflecting our curiosity, our creativity, and our commitment to progress. As we continue our voyage into the heart of this exciting frontier, let us embrace the promise, the challenges, and the transformative power of AI. As we immerse ourselves in this captivating exploration, we become not just observers, but active participants in the thrilling narrative of AI, shaping and being shaped by the trajectory of this technological marvel.

# Chapter 2: The Significance of AI in the Modern World

The advent of the 21st century has witnessed the proliferation of an intricate digital ecosystem, a world that thrives on information, innovation, and interconnectivity. At the nexus of this revolutionary transformation stands the commanding figure of Artificial Intelligence (AI). As we advance into this chapter, our exploration delves into the profound implications of AI, articulating its transformative potential and its multifaceted role in shaping our modern world.

### The New Industrial Revolution

Often lauded as the catalyst for the fourth industrial revolution, AI weaves a compelling narrative of technological advancement and societal transformation. Just as steam power, electricity, and digitization altered the trajectory of human progress in the previous revolutions, AI, with its inherent potential to emulate and even surpass human cognitive abilities, signals an unprecedented paradigm shift. This metamorphosis is not confined to a single sector or discipline but permeates the breadth and depth of human enterprise, from industry to academia, from healthcare to finance, from e-commerce to entertainment.

Take, for example, the realm of e-commerce. AI-powered systems are progressively becoming the backbone of this industry, driving personalized customer experiences, streamlining supply chains, and enabling sophisticated predictive analytics. Whether it's a chatbot assisting you with a purchase or an algorithm predicting your shopping preferences, AI's footprint is distinctly visible.

In the sphere of healthcare, AI's application extends beyond administrative functions to critical areas such as disease detection, drug discovery, and personalized medicine. AI tools that can analyze complex medical data to predict health risks or recommend treatments herald a new era of precision medicine, radically transforming patient care and outcomes.

### The Evolution of the Workforce

As AI continues to evolve, it is not only transforming industries but also redefining the nature of work and the workforce. AI's capacity to automate routine tasks is triggering a shift in labor dynamics, necessitating a fresh perspective on employment and skills development. On one hand, it is fostering a demand for professionals adept in AI and related technologies. On the other hand, it compels us to reimagine roles and responsibilities in a world where AI systems coexist with humans in the workplace. Thus, the story of AI is also a narrative of human adaptation and resilience, a testament to our ability to harness change for progress.

Privacy, Ethics, and Governance

As AI permeates our lives, it inevitably brings to the forefront complex issues related to privacy, ethics, and governance. AI systems, dependent on vast quantities of data for their functioning, evoke legitimate concerns about data privacy and security. Furthermore, decisions made by AI algorithms, while being efficient, can potentially be biased or opaque, raising ethical dilemmas. These challenges necessitate robust frameworks for AI governance, emphasizing transparency, accountability, and inclusivity. Therefore, the journey of AI is not merely a technological voyage but also a philosophical and moral quest, a delicate balancing act between leveraging AI's benefits and safeguarding societal values.

The Road Ahead

As we delve deeper into the significance of AI in our modern world, we will explore its manifold facets and nuances. Each narrative thread of AI - its transformative impact, its role in workforce evolution, its ethical implications - offers a unique lens to understand this technological marvel. But the AI tapestry is far from complete. The road ahead is filled with challenges to be surmounted and opportunities to be seized. As we navigate the contours of the AI landscape, we are, in essence, shaping our collective future, a future interwoven with the limitless potential of Artificial Intelligence.

This exploration encourages us to not just comprehend AI but to engage with it critically, to question, to reflect, and to participate in the ongoing discourse. As we venture further into this exciting realm, we extend an invitation to the reader to join us in this fascinating journey, a journey that holds the promise of unravelling the expansive vistas of AI and its significant role in the canvas of our modern world.

## AI's Pervasive Influence: Highlighting how AI has become an integral part of everyday life, from smartphones to virtual assistants.

One of the most striking aspects of the modern world is the ubiquity of Artificial Intelligence (AI). It has stealthily permeated the fabric of our everyday lives, casting an influence that ranges from the subtle to the transformative. From the virtual assistants that organize our schedules to the algorithms that curate our digital content, from the smartphones that accompany us throughout the day to the smart homes that optimize our comfort and safety, AI's imprint is everywhere.

## Smartphones: The Everyday Companion

In the context of the smartphone, the pervasive influence of Artificial Intelligence is undeniably profound. These devices, which have transitioned from being a luxury to an indispensable part of our lives, are a testament to the transformative power of AI. The sophisticated capabilities that we now associate with smartphones - from voice recognition and image analysis to predictive text input - are the result of years of advancement in AI technologies.

One of the most noteworthy of these AI-driven features is voice recognition. This technology, bolstered by AI's ability to understand and interpret human speech, allows users to interact with their devices using their voice. Tasks that once required tactile interaction, such as dialing a phone number or typing a search query, can now be executed through voice commands, making the user experience more intuitive and hands-free.

Complementing voice recognition is image analysis or computer vision. AI algorithms enable smartphones to recognize and interpret visual data. For instance, when you point your smartphone camera at a landmark, AI can identify it and provide relevant information. Similarly, in social media apps, image analysis is used for features like tagging friends in photos or applying filters that can detect and enhance human faces.

Predictive text input, another AI-driven feature, has revolutionized the way we communicate using smartphones. AI learns from your typing habits and vocabulary to predict the next word in a sentence, speeding up text input and making it more efficient. Furthermore, this technology can learn and adapt to different languages and colloquial phrases, making digital communication more personalized and efficient.

The integration of AI into smartphones has also given rise to virtual personal assistants, such as Apple's Siri, Google's Assistant, and Amazon's Alexa. These assistants are powered by sophisticated AI technologies, notably natural language processing (NLP) and machine learning. NLP enables these assistants to understand, interpret, and generate human language, facilitating interactive and natural communication. Machine learning, on the other hand, allows these assistants to learn from user interactions and improve their performance over time, making them more accurate and personalized.

Virtual assistants, owing to these underlying AI technologies, have transformed the way we interact with our smartphones. They are capable of setting reminders, providing real-time traffic updates, answering trivia, controlling smart home devices, and much more. Through their versatility and intelligence, virtual assistants have evolved from being a novelty feature to a productivity tool that shapes our interactions with technology.

Thus, in the realm of smartphones, AI has shifted from being a supporting technology to a central one. It has not only enhanced existing features but has also facilitated new forms of interaction and functionality, redefining the smartphone experience and setting the stage for future innovations.

Virtual Assistants: The Digital Aide

The reach of Artificial Intelligence extends far beyond the realm of smartphones and personal devices. One of the most significant manifestations of this expansion is the prevalence of standalone virtual assistants that have started to pervade our homes and workplaces. These devices, such as Amazon Echo or Google Home, represent a noteworthy evolution of the AI technologies that have revolutionized the smartphone experience.

These standalone virtual assistants are the embodiment of a new era of AI-driven home automation. Echoing the intelligence and versatility of their smartphone counterparts, these devices integrate seamlessly into everyday life, transforming routine tasks into convenient, voice-controlled activities. Whether it is streaming music, making phone calls, setting alarms, or managing to-do lists, these tasks can now be accomplished with a simple voice command, thereby streamlining our interactions with technology and enhancing our productivity.

Beyond mere task execution, these virtual assistants leverage the power of AI to offer more personalized and predictive experiences. Through the continuous analysis of user behavior and preferences, these devices can tailor their responses and actions to individual users. For instance, a virtual assistant can suggest a playlist based on your music preferences or recommend a recipe based on your dietary habits. They can remind you of a meeting based on your calendar events or alert you about potential traffic on your regular commute. In essence, they can anticipate your needs and adapt their functionality accordingly, offering a level of personalization that was previously unattainable.

Moreover, these virtual assistants are increasingly being integrated with other smart devices, contributing to the creation of fully connected smart homes. They can control lighting systems, regulate thermostats, manage security systems, and even communicate with smart appliances. As such, they are evolving into central control units for smart homes, underlining AI's pivotal role in the future of home automation.

Thus, standalone virtual assistants represent a significant stride in AI's journey, demonstrating the technology's potential to transcend device boundaries and become an integral part of our living spaces. They are transforming our interaction with technology, from a tool that responds to commands to an intelligent aide that understands, anticipates, and complements our lifestyle. This pervasive influence of AI marks a significant shift in how technology is integrated into our lives, setting the stage for future advancements that will continue to reshape our experiences in unprecedented ways.

## Home Automation: The AI-enabled Habitat

The burgeoning domain of home automation stands as a testament to the pervasiveness of Artificial Intelligence. An idea that was once confined to the realm of science fiction, the concept of smart homes, is today's reality, thanks largely to advancements in AI. From regulating ambient conditions to enhancing security, and from optimizing energy consumption to automating mundane chores, AI systems have introduced a new dimension of convenience, safety, and efficiency in our homes.

Consider, for example, the implementation of AI in managing lighting and temperature. Intelligent systems can adjust these parameters based on various factors such as time of the day, occupancy, and even individual preferences, thereby ensuring a comfortable living environment while simultaneously promoting energy efficiency. Further, advanced security systems, employing AI, can detect unusual activity or recognize familiar faces, providing enhanced security while reducing false alarms.

The narrative of AI's impact on home automation, however, extends beyond just comfort and security. A new generation of AI-powered home appliances is changing the way we accomplish routine tasks. Take, for instance, robotic vacuum cleaners. These devices leverage AI to map and navigate homes, adapt to different floor types, and even return to their docking station for recharge when low on battery. They can be scheduled to clean at specific times and can avoid obstacles, making them a reliable tool for maintaining cleanliness with minimal human intervention.

Another significant development is the advent of smart refrigerators. These sophisticated appliances can monitor their contents, alert users when they are running low on certain items, suggest recipes based on available ingredients, and even order groceries online. Through AI, they are able to learn user habits, anticipate needs, and provide predictive maintenance alerts, ensuring optimal performance and longevity.

These examples of robotic vacuum cleaners and smart refrigerators represent just the tip of the iceberg. From AI-powered washing machines that can optimize washing cycles, to smart ovens that can suggest cooking times based on the type of food being cooked, the possibilities are seemingly endless.

By automating routine tasks and offering personalized, predictive features, these appliances are not just enhancing convenience, but also freeing up valuable time for users, thereby improving their quality of life. They underscore the power of AI to transform ordinary homes into intelligent spaces that are attuned to the needs of their inhabitants, and this is a reality that is no longer a distant future but an unfolding present. As we continue to unlock AI's potential, it is certain that we will witness further advancements that will redefine the concept of home and lifestyle in profound ways.

## In Closing

In the contours of our contemporary existence, Artificial Intelligence stands as a ubiquitous presence, akin to an unseen yet crucial architect that is constantly shaping, enhancing, and transforming our experiences. Be it the smartphones that have become extensions of our personalities or the virtual assistants that have become our companions, AI's imprint on our lives is both profound and wide-ranging. It lends sophistication to our tasks, intuitively personalizes our interactions, and adds an unprecedented layer of convenience, weaving an intricate and indelible pattern of influence in the fabric of our daily lives.

Consider, for example, how seamlessly AI-enhanced features have become intertwined with our smartphone usage. Predictive text input, voice recognition, and image analysis have transformed from being novelties to necessities, augmenting our communications, productivity, and creativity. Moreover, the advent of personal assistants such as Siri, Google Assistant, and Alexa have taken this interplay of AI and smartphones a step further, blurring the boundaries between human and technological interactions and creating a more intuitive, personalized, and efficient interface with technology.

Our homes and workplaces, too, have witnessed the transformative touch of AI. Smart devices and appliances, equipped with AI capabilities, have started to infuse intelligence into our living and working spaces. Our commands are heard and executed, our preferences are learned and catered to, and our needs are anticipated and fulfilled, all facilitated by AI's invisible hand. This ongoing evolution of our immediate environments is more than just a testament to technological progress. It is a mirror reflecting how AI, as a pervasive force, is redefining the way we live, work, and interact with the world.

Yet, the current manifestation of AI, as transformative as it is, represents merely the nascent stages of its potential. As AI continues to evolve, driven by relentless research and ceaseless innovation, its integration into our lives will deepen, its applications will broaden, and its influence will amplify. AI is poised to transcend its role as an enabling technology and become an integral part of our social, economic, and cultural fabric.

In the realm of healthcare, for instance, we can envision AI-powered systems not only diagnosing diseases and prescribing treatments but also predicting health risks and promoting preventive care. In education, AI could transcend geographical barriers, personalizing learning experiences for each student, and even identifying and nurturing their unique talents. The world of commerce could see AI creating hyper-personalized shopping experiences, predicting market trends, and revolutionizing supply chains. The possibilities are as myriad as they are tantalizing.

Thus, as we stand on the cusp of this AI-enhanced future, it is incumbent upon us to foster a deeper understanding of this transformative technology, to engage with it critically and ethically, and to harness its potential in a manner that augments our collective good. The journey ahead with AI promises to be as exciting as it is transformative, and we are but explorers on this path, limited only by the boundaries of our imagination.

## AI in Business: Exploring the role of AI in modern enterprises, from startups to multinational corporations.

Artificial Intelligence, with its transformative capabilities and unparalleled potential, has emerged as a defining force in the realm of business. It has insinuated itself into the very foundation of modern enterprises, acting as a cornerstone for innovation, driving growth, and redefining business landscapes across the spectrum, from fledgling startups to multinational corporations. The interplay of AI and business, marked by disruption and evolution, opens up new vistas of possibilities and engenders unprecedented challenges, shaping a dynamic and compelling narrative worth examining.

To begin with, consider the realm of customer interactions, a critical area for any business. AI-driven technologies such as chatbots and virtual assistants are revolutionizing this sphere. They handle customer queries in real-time, ensure round-the-clock availability, provide personalized assistance, and learn from each interaction to improve future responses. Companies like Amazon and Netflix leverage AI to offer personalized product recommendations based on customer behavior and preferences, thereby enhancing the user experience and boosting sales.

In the sphere of operations and supply chain management, AI offers innovative solutions that enhance efficiency and productivity. Predictive analytics, underpinned by machine learning, allows businesses to forecast demand with higher accuracy, optimize inventory levels, and improve supply chain efficiency. Tech giant IBM, for instance, has been using its AI platform Watson to predict demand for products and optimize its supply chain, thereby reducing costs and improving customer service.

Another arena where AI's impact is significant is in strategic decision-making. Business intelligence tools powered by AI algorithms analyze vast amounts of data to unearth valuable insights, trends, and patterns that assist executives in making informed and strategic decisions. These insights are invaluable in enhancing competitive advantage and driving business growth.

Furthermore, in the finance sector, the influence of AI is pervasive. Robo-advisors, powered by AI, have democratized access to financial advice, making it affordable and accessible. High-frequency trading firms use AI to analyze market trends and execute trades at lightning speeds, significantly outpacing human traders.

AI's influence extends to the realm of human resources as well. AI-driven platforms help automate repetitive tasks such as resume screening, skill assessment, and initial rounds of interviews. They also provide data-driven insights to inform talent management and employee engagement strategies, thereby promoting a more productive and engaged workforce.

Even as we celebrate AI's contributions to business, it is imperative to engage with its challenges and implications critically. As AI continues to automate tasks, the question of job displacement comes to the fore. Businesses need to navigate this complex issue responsibly, ensuring that the transition towards AI-driven operations is just and inclusive.

Moreover, as businesses leverage AI to handle sensitive data, issues of privacy and data security gain prominence. Regulatory frameworks need to evolve in tandem with technological advancements to ensure data integrity and privacy.

Despite these challenges, the integration of AI in business represents a compelling opportunity. As we look towards the future, AI's role in reshaping business landscapes promises to be transformative and disruptive in equal measure. As businesses adapt to this AI-driven paradigm, the potential for innovation, efficiency, and growth is unprecedented.

In essence, AI's symbiosis with business is a testament to its transformative potential and a glimpse into a future where this interplay could redefine the very contours of commerce and enterprise. For startups and corporations alike, embracing AI signifies not only a commitment to innovation and growth but also a willingness to navigate the complex challenges it presents, marking the beginning of an exciting and transformative journey in the world of business.

## Impact on the Workforce: Discussing how AI is reshaping the job market and workforce dynamics.

Artificial Intelligence, in its relentless march towards progress and innovation, is transforming the fundamental structures of the workforce and the job market. The rise of AI has catalyzed a paradigm shift in employment dynamics, heralding a future marked by automation, enhanced productivity, and evolving job roles. This transformative journey, however, is not without its challenges. The discourse around AI's impact on the workforce necessitates a nuanced understanding of both its potential to create new opportunities and its implications for job displacement and skills realignment.

The notion of AI-powered automation, while not entirely new, has accelerated in recent years. From manufacturing and logistics to customer service and data analysis, AI-driven automation is revolutionizing various sectors. Companies like Amazon are deploying robots to automate warehouse operations, thereby significantly increasing efficiency and reducing the scope of human error.

Simultaneously, AI-driven software and algorithms are performing complex tasks such as data analysis, which previously required human intervention. This has resulted in enhanced productivity, more accurate predictions, and improved decision-making capabilities within organizations.

AI's automation capabilities, however, raise concerns about job displacement. The fear that machines might render certain jobs obsolete is not unfounded. McKinsey Global Institute's report suggests that by 2030, between 400 million and 800 million individuals could be displaced by automation and would need to find new jobs.

While this narrative might seem alarming, it is crucial to delve deeper and recognize that the impact of AI on the workforce is not a zero-sum game. History offers reassurance that technological advancements, while displacing certain jobs, often create new ones. The invention of the automobile, for instance, ended many jobs tied to horse-drawn carriages but created far more in the automobile industry.

In this context, AI is expected to spawn an array of new job roles and opportunities. Roles such as AI Specialist, Data Scientist, and Machine Learning Engineer are increasingly in demand, commanding attractive compensation and offering significant growth prospects. In fact, LinkedIn's 2020 Emerging Jobs Report identified artificial intelligence as the top emerging job category.

AI also holds the promise of enhancing job quality by automating mundane and repetitive tasks, thereby freeing up workers to focus on more complex and creative aspects of their work. This shift towards higher-value tasks can lead to more engaging and fulfilling job roles and can potentially boost job satisfaction and productivity.

However, the transition to an AI-driven job landscape is not automatic and demands a concerted effort towards reskilling and upskilling the workforce. Governments, educational institutions, and businesses have a shared responsibility to create a robust ecosystem for lifelong learning and skills development.

For instance, businesses could invest in training programs to help their workforce acquire AI-related skills. Governments could incentivize such initiatives and could also reform education systems to incorporate AI and other relevant technologies into curricula. This would help prepare the younger generations for the AI-dominated job market of the future.

Moreover, attention should also be given to those at the risk of displacement, particularly workers in lower-skilled jobs. Adequate social protection measures, robust retraining programs, and policy interventions would be crucial to ensure a just transition to the new AI-driven job landscape.

In conclusion, AI's impact on the workforce presents a complex narrative of opportunities and challenges. While AI holds immense potential to revolutionize the job market and create new avenues of employment, it also demands a proactive approach towards managing displacement risks and facilitating skills realignment. As we navigate this transformative phase, a balanced, inclusive, and forward-thinking approach would be instrumental in harnessing AI's potential while mitigating its challenges, paving the way for a future where AI and human ingenuity coexist and thrive.

# Chapter 3: Unleashing the Potential: How AI Transforms Industries

The dawn of the twenty-first century heralded an age of unprecedented technological innovation and progress, characterized by the rapid advancement of digital technologies and, more notably, the rise of Artificial Intelligence (AI). This profound technological revolution, powered by AI, is permeating every facet of society, economy, and industry, engendering a paradigm shift in the way we live, work, and interact with the world around us.

AI's ascendancy is not an isolated phenomenon; rather, it is part of a broader tapestry of digital transformation sweeping across industries, reshaping traditional business models, and fostering new forms of value creation. The pervasive influence of AI transcends the confines of technology-centric sectors and permeates a myriad of industries, from e-commerce and healthcare to banking and finance.

In the realm of e-commerce, AI is propelling a new era of personalized shopping experiences, harnessing the power of data analysis to offer tailored product recommendations, optimize pricing strategies, and streamline supply chains. From predictive algorithms that anticipate customer preferences to chatbots that enhance customer engagement, AI's transformative role in e-commerce is palpable.

In the healthcare sector, AI's impact is equally profound, marking a seismic shift in patient care, diagnostics, and therapeutics. The advent of AI-powered technologies such as predictive analytics, machine learning algorithms, and computer vision, among others, is enabling early disease detection, personalized treatment plans, and enhanced patient outcomes, thus setting the stage for a future where healthcare is not only reactive but also proactive and personalized.

Similarly, in banking and finance, AI's role is transformative, ushering in a new age of financial services marked by automation, personalization, and enhanced risk management. AI-powered applications such as robo-advisors, fraud detection algorithms, and automated customer service platforms are transforming the financial landscape, offering unparalleled convenience, security, and efficiency.

Yet, as we delve into the transformative potential of AI, it is imperative to acknowledge the challenges and ethical considerations this potent technology presents. The implications of AI for data privacy, job displacement, and social inequality necessitate rigorous scrutiny and thoughtful deliberation. Balancing the promise of AI-driven progress with the need for equitable, responsible, and ethical AI applications is a critical task that lies before policymakers, business leaders, and society at large.

In this chapter, we shall explore the transformative potential of AI across various industries, highlighting not only the opportunities it presents but also the challenges it poses. Through this analytical and balanced discourse, we aim to provide a comprehensive understanding of AI's industry-transforming capabilities, as well as the strategies required to harness this potential responsibly and ethically.

As we embark on this journey, we invite you to delve into the fascinating world of AI and its industry-transforming potential, exploring its manifold applications, understanding its challenges, and envisaging a future shaped by the intelligent and responsible use of this revolutionary technology.

## AI in E-commerce: Examining the use of AI-powered recommendation engines, personalized marketing, and supply chain optimization in the e-commerce sector.

Artificial Intelligence (AI) has increasingly become a powerful force within the realm of e-commerce, fundamentally reshaping the landscape of online retail. The synthesis of vast amounts of data with sophisticated AI algorithms has unlocked a new era of e-commerce characterized by personalization, efficiency, and enhanced user engagement. Central to this transformation are AI-powered recommendation engines, personalized marketing strategies, and supply chain optimization.

Recommendation engines are among the most prevalent and impactful manifestations of AI within e-commerce. These sophisticated algorithms sift through vast troves of user data – including browsing history, purchase history, and customer demographics – to generate personalized product recommendations. Amazon's recommendation engine, for instance, is famously responsible for a substantial proportion of the company's sales. This level of personalization enhances the customer experience by providing relevant product suggestions, which in turn drives sales and increases customer loyalty. In essence, recommendation engines serve as a prime example of how AI can create a win-win scenario: enhancing customer satisfaction while concurrently driving business growth.

Personalized marketing is another potent application of AI in e-commerce. Leveraging AI's capability to analyze and learn from large volumes of data, businesses can deliver tailored marketing messages and offers that resonate with individual customers' preferences and behavior. For instance, AI can be used to predict which customers are most likely to respond to specific promotional offers or which products a customer may be interested in based on their past behavior. This level of personalization not only improves the effectiveness of marketing campaigns but also enhances the overall customer experience by making marketing communication more relevant and less intrusive.

AI's transformative influence extends to the realm of supply chain management, where it is being leveraged to optimize logistics and streamline operations. Through machine learning algorithms, AI can analyze vast amounts of supply chain data to predict trends, identify potential disruptions, and make recommendations for optimization. For instance, AI can be used to predict demand for specific products, allowing businesses to adjust their inventory levels accordingly, thereby reducing the cost of holding excess inventory and minimizing the risk of stockouts. Furthermore, AI can also assist in identifying the most efficient delivery routes, thus reducing shipping times and costs.

In summary, AI is exerting a transformative influence on the e-commerce sector, reshaping the ways businesses engage with customers, market their products, and manage their supply chains. Through AI-powered recommendation engines, personalized marketing, and supply chain optimization, businesses are not only enhancing their operational efficiency but also providing a more personalized and satisfying customer experience. However, as businesses increasingly harness the power of AI, they must also grapple with related ethical considerations, such as data privacy and security, ensuring they responsibly manage and protect the vast amounts of customer data they collect. As we move forward, it will be fascinating to observe how the relationship between AI and e-commerce continues to evolve and shape the future of online retail.

## Revolutionizing Healthcare: Investigating AI's impact on medical imaging, diagnostics, drug discovery, and telemedicine, among other aspects of healthcare.

The advent of artificial intelligence (AI) has ushered in a new era in many fields, none more so than healthcare. With its potential for enhancing accuracy, efficiency, and personalized care, AI is poised to revolutionize several facets of the healthcare industry, spanning from medical imaging and diagnostics to drug discovery and telemedicine.

Let's first turn our attention to medical imaging, an essential component of diagnostics. The application of AI, particularly deep learning, has proven beneficial in this domain. Deep learning, a subset of AI, utilizes complex neural networks to mimic human intelligence. AI algorithms can be trained to recognize patterns and abnormalities within medical images, thereby assisting in early detection of diseases. For instance, AI is being leveraged to improve the accuracy and speed of interpreting mammograms, CT scans, and other types of medical images, thereby enhancing the detection of cancers and other diseases at their earliest stages when they are more treatable.

AI also has a profound impact on diagnostics. AI algorithms can comb through extensive databases of patient data to detect patterns that might elude human clinicians. IBM's Watson, for example, has been employed in oncology to recommend treatment plans by analyzing patient data against a multitude of oncology literature. The potential here is not only to augment physicians' diagnostic accuracy but also to significantly reduce diagnostic times.

In the realm of drug discovery, AI provides a paradigm shift. Traditionally, discovering a new drug is a time-consuming and costly endeavor, often taking years, if not decades, and billions of dollars to accomplish. AI has the potential to expedite this process significantly. By analyzing large sets of biochemical data, AI can predict how different compounds will behave and how likely they are to make an effective drug, allowing researchers to focus on the most promising candidates. This innovative approach can significantly shorten the drug discovery timeline, thereby bringing essential medications to patients more swiftly and cost-effectively.

Moreover, the surge of telemedicine, particularly in the wake of the COVID-19 pandemic, has opened new frontiers for AI application. AI chatbots, for instance, can handle initial patient screenings, gather preliminary information, and guide patients to the appropriate care, thereby increasing the healthcare system's efficiency. AI's role in remote patient monitoring is also burgeoning. Algorithms can analyze data from wearable devices to predict potential health crises before they occur, allowing for timely intervention.

However, it is crucial to acknowledge that despite these potential benefits, the application of AI in healthcare is not without challenges and ethical considerations. Questions about data privacy, the potential for AI to reinforce existing healthcare disparities, and the necessity for explainability and transparency in AI decision-making processes are all pressing issues that must be addressed as we navigate the future of AI in healthcare.

In conclusion, the incorporation of AI into healthcare has shown immense promise, with its ability to enhance diagnostic accuracy, expedite drug discovery, personalize treatments, and optimize healthcare delivery. As we continue to explore and refine AI's capabilities, we can anticipate a future where healthcare is increasingly predictive, personalized, and accessible. But as we move forward, we must also ensure that this is done in a way that is ethically sound, inclusive, and transparent. The potential of AI in healthcare is enormous, but it is up to us to harness it responsibly.

# AI in Banking and Finance: Discussing AI's role in fraud detection, risk assessment, customer service, and algorithmic trading, driving innovation in the financial industry.

Artificial Intelligence (AI) has pervaded numerous sectors, with the financial industry being a prominent adopter of this transformative technology. It is utilized in a multitude of applications, ranging from fraud detection and risk assessment to enhancing customer service and facilitating algorithmic trading. This section explores these myriad applications and discusses how AI is driving unprecedented innovation in the banking and finance landscape.

Starting with fraud detection, the advent of AI has been nothing short of a game-changer. Traditional fraud detection systems relied on set rules and heuristics, which were not always effective in identifying sophisticated fraudulent patterns. AI, with its ability to learn from large datasets and identify anomalous behavior, has significantly enhanced the efficiency and effectiveness of fraud detection systems. Machine learning algorithms, a subset of AI, can analyze patterns across a vast number of transactions in real-time, flagging any transactions that deviate from the norm. This not only enhances accuracy but also reduces the number of false positives, leading to more efficient fraud management systems.

Risk assessment is another critical area in the banking and finance industry where AI has made substantial inroads. Credit scoring, for instance, is an essential component of risk management. AI algorithms can analyze vast and diverse datasets, including social media activity and online transactions, to predict a customer's creditworthiness more accurately. Furthermore, AI is increasingly being used to assess market risks, analyzing vast quantities of data from various sources to predict market trends and identify potential risks more accurately and swiftly than traditional methods.

When it comes to customer service, AI's role is becoming increasingly prominent. AI-powered chatbots and virtual assistants have become common features on banking websites and apps, providing 24/7 customer support, answering queries, and performing simple tasks such as balance inquiries and money transfers. By automating these tasks, banks can provide a higher level of service more efficiently.

Algorithmic trading is another application of AI that is revolutionizing the financial industry. This involves the use of AI algorithms to make trading decisions at speeds and frequencies that are beyond human capabilities. These algorithms analyze multiple market conditions in real-time, making trades based on predefined instructions, and adapting to market changes. This can lead to more profitable trading outcomes and a more efficient market.

However, while AI offers numerous advantages, it is not without challenges. One of the most prominent is the 'black box' problem, which refers to the lack of transparency in AI decision-making processes. This is particularly concerning in the financial industry, where decisions can have substantial economic implications. Further challenges include data privacy issues and the risk of AI systems being targeted by cybercriminals.

In conclusion, AI is undeniably transforming the banking and finance industry, offering improved efficiency, accuracy, and personalized services. However, as we continue to navigate this brave new world of AI-driven finance, it is critical to address the associated challenges head-on, ensuring that this powerful technology is used responsibly and ethically. The potential of AI in banking and finance is immense, but it is the onus of us, the users and developers, to harness it in a way that brings the maximum benefit to all stakeholders.

# Blockchain and AI Integration: Exploring the convergence of blockchain technology and AI, with the potential to disrupt various industries beyond finance.

The fusion of Artificial Intelligence (AI) and blockchain represents a profound development in the realm of technology. While each has independently brought about significant changes across various industries, their integration promises to create a confluence of enhanced trust, transparency, and automated intelligence capable of disrupting many sectors beyond finance. The synergy between AI and blockchain has the potential to revamp the data economy, improve decision-making processes, and heighten security measures, among other benefits.

At the heart of the interplay between AI and blockchain is the foundational premise of each technology. Blockchain provides a transparent and immutable record of transactions, enabling an unprecedented level of data accuracy and security. AI, on the other hand, thrives on vast datasets, employing intricate algorithms to learn from the data, identify patterns, and make predictions or decisions. The merging of these two technologies, therefore, not only amplifies the capabilities of each but also addresses their individual limitations.

Blockchain, for instance, can mitigate the 'black box' problem inherent to many AI systems. By maintaining a transparent record of an AI's decision-making process, blockchain can provide valuable insights into why a certain decision was made, thereby increasing the system's transparency and fostering greater trust. This feature could be particularly impactful in sectors like healthcare, where understanding an AI's decision-making process could improve diagnosis and treatment strategies.

On the other hand, AI can address one of the most significant challenges faced by blockchain - the issue of scalability. AI algorithms can be used to streamline the validation process of transactions, significantly reducing the time and computational resources required. Moreover, AI can enhance the security of a blockchain network by identifying and countering potential threats more effectively and efficiently than traditional security measures.

Beyond these direct advantages, the convergence of AI and blockchain can potentially lead to a shift in how data is managed and valued across sectors. As data becomes increasingly important, ensuring its integrity, privacy, and usability is paramount. Blockchain can provide the foundation for data integrity and privacy, while AI can unlock its value through advanced analytics and predictive capabilities. This fusion can lead to the creation of a secure, transparent, and value-driven data economy that can benefit industries ranging from advertising to healthcare, supply chain management to governance.

Consider, for example, a supply chain management system powered by AI and blockchain. The blockchain could provide a transparent and immutable record of each product's journey from production to delivery, while an AI system could analyze this data in real-time to identify inefficiencies, predict potential disruptions, and recommend improvements. This would not only increase efficiency but also improve accountability and sustainability.

Despite the promising potential, the integration of AI and blockchain is not without its challenges. These include technical issues related to interoperability and standardization, regulatory uncertainties, and ethical considerations surrounding data privacy and AI decision-making. As these technologies continue to evolve, it is incumbent upon policymakers, technologists, and society at large to address these challenges in a balanced and forward-looking manner.

In conclusion, the convergence of blockchain and AI heralds a new era of technological innovation with the potential to transform various industries beyond finance. By bringing together blockchain's transparency and immutability with AI's learning and predictive capabilities, this integration can unlock new opportunities, drive efficiencies, and create a more secure and value-driven data economy. However, realizing this potential requires a thoughtful approach that addresses the associated challenges and leverages the unique strengths of each technology.

# Chapter 4: The Disruptive Force of AI

The phenomenon of artificial intelligence (AI) is not a new concept. Its theoretical foundations were laid down by the seminal work of Alan Turing and others in the mid-twentieth century, with the vision of creating an artificial brain. Turing's famous question, "Can machines think?", set in motion an intellectual journey, which has led us to the current age of AI disruption.

However, AI, as we know it today, is a product of developments that have taken place over the last few decades. It is the culmination of advancements in machine learning, computational power, data storage, and analytics capabilities, which have fueled AI's transition from a theoretical concept to a transformative technology. The narrative of AI's journey is filled with peaks of inflated expectations and troughs of disillusionment. Nevertheless, the past decade has been marked by an explosion of interest, investment, and deployment of AI, making it a cornerstone of modern technological innovation.

The impact of AI extends beyond mere augmentation of computational abilities. It is a disruptive force that is reshaping economies, altering industry dynamics, and transforming social structures. From autonomous vehicles disrupting transportation and logistics to predictive analytics redefining healthcare, the influence of AI has been all-encompassing. However, disruption, in this context, is not merely a destructive force. It paves the way for innovation, engenders new business models, and spurs societal progression.

The vast scope of AI's application has given rise to its disruptive potential. The ability of AI algorithms to parse through voluminous data sets, recognize patterns, learn from experiences, and make informed decisions positions it as a game-changer in a myriad of sectors. Industries that were traditionally seen as manual or human-intensive are now witnessing a paradigm shift with the incorporation of AI capabilities.

However, as with any profound technological shift, the advent of AI is also fraught with challenges and concerns. It raises pertinent questions about data privacy, job displacement, security, and ethical implications, all of which require careful examination and robust policy interventions.

This chapter aims to traverse the disruptive journey of AI, explore its implications across various sectors, and delve into the challenges it poses. It is intended to provide a balanced perspective on AI disruption, presenting an amalgamation of its awe-inspiring potential and the cautionary tales that we must heed to navigate our AI-driven future responsibly and effectively. As we delve into this exploration, it is important to keep in mind that AI is not a panacea for all societal or economic challenges, nor is it an unbridled threat. It is a tool, whose utility or detriment is determined by the intention and wisdom behind its application.

# Disrupting Traditional Business Models

The digital age, marked by the pervasive use of technology and connectivity, has eroded the geographical and temporal boundaries that once defined business operations. Amid this landscape, artificial intelligence has emerged as a significant driving force, instigating a disruption in traditional business models. By enabling intelligent automation, enhancing customer experiences, and offering valuable insights through data analysis, AI has engendered a culture of innovation, adaptation, and competitiveness.

A. Impact on Various Industry Verticals

AI's influence has permeated various sectors, leading to the advent of novel business models. In the retail industry, for instance, the concept of physical stores is being increasingly supplanted by the rise of e-commerce, a transformation hastened by AI's capabilities in customer behavior prediction, personalized recommendations, and efficient supply chain management. Similarly, in the financial sector, AI's prowess in handling complex calculations and large datasets has expedited the shift towards algorithmic trading and automated financial advisory, heralding the era of 'fintech'.

The healthcare sector too is being reshaped by AI. From predictive diagnostics and precision medicine to robot-assisted surgeries, AI is revolutionizing the way healthcare is delivered, making it more accurate, personalized, and accessible. In the media industry, AI is augmenting content curation and delivery, tailoring it to the individual preferences of each user.

B. Emergence of Data as a Key Asset

The disruption induced by AI has underscored the importance of data as a key business asset. As businesses are gradually transitioning from a product-oriented to a service-oriented model, data has become the crux of value creation. Companies such as Google, Amazon, and Facebook, often referred to as 'data behemoths', are testament to the business potential of data.

C. Enabling Greater Customer Centricity

The application of AI has also fostered a shift from a product-centric to a customer-centric business paradigm. AI-powered systems enable businesses to analyze customer preferences, predict future behavior, and deliver personalized experiences, thus creating a more direct, enduring, and profitable relationship with customers.

D. Optimizing Operations and Enhancing Efficiency

AI and its related technologies have also brought about significant changes in internal business operations, promoting efficiency and cost-effectiveness. For instance, AI-powered process automation can perform routine tasks more quickly and accurately than humans, thereby reducing operational costs, minimizing errors, and freeing up human employees for more strategic tasks.

E. Challenges in the Path of Disruption

While AI's potential to disrupt traditional business models is evident, this journey is not devoid of challenges. For one, implementing AI necessitates significant investments in technology and skills. Additionally, the use of AI raises critical issues related to data security, privacy, and ethics, necessitating careful handling and robust governance frameworks.

Furthermore, the transition from traditional to AI-enabled models requires a cultural shift within organizations. Employees must be upskilled to work in tandem with AI, and there should be a willingness at the managerial and executive levels to embrace change and experiment with novel business approaches.

In conclusion, as AI continues to evolve and become more sophisticated, it is likely to instigate further disruptions in traditional business models, thereby fostering a culture of continuous adaptation and innovation. Businesses that manage to harness the power of AI, navigate its challenges, and align it with their strategic objectives will be well-positioned to thrive in the increasingly competitive business landscape.

**Definition of traditional business models and their limitations.**

A. Defining Traditional Business Models

Traditional business models represent structured strategies for an organization to deliver value to its customers and sustain itself financially. These models, primarily linear in their approach, have historically focused on the transactional exchange of goods and services for economic compensation. Some prevalent forms include brick-and-mortar retail models, subscription-based services, or commission-based structures, among others.

Traditional business models have been widely adopted due to their simplicity and proven track record. They revolve around tangible products or services and exhibit clear separation between different stages of the value chain – production, marketing, sales, and customer service.

B. Limitations of Traditional Business Models

While traditional business models have served businesses well for many decades, they have inherent limitations, particularly in the contemporary digital and data-driven age.

Limited Scalability: Traditional models often face scalability constraints due to high operational costs and logistical complexities. Physical businesses, for instance, must incur substantial expenditure for expanding their geographic reach.

Lack of Personalization: These models typically adopt a 'one-size-fits-all' approach, lacking the ability to tailor offerings to individual customer preferences. In today's consumer landscape, where personalization is key to customer retention, this can be a significant drawback.

Slow Adaptation to Market Changes: Traditional business models can be less agile in responding to market shifts. Their entrenched processes and systems may impede rapid adaptation to evolving consumer behavior, emerging technologies, or changes in the regulatory environment.

Insufficient Utilization of Data: In the era where data is the 'new oil', traditional business models often underutilize the potential of data. They may not have the requisite infrastructure or expertise to collect, process, and analyze large volumes of data for actionable insights.

C. Necessity for Transformation

These limitations necessitate a transformation towards more flexible, scalable, and data-driven business models. With the advent of advanced technologies like AI, businesses have the opportunity to redefine their models, making them more customer-centric, efficient, and adaptive to market changes. However, it is critical to approach this transformation thoughtfully, considering factors like technological readiness, organizational culture, market dynamics, and regulatory constraints.

# Case Studies: Examples of industries and businesses that AI has disrupted.

### A. E-commerce: Amazon's Unprecedented Personalization

The e-commerce titan, Amazon, is an epitome of AI's transformative power. It has fundamentally disrupted traditional retail models through its sophisticated use of AI. One of the keys to Amazon's extraordinary success has been its ability to personalize customer experiences. By leveraging machine learning algorithms, Amazon analyzes vast amounts of customer data to predict preferences, recommend products, and tailor its marketing strategies, thus dramatically boosting customer engagement and sales.

### B. Healthcare: Aidoc's AI-Powered Diagnostics

Moving to the healthcare sector, a paradigm shift is evident with AI's integration. Companies like Aidoc are at the forefront of this revolution. Aidoc uses advanced AI algorithms to analyze medical imaging data and detect critical conditions such as strokes, pulmonary embolisms, and cervical spine fractures in near real-time. The technology offers a clear demonstration of how AI can enhance diagnostics, improve patient outcomes, and reduce healthcare costs, disrupting the traditional diagnostic process.

### C. Banking and Finance: Ant Financial's AI-Driven Services

Ant Financial, an affiliate company of the Alibaba Group, has revolutionized the financial industry in China and beyond. Its AI-driven technology allows the company to assess credit risk, detect fraudulent transactions, and deliver personalized financial services at an unprecedented scale and speed. For instance, Ant's MYbank, an AI-powered online bank, has provided small and micro-sized businesses, which were previously underserved by traditional banking institutions, with much-needed access to credit.

### D. Manufacturing: Siemens' AI-Enhanced Processes

In the realm of manufacturing, Siemens provides a compelling example of AI-driven transformation. The company's AI systems analyze a plethora of data from machinery and equipment, predict maintenance needs, optimize production processes, and even design new products. Such AI-enabled predictive maintenance and process optimization significantly enhance operational efficiency and product quality, challenging conventional manufacturing practices.

Each of these cases demonstrates the disruptive potential of AI in divergent fields, providing a tangible understanding of how the technology can redefine traditional business models. As AI continues to evolve and permeate more sectors, such transformations are likely to become even more profound and widespread. This widespread disruption presents both an opportunity and a challenge for businesses, policymakers, and society at large, underscoring the need for proactive adaptation and thoughtful regulation.

**The role of AI in fostering a culture of innovation and adaptation.**

Artificial intelligence has acted as a catalyst, inspiring a culture of innovation and adaptation across industries. It is the engine behind a new era, one where conventional methods are giving way to intelligent systems that improve efficiency, accuracy, and profitability.

A. AI as an Innovation Catalyst

One of AI's most significant contributions to the modern business landscape is its role as an innovation catalyst. It has created opportunities for novel product and service offerings that were previously unimaginable.

For example, autonomous vehicles have emerged from the integration of AI with automotive engineering. Companies like Tesla and Waymo have pushed the boundaries of what is possible in transportation, creating vehicles that can navigate complex traffic situations with minimal or no human intervention. This innovation has the potential to redefine our transport systems, improving safety and efficiency.

In the healthcare industry, AI's transformative potential is also manifest. AI algorithms have facilitated the development of precision medicine, where treatments are tailored to individual patients based on their genetic makeup and other unique health factors. These innovations have not only improved patient outcomes but also challenged the one-size-fits-all approach of traditional medicine.

B. AI and Organizational Adaptation

Simultaneously, AI's disruptive force necessitates adaptation. Organizations must be agile in their operations and strategies to harness AI's benefits and mitigate potential downsides.

Organizations are increasingly investing in AI skills, either through recruitment or training programs, to build competencies required for the AI-powered future. This skill acquisition is accompanied by a reshaping of organizational structures and processes to accommodate the changes brought by AI. For instance, data-driven decision-making is becoming central to strategy development and execution in many firms.

Moreover, ethics and governance considerations are gaining prominence as AI becomes increasingly integrated into operations. Organizations are grappling with issues around data privacy, algorithmic bias, and transparency, necessitating the development of comprehensive AI ethics and governance frameworks.

C. Cultivating an AI-Ready Culture

To thrive in this dynamic environment, companies must foster a culture that embraces AI. This involves nurturing a mindset of continuous learning among employees, encouraging experimentation, and promoting cross-functional collaboration.

Business leaders must also understand that AI integration is not merely a technological shift but also a cultural one. Therefore, managing the human side of the AI transformation – including addressing fears about job displacement and promoting the benefits of AI – is crucial.

In conclusion, AI's role in driving innovation and requiring adaptation has far-reaching implications. While it offers immense opportunities, it also necessitates significant organizational changes. Thus, embracing AI means embracing a culture of innovation, continuous learning, and adaptability. This cultural shift, accompanied by strategic investment in AI technologies and skills, will position organizations to capitalize on the AI revolution's benefits.

**Analysis: How AI challenges the status quo and encourages new business models.**

Artificial intelligence's incursion into the business world has brought about a paradigm shift, challenging established norms and encouraging innovative business models. In essence, AI is not just a new tool for doing business but a catalyst altering the fundamental ways we conduct business.

A. Disrupting the Status Quo

At the core of AI's disruptive influence is its capacity to automate cognitive tasks, traditionally seen as the domain of human intelligence. It transcends the boundaries of routine mechanization and enters a realm where it can process information, make decisions, and learn from those decisions.

For instance, in the financial industry, robo-advisors have challenged traditional wealth management's status quo. These AI-powered platforms provide personalized investment advice at a fraction of a human financial advisor's cost. The automated, low-cost model has not only threatened established wealth management firms but also democratized access to investment advice, which was previously accessible mostly to high net worth individuals.

In the retail sector, AI-powered recommendation engines have revolutionized customer experiences. Companies like Amazon and Netflix use AI to provide personalized product and content recommendations, challenging the mass marketing approach of traditional retail and entertainment industries.

B. Fostering New Business Models

As AI disrupts the status quo, it simultaneously fosters new business models, often centered around data and network effects.

AI's value increases with the volume and quality of data it can access, leading to business models where data acquisition and utilization are key. For example, Google has built an advertising empire leveraging user data to provide highly targeted ads.

Furthermore, AI can harness network effects, where a service's value increases as more people use it. Ride-sharing companies like Uber and Lyft exemplify this. They use AI algorithms to match drivers with riders, optimize routes, and set dynamic pricing. As more drivers and riders join the network, the service improves, creating a positive feedback loop.

These new business models often challenge traditional notions of value creation and competition. In an AI-driven world, the competitive advantage is not necessarily derived from physical assets or human capital but from proprietary algorithms, data, and network effects.

C. Balancing AI's Promise and Perils

While AI fosters innovation and new business models, it also brings challenges that businesses need to address. Issues around data privacy, algorithmic transparency, and job displacement due to automation require careful navigation. Firms need to balance the drive for AI-powered innovation with ethical, societal, and regulatory considerations.

In conclusion, AI challenges the status quo, necessitating that businesses adapt or risk obsolescence. It simultaneously fosters innovative business models centered around data and network effects. To navigate this AI-powered business landscape, firms need to innovate, adapt, and carefully consider AI's societal implications. By doing so, they can harness AI's full potential and pave the way for sustainable, responsible, and inclusive growth.

## The Era of Data-Driven Decisions

We stand on the cusp of a transformative era, an epoch underpinned by data, powered by artificial intelligence, and characterized by a new paradigm of decision-making. As the digital revolution matures, we find ourselves increasingly reliant on vast repositories of data and sophisticated computational models. This is the dawn of the era of data-driven decisions.

A. The Data Deluge and the Rise of AI

The proliferation of digital technologies, ranging from the Internet of Things (IoT) to ubiquitous mobile devices, has led to an unprecedented explosion in data generation. According to estimates by the International Data Corporation, the global datasphere will grow to an astounding 175 zettabytes by 2025, signifying a proliferation of digital data at a scale hitherto unfathomable.

In tandem with the proliferation of data, we are witnessing advancements in artificial intelligence and machine learning, tools that enable us to extract meaning from this sea of information. The rise of AI in recent years is fundamentally tied to the availability of large data sets. These vast volumes of data, combined with increasingly powerful computational resources and sophisticated algorithms, are what make the current wave of AI possible.

B. The Advent of Data-Driven Decision Making

Within this landscape of abundant data and AI capabilities, the paradigm of decision-making in business is being remodeled. Gone are the days when decisions were purely reliant on human intuition, experience, or qualitative factors. The age of data-driven decision making has arrived, with AI enabling businesses to harness their data's full potential to generate actionable insights, optimize operations, and guide strategic decisions.

In this context, data-driven decision making involves analyzing large amounts of data—often in real-time—to inform business decisions. This approach empowers companies to move away from instinct-based decisions towards data-supported ones. The spectrum of its applications is vast, from optimizing supply chains in manufacturing to personalized product recommendations in e-commerce, and risk assessment in finance.

C. Pioneering a New Era

As we stand at the beginning of this new era, it is paramount to comprehend the scale of the shift and its implications. Businesses that adapt to this data-centric landscape will not only survive but thrive, innovating new products, optimizing their operations, and creating value in ways we are just beginning to understand. This era promises not only a renaissance in how decisions are made but also how businesses operate, innovate, and compete.

At the same time, it is crucial to understand the risks and challenges associated with the era of data-driven decisions. Issues related to data privacy, data quality, and the risk of over-reliance on data at the expense of human judgement need careful consideration. As we journey further into this era, it will be incumbent upon us to navigate these challenges thoughtfully and responsibly.

The era of data-driven decisions, underpinned by the symbiosis of vast data and powerful AI, has just begun. It is an era that will reshape the landscapes of business, society, and even human cognition. We must strive to harness its potential while judiciously addressing its challenges, marking the path to a future characterized by informed decision-making, increased efficiency, and transformative innovation.

**The increasing importance of data in the contemporary business landscape.**

A. Data as a Catalyst for Innovation

In the contemporary digital age, data has emerged as a resource of immense significance. It serves as the lifeblood of modern businesses, acting as a catalyst for innovation, driving efficiencies, and creating new avenues for revenue generation. The transformative potential of data transcends traditional boundaries and permeates all aspects of business, from strategy formulation to customer engagement, operational efficiency, and even business model design.

The e-commerce industry provides a cogent example. Giants like Amazon and Alibaba leverage vast troves of customer data, including purchasing habits, search histories, and demographic information, to generate personalized product recommendations, optimise logistics, and predict demand trends. Their ability to convert raw data into actionable insights has proved pivotal in redefining the retail landscape.

## B. Data and Informed Decision Making

Data has also elevated the art of decision-making into a more exacting science. By analysing historical data and discerning patterns, businesses can predict future trends, make proactive decisions, and manage risks effectively. For instance, in finance, companies use sophisticated machine learning algorithms to assess credit risk by analysing hundreds of data points that would be impractical for a human to process.

## C. Data-Driven Operational Efficiency

Furthermore, data aids in enhancing operational efficiency, a key determinant of competitiveness in today's fast-paced business landscape. The manufacturing sector demonstrates this well, where data from sensors embedded in machinery is analysed to predict equipment failure, improve maintenance schedules, and optimise production processes. This predictive maintenance approach minimises unplanned downtime, thereby improving productivity and reducing costs.

## D. Data in Healthcare and Beyond

Beyond the realm of business, data's transformative potential is impacting various sectors, including healthcare. Hospitals and research centres analyse vast datasets to make diagnostic processes more accurate, predict disease outbreaks, and design personalised treatment plans. This emerging field, known as precision medicine, would not be feasible without the power of data.

## E. The Flip Side: Challenges and Considerations

Notwithstanding the clear advantages, it is crucial to understand the challenges associated with the expanding role of data in business. Privacy and security concerns are paramount, given the sensitive nature of much of the data collected and analysed by companies. Regulatory compliance is another critical issue, with legislation such as the European General Data Protection Regulation imposing stringent rules on data handling.

Additionally, businesses must grapple with issues related to data quality. Poor quality data, riddled with inaccuracies or inconsistencies, can lead to flawed insights and ill-informed decisions. Therefore, companies must invest in robust data management and governance practices to ensure the integrity of their data.

F. Charting the Course Ahead

In this era of big data, companies that effectively harness the potential of their data will hold a significant competitive advantage. However, this requires more than just collecting vast amounts of data. It necessitates a commitment to building a data-centric culture, investing in the necessary tools and skills, and addressing the associated challenges responsibly.

The increasing importance of data in the contemporary business landscape is an irrefutable reality. Companies must therefore recognise data as a vital asset, essential to innovation, competitiveness, and growth. As we move further into this data-driven age, those who can master the art of extracting value from their data will be the ones who shape the future of business.

## The transformational role of AI in data interpretation and analysis.

A. AI: The New Age Oracle of Data

Artificial Intelligence (AI) has radically altered the landscape of data interpretation and analysis. In a world inundated with data, AI, with its capability to process and analyse vast quantities of information, has emerged as a new age oracle, delivering valuable insights from the chaotic cacophony of data.

B. The AI-Powered Analytical Engine

AI-based algorithms, such as machine learning, have revolutionized data analysis, replacing traditional statistical techniques that often struggled to manage the volume, variety, and velocity of modern data. Machine learning algorithms can identify complex patterns within massive datasets, facilitating predictive analytics, anomaly detection, and decision-making processes.

For instance, financial institutions leverage AI for fraud detection. Unusual patterns of transactions that may signify fraudulent activities can be identified in real time by employing AI-driven anomaly detection techniques, thereby securing the finances of individuals and corporations.

C. AI and Business Intelligence

AI has also breathed new life into the realm of business intelligence. Natural language processing (NLP), an AI technique, has made data analytics more accessible and intuitive, allowing users to interact with data using everyday language. Moreover,

AI-powered predictive analytics helps businesses anticipate market trends and consumer behaviours, thereby informing strategy and decision-making processes.

A noteworthy example can be found in the e-commerce industry, where platforms use AI algorithms to analyse customer behaviour and preferences, enabling personalized marketing and effective inventory management. By accurately predicting demand, businesses can maintain optimal stock levels and reduce holding costs, resulting in improved profitability.

D. AI in Healthcare Data Analysis

The transformative role of AI is particularly pronounced in the healthcare sector, where it helps decode the mysteries within vast medical datasets. AI-driven techniques analyse patient data to diagnose diseases, predict treatment outcomes, and even develop personalized medicine. AI has also shown promise in analysing medical imaging data, reducing the burden on radiologists and enhancing diagnostic accuracy.

E. Data Interpretation and the AI Challenge

While AI's role in data interpretation and analysis is transformative, it is not devoid of challenges. AI's reliability is contingent upon the quality of the data it analyses. Erroneous or biased data can lead to flawed insights, propagating what is often termed as 'garbage in, garbage out'. Ensuring data integrity and reducing bias in AI models is therefore critical.

Furthermore, the 'black box' nature of many AI algorithms, where the decision-making process is opaque, poses challenges, particularly when AI is used in sensitive areas like healthcare or finance. There is an emerging consensus on the need for explainable AI, where the reasoning behind AI decisions is clear and understandable to humans.

F. The Future of AI in Data Analysis

As we forge ahead into an increasingly data-centric era, the role of AI in data interpretation and analysis will only grow more crucial. Whether in transforming businesses or enhancing healthcare, AI promises to unlock the full potential of data, heralding a new era of data-driven decision-making.

However, realizing this potential will require addressing AI's challenges and ensuring its responsible use. A thoughtful and balanced approach, embracing AI's capabilities while acknowledging its limitations, will be essential as we navigate this transformative journey. In the final analysis, it is the human-AI partnership, combining AI's analytical prowess with human intuition and ethical judgement, that will chart the path towards a data-rich future.

**Case Studies: AI's impact on decision-making and predictive insights across industries.**

A. AI in E-commerce: Predicting Consumer Behaviour

The world of e-commerce provides an ideal setting for studying the impact of Artificial Intelligence (AI) on decision-making and predictive insights. Consider the case of Amazon, a titan in the e-commerce industry. Amazon utilizes AI to create highly personalized shopping experiences. It accomplishes this through predictive algorithms that analyze a customer's browsing history, purchase history, and other online behaviour to anticipate what a customer may wish to purchase next.

These algorithms can detect patterns not readily apparent to humans. For instance, a customer who has recently purchased running shoes may next be in the market for running attire, a water bottle, or a fitness tracker. Recognizing these patterns, the AI can tailor product recommendations, aiding Amazon in decision-making relating to marketing and inventory management. By delivering relevant suggestions, Amazon has been able to significantly increase its revenue, showcasing the significant potential of AI in enhancing business performance.

B. AI in Banking: Mitigating Financial Risks

Artificial Intelligence's transformative power extends beyond retail. In banking, AI is reshaping the decision-making process, particularly in the realm of risk management. JP Morgan Chase, one of the world's leading financial institutions, provides a compelling example of this. The bank employs AI to identify potentially fraudulent transactions and assess credit risks, thereby enhancing the security of its operations and improving its risk management.

AI's advanced algorithms scrutinize banking transactions in real-time, identifying anomalous patterns that may signal fraudulent activity. This proactive detection aids in preventing fraud before it impacts the customer or the bank. Moreover, AI algorithms sift through vast amounts of data, including a client's credit history, income level, and market trends, to evaluate credit risks, thereby assisting the bank in making more informed lending decisions. By optimizing decision-making and risk analysis, AI is fostering a more secure and efficient banking landscape.

C. AI in Healthcare: Revolutionizing Diagnosis and Treatment

Healthcare is another domain that has been profoundly influenced by AI. IBM's Watson Health serves as a prime example of AI's capacity for decision-making and predictive insights. Watson Health applies AI to analyze vast databases of medical literature, clinical guidelines, and patient data. It can predict the probability of disease, propose a suitable treatment plan, and even anticipate patient responses to treatments.

For instance, Watson Health was deployed in oncology to assist doctors in diagnosing and treating cancer. By comparing a patient's medical records with a vast array of oncological literature, Watson Health can suggest a personalized treatment plan that considers the most up-to-date research and guidelines. This ability of AI to deliver predictive insights and aid in decision-making has immense implications for patient care, potentially revolutionizing diagnosis, treatment, and prognosis.

D. Reflections on AI Across Industries

These case studies from diverse industries demonstrate AI's transformative impact on decision-making and predictive insights. By applying AI, organizations are gaining unparalleled insight into patterns and trends, empowering them to anticipate the future and make informed decisions. However, the benefits of AI must be balanced against challenges, including data privacy concerns and the risk of algorithmic bias. As the use of AI continues to expand, ongoing vigilance will be necessary to ensure that AI's predictive powers are harnessed responsibly and ethically.

**The future of data-driven decisions: Opportunities and challenges in the era of AI.**

Artificial Intelligence (AI) has undeniably reshaped the landscape of decision-making across various sectors, ushering in an era characterized by predictive insights and data-driven choices. As we peer into the future, we must consider both the opportunities AI presents and the challenges it poses to truly harness its transformative potential.

A. Opportunities in the Era of AI

Accelerated Innovation: The use of AI in decision-making can lead to a surge in innovation. As AI continues to evolve and learn, it has the potential to uncover insights humans might overlook, leading to the creation of novel products, services, or solutions. For example, pharmaceutical companies are using AI to accelerate drug discovery by identifying potential therapeutic compounds more efficiently than traditional methods.

Enhanced Efficiency: The implementation of AI can streamline processes, reducing the time and resources spent on tasks that can be automated. In the finance sector, AI is being used to automate manual tasks such as data entry and report generation, freeing up employees to focus on more strategic work.

Personalized Experiences: AI's ability to analyze large data sets enables the creation of highly personalized experiences. In the realm of e-commerce, businesses can provide custom-tailored product recommendations based on a customer's past behavior, improving customer satisfaction and boosting revenue.

B. Challenges in the Era of AI

Ethical and Privacy Concerns: With the vast amounts of data AI systems require, questions of privacy and ethics inevitably arise. Ensuring transparency in how data is collected, stored, and used is paramount. Furthermore, the potential for AI systems to make decisions that affect individuals raises concerns about fairness and bias, necessitating careful oversight.

Security Risks: The reliance on AI and data also introduces new security vulnerabilities. Protecting the data that powers AI from malicious actors is an ongoing challenge, requiring robust cybersecurity measures.

Skills Gap: As AI becomes increasingly integral to decision-making processes, there is a growing need for skilled workers who can manage and interpret AI systems. This skills gap can impede the adoption of AI and underscores the importance of education and training in this burgeoning field.

C. Balancing the Scale: The Future of Data-Driven Decisions

In the era of AI, data-driven decisions hold the promise of transformative change across all sectors. However, realizing this potential requires a balanced approach that harnesses the opportunities AI presents while thoughtfully addressing its challenges.

From a positive perspective, AI offers the prospect of unprecedented innovation, efficiency, and personalization. However, we must also consider the ethical, privacy, security, and skill challenges that the use of AI introduces. By doing so, we can ensure that data-driven decisions in the era of AI not only enhance our economic and social potential but also uphold our values and principles.

In conclusion, the future of data-driven decision-making in the era of AI is teeming with opportunities and challenges. As we navigate this exciting yet complex landscape, a measured approach will be essential, one that embraces the transformative potential of AI while diligently addressing the ethical, privacy, and security concerns it brings to the fore. In this way, we can responsibly harness the power of AI to shape a future characterized by innovation, efficiency, and fairness.

## Embracing Change: AI as a Catalyst for Growth and Competitiveness

As we venture into the future, Artificial Intelligence (AI) is proving to be an indispensable agent of growth and competitiveness in the business landscape. Its transformative potential is apparent in how it reshapes traditional business models, accelerates innovation, and provides a distinct competitive edge. However, embracing this change requires acknowledging the inherent challenges while simultaneously exploiting the unprecedented opportunities AI offers.

A. AI and the Reshaping of Business Models

AI plays an instrumental role in transforming traditional business models into more streamlined and efficient entities. By automating repetitive tasks, AI allows organizations to redirect human resources towards more complex and strategic roles. Furthermore, AI's predictive capabilities facilitate proactive decision-making based on insights gleaned from the vast sea of data.

For instance, in the healthcare industry, predictive algorithms are being deployed to anticipate patient needs and optimize healthcare delivery. This predictive approach restructures the traditional reactive model, ensuring better patient outcomes and more efficient resource allocation.

B. AI as a Driver of Innovation

In the innovation paradigm, AI serves as a potent catalyst. By processing and analyzing vast data quantities, AI can identify patterns, trends, and relationships that might remain concealed from the human eye. This capability empowers organizations to innovate in their product offerings, operational processes, and customer engagement strategies.

A case in point is the finance sector, where AI-driven robo-advisors provide customized financial advice based on a user's financial history, goals, and risk tolerance. This innovative service transforms the traditional advisory model by providing a highly personalized and scalable solution.

C. AI as a Source of Competitive Advantage

AI also acts as a powerful source of competitive advantage. Companies that harness the potential of AI can gain a significant edge over competitors by leveraging insights from data analysis, offering personalized experiences, and improving operational efficiency.

An example is the e-commerce sector. Companies like Amazon harness AI to deliver highly personalized shopping experiences, predicting consumer behavior and preferences. These personalized experiences provide a competitive advantage by fostering customer loyalty and increasing sales.

D. Navigating the Challenges

While AI offers profound opportunities, it is incumbent upon us to address the associated challenges responsibly. These include issues surrounding data privacy and security, ethical dilemmas about AI decision-making, and the risk of deepening digital divide and social inequalities.

E. The Future: Harnessing AI's Potential

As we look to the future, AI presents an exciting prospect as a catalyst for growth and competitiveness. Businesses that can adeptly navigate the challenges and embrace the opportunities AI offers will find themselves at the vanguard of their industries.

AI's potential as a transformative force is not in dispute. However, its responsible and effective implementation will require strategic planning, continuous learning, and a commitment to ethical and inclusive practices. By embracing these principles, businesses can leverage AI as a potent tool for growth and competitiveness, catalyzing a future defined by innovation, efficiency, and equitable prosperity.

**The strategic imperative for businesses to embrace AI.**

As we stand at the precipice of an unprecedented digital revolution, it has become an incontestable strategic imperative for businesses to embrace Artificial Intelligence (AI). The ubiquity of AI's transformational influence — ranging from operational efficiencies to customer engagement and strategic decision-making — calls for an immediate and unequivocal commitment from businesses to embed AI into their strategic plans.

### A. The AI-Enabled Efficiency

One of the primary justifications for the integration of AI is the substantial efficiency gains it promises. The automation potential of AI can help businesses eliminate time-consuming and repetitive tasks, freeing up human resources for more strategic and high-value tasks. For example, in the banking sector, AI-powered chatbots are addressing routine customer inquiries, thereby allowing human employees to concentrate on complex problem-solving and personalized customer service.

### B. Enhancing Customer Engagement

AI's prowess also extends to improving customer engagement. By harnessing the predictive analytics capabilities of AI, businesses can gain deep insights into customer behaviors, preferences, and trends, thus enabling personalized interactions. E-commerce businesses such as Amazon are leveraging AI to curate personalized product recommendations, demonstrating the potential of AI in enhancing customer engagement and driving sales.

### C. Decision-Making and Strategic Planning

AI has also introduced a paradigm shift in decision-making and strategic planning. By processing and interpreting vast quantities of data, AI can provide predictive insights and foresight that would be beyond the scope of human analysis. This data-driven approach empowers businesses to make proactive strategic decisions. For instance, healthcare providers are leveraging AI to predict disease patterns, enabling strategic planning for healthcare delivery and disease prevention.

### D. Staying Competitive in the Digital Economy

In the contemporary digital economy, the refusal to embrace AI can be a strategic blunder. With the growing number of businesses integrating AI into their operations, those failing to do so risk being left behind. To maintain and enhance competitiveness, businesses must invest in AI and harness its potential effectively.

### E. Anticipating the Challenges

While the strategic imperatives for adopting AI are clear, it is also crucial to acknowledge and prepare for the challenges this adoption might entail. Issues such as data privacy, security, ethical considerations, and the need for technical skills necessitate thoughtful strategies and policies.

F. Conclusion: An Imperative, Not an Option

In sum, it is no longer a question of whether businesses should adopt AI, but rather a question of how and when. The strategic imperative to embrace AI is clear and compelling. By integrating AI into their strategic planning, businesses can harness its transformative potential, driving efficiency, enhancing customer engagement, improving decision-making, and ensuring competitiveness in the digital economy. The future belongs to those who can adeptly navigate the challenges, seize the opportunities, and strategically integrate AI into their business models.

## Case Studies: Successful adoption and integration of AI in businesses.

A comprehensive examination of the successful adoption and integration of Artificial Intelligence (AI) in businesses would not be complete without discussing concrete examples from a variety of industries. The case studies presented in this chapter demonstrate the transformative potential of AI, providing readers with insights into how these pioneering organizations have utilized AI to revolutionize their operations, customer engagement, decision-making processes, and overall competitive standing.

A. Amazon: Revolutionizing E-commerce with AI

Amazon, the global e-commerce behemoth, provides a paradigmatic example of successful AI integration. Through AI, Amazon has personalized the online shopping experience, leveraging predictive analytics to recommend products based on consumers' browsing and purchasing histories. Further, Amazon's AI-powered virtual assistant, Alexa, has revolutionized customer interactions, creating a seamless and interactive experience that extends beyond traditional online shopping. The company's AI endeavors have yielded considerable benefits, including improved customer engagement, increased sales, and an enhanced competitive position.

B. Bank of America: Enhancing Customer Service with AI

In the banking industry, Bank of America stands out as an example of successful AI integration. Their virtual assistant, Erica, uses AI and predictive analytics to offer personalized financial advice to customers, thereby transforming the customer service experience. Erica can analyze individual customers' spending habits, providing personalized recommendations for budgeting and saving. This proactive approach to customer service exemplifies how AI can enhance customer engagement and satisfaction in the banking sector.

### C. Johnson & Johnson: AI in Healthcare

In the healthcare sector, Johnson & Johnson's adoption of AI for surgical procedures serves as a benchmark. The company developed the Sedasys system, an AI-powered machine that automates the delivery of anesthesia in minor surgeries. The system monitors patient vital signs and adjusts the anesthesia dosage in real-time, increasing the safety and efficiency of procedures. This case demonstrates AI's potential to improve operational efficiencies and patient outcomes in healthcare.

### D. Google: Using AI to Drive Sustainability

Google's use of AI extends to its sustainability efforts. The company used DeepMind, its AI platform, to reduce the energy used for cooling its data centers by 40%. DeepMind's machine learning algorithms analyzed various factors such as weather and cooling configurations to predict and optimize energy usage, demonstrating how AI can contribute to operational efficiency and sustainability goals.

### E. Conclusion: AI as a Transformational Force

These case studies exemplify the transformative potential of AI when successfully integrated into business operations. They underline the ways in which AI can revolutionize customer engagement, enhance operational efficiency, improve decision-making processes, and even contribute to sustainability efforts. As more businesses continue to recognize and embrace the immense potential of AI, we can expect to see more innovative applications and success stories in the near future. The strategic adoption and integration of AI, as these case studies show, is not merely a path to incremental improvement but a potent catalyst for business transformation.

**Analysis: AI's impact on business growth, competitiveness, and sustainability.**

Artificial Intelligence (AI) has undeniably emerged as a key driver of business growth, competitiveness, and sustainability. Its pervasive influence traverses various industries and sectors, instigating a radical transformation in traditional business operations, strategies, and philosophies. In this chapter, we shall delve into a profound analysis of the impact of AI on business growth, competitiveness, and sustainability, employing theoretical discussions, real-world examples, and comprehensive data.

## A. AI and Business Growth

Artificial Intelligence has ushered in a new era of business growth. By automating routine tasks, businesses can increase operational efficiency, reduce human error, and refocus human labor on strategic and creative tasks. This capability is evidenced in the manufacturing industry, where robotics and AI systems are being leveraged to automate assembly lines, resulting in increased productivity and cost savings.

AI's ability to analyze large data sets and derive actionable insights enables businesses to better understand their customer base and the market at large. Such insights can guide decision-making processes and strategy formulation, providing companies with a competitive edge and fueling business growth.

## B. AI and Competitiveness

AI is increasingly being recognized as a crucial element in business competitiveness. In a data-driven economy, AI's superior data processing and analytical capabilities enable businesses to draw deep insights, thereby informing strategic decisions.

Moreover, AI's ability to provide personalized experiences is redefining competitiveness. For instance, Netflix's AI algorithms recommend shows based on user preferences, providing a personalized customer experience that distinguishes Netflix from competitors.

AI also provides businesses with tools to enhance their products or services. Tesla's AI-driven Autopilot system exemplifies this potential, providing the company with a competitive advantage in the electric vehicle market.

## C. AI and Sustainability

Lastly, AI's impact extends to business sustainability. AI can optimize resource allocation, reducing waste and improving efficiency. For example, Google used its AI platform, DeepMind, to reduce the energy used for cooling its data centers by 40%. AI can also facilitate sustainable decision-making by providing insights into the environmental impact of different business practices.

AI can also play a pivotal role in corporate social responsibility (CSR) initiatives. AI-powered solutions can be used to tackle various social and environmental challenges, thereby contributing to the achievement of a company's CSR objectives and bolstering its reputation.

D. Conclusion: The Triumvirate of Growth, Competitiveness, and Sustainability

In sum, AI significantly impacts business growth, competitiveness, and sustainability, driving operational efficiency, enabling informed decision-making, fostering product or service innovation, facilitating resource optimization, and contributing to corporate social responsibility initiatives.

As AI continues to evolve and permeate various aspects of business, its influence on these three crucial domains is set to amplify further. Hence, businesses aspiring to grow, compete, and sustain in the modern economy must strategically adopt and leverage AI technologies, thereby embracing the transformative potential that AI embodies.

**Guidance: Steps for businesses to successfully embrace AI and leverage it for growth.**

The transition towards a business landscape dominated by Artificial Intelligence (AI) requires strategic integration and a comprehensive understanding of its implications. Hence, for businesses to exploit AI's growth potential effectively, there are several steps that they ought to consider, encapsulated in the following directives.

Grasp AI Fundamentals: The first step for any organization aiming to adopt AI is to develop an understanding of AI's core concepts, capabilities, and limitations. This understanding serves as a foundation for informed decisions about AI integration and aids in navigating the complexities associated with AI technologies.

Develop a Clear Strategy: Businesses must align AI adoption with their strategic objectives. They need to delineate how AI can help achieve these goals and create a roadmap detailing the integration process. A clear AI strategy should address potential challenges, investment plans, and KPIs to measure AI's effectiveness.

Invest in Data Infrastructure: AI technologies thrive on data. Therefore, building robust data infrastructure is vital. Businesses should consider how to collect, store, manage, and analyze data while adhering to regulatory requirements and ethical considerations related to data privacy and security.

Cultivate Talent and Expertise: The successful implementation of AI necessitates a skilled workforce adept in AI technologies. Organizations must invest in upskilling existing staff and consider hiring new talent with AI expertise. Collaborations with academic institutions and technology firms can also be beneficial in this regard.

Leverage AI for Customer-centric Services: AI can help businesses understand their customers better and deliver personalized experiences. Therefore, businesses should consider implementing AI in customer-facing roles such as customer service, sales, and marketing.

Employ AI for Operational Efficiency: AI can automate various business processes, leading to cost savings and enhanced productivity. Businesses should identify areas where automation can yield significant benefits and deploy AI technologies accordingly.

Experiment and Learn: The AI landscape is continually evolving, and businesses must adopt a culture of continuous learning. They should not be afraid to experiment with new AI applications and learn from both successes and failures. Regular assessments and revisions of the AI strategy are also integral to this process.

Engage in Ethical AI Practices: As businesses integrate AI, they must consider ethical dimensions, including bias, fairness, transparency, and accountability. Businesses should establish ethical guidelines for AI usage, keeping in mind both regulatory requirements and societal expectations.

To encapsulate, successful integration of AI into business operations requires a comprehensive understanding of AI, strategic planning, robust data infrastructure, talent cultivation, customer-centric application, operational efficiency, continual experimentation, and an ethical approach. Each step is a significant contributor to the overall process of leveraging AI for growth, demanding thoughtful consideration and diligent execution. As such, businesses that follow these steps with tenacity and foresight will undoubtedly be better poised to exploit AI's immense potential and reap the benefits of enhanced growth and competitiveness.

# The Downside of AI Disruption: Addressing Concerns and Challenges

Artificial Intelligence, despite its numerous advantages and transformative potential, presents a complex tableau of challenges. By unveiling these concerns, we allow for a more comprehensive understanding of AI, paving the way for responsible and effective utilization of the technology.

One of the major areas of concern is the Impact on Employment. As AI becomes increasingly proficient in performing routine tasks, the displacement of human labour in certain sectors is a foreseeable consequence. Jobs with repetitive tasks, such as manufacturing and data entry, are particularly vulnerable. However, AI also has the potential to create new job categories and stimulate demand for high-skilled labour, a phenomenon previously observed during the Industrial Revolution.

The impact of AI is also closely tied to Privacy and Security concerns. AI's capacity for data analysis, while beneficial in many contexts, could be misused to breach privacy boundaries. Further, as AI systems become more integral to our infrastructure, they become attractive targets for cyber-attacks.

An additional challenge is the Bias and Fairness in AI systems. AI algorithms are trained on vast datasets, and if these datasets reflect societal biases, the AI system can inadvertently perpetuate or even amplify these biases. This issue is not merely a technical problem; it requires addressing underlying social and ethical questions about what constitutes fairness and how it should be encoded into our systems.

The problem of Transparency and Explainability is another hurdle in AI's path. The decision-making process of complex AI models, particularly deep learning networks, is often opaque. This "black box" problem makes it difficult to understand how the AI system arrived at a particular decision, which is particularly problematic in high-stakes areas like healthcare or criminal justice.

Moreover, the Regulatory Challenges for AI are substantial. Existing legal and regulatory frameworks are not fully equipped to address the unique challenges posed by AI. For instance, who is responsible if an AI system makes a mistake or causes harm? The development of comprehensive and effective regulations for AI is an ongoing global challenge.

Last but not least, the issue of AI and Inequality merits attention. As AI continues to transform industries, there is a risk that the benefits will disproportionately accrue to those already advantaged, thereby exacerbating existing socio-economic inequalities.

Addressing these challenges necessitates a multi-disciplinary, inclusive, and forward-looking approach. Businesses, policymakers, educators, and civil society must collaborate to shape AI's trajectory. This involves proactively shaping policies to manage job transitions, enforce privacy and security, address biases, increase transparency, create regulatory frameworks, and ensure the equitable distribution of AI's benefits. As such, the conversation about AI's downside is not a deterrent but a crucial aspect of our journey towards a future where AI is used responsibly and to the benefit of all.

# Discussion on the potential drawbacks and challenges posed by AI, such as job displacement, privacy concerns, and security risks.

Artificial Intelligence, while a catalyst for transformative change across sectors, also introduces profound challenges that are critical to address for the holistic incorporation of this technology in our societal fabric.

A. Job Displacement

The potential of AI to automate a multitude of tasks brings forth the specter of job displacement, a concern that transcends sectors and geographical boundaries. The susceptibility of various vocations to automation is highly dependent on the nature of their tasks. Professions replete with routine, predictable activities are prime candidates for automation. This applies not just to manual labour, but also to certain white-collar jobs. For instance, positions in data entry, telemarketing, and even some aspects of accounting are under threat.

However, this narrative of job displacement must be counterbalanced with the potential of AI to spur job creation. Like the advent of the personal computer in the 20th century that birthed an entirely new industry, AI is expected to generate novel roles requiring specialized skills. This perspective suggests that AI will not so much eliminate jobs as it will transform them, necessitating a shift in skills and retraining efforts on an unprecedented scale.

B. Privacy Concerns

The extensive data-processing capabilities of AI generate substantial privacy concerns. AI's efficacy is often tied to the volume of data it can access, and this dependency on data can lead to invasive practices. Organizations may exploit data under the guise of personalization or enhancing user experience, leading to unethical use or exposure of personal information.

Additionally, AI's potential to recognize patterns and make predictions could be used to deduce sensitive information, even when such information has not been explicitly provided. This aspect raises a plethora of ethical and legal questions regarding consent, the right to privacy, and how personal data ought to be handled.

C. Security Risks

AI's integration into various aspects of society also presents significant security risks. As AI systems become critical components in fields like finance, healthcare, and national security, they become attractive targets for cyber-attacks. Adversaries could manipulate AI systems, causing them to behave unpredictably or making them hostile. Further, the use of AI in cyber-attacks could lead to more sophisticated and challenging-to-counter threats, transforming the landscape of cyber warfare.

It's also important to recognize that these risks are not standalone issues but are intricately interlinked. For example, a security breach could lead to massive privacy violations, and widespread job displacement could lead to socio-economic instability. Therefore, the strategies to address these challenges require a holistic and multi-disciplinary approach, taking into account the technical, ethical, and societal dimensions of AI.

These potential drawbacks of AI are not indicators to halt progress but to proceed with caution, with due consideration to the societal ramifications of technological advancements. These challenges, when addressed, could pave the way for an AI-integrated society that respects human dignity, fosters economic growth, and enhances the quality of life.

## Balancing AI's disruptive potential with ethical considerations and regulatory frameworks.

AI's disruptive potential is undoubtedly significant, providing unprecedented opportunities for economic growth, societal advancement, and the resolution of complex problems. However, the very attributes that lend AI its transformative power also present unique challenges, necessitating careful considerations of ethics and the development of comprehensive regulatory frameworks.

A. Ethical Considerations

AI's ethical implications emerge largely from its autonomous decision-making capabilities, vast data processing abilities, and potential for pervasive integration into daily life.

Firstly, AI's decision-making autonomy engenders questions of accountability and transparency. For instance, if an AI system causes harm, where does the accountability lie? This question becomes particularly significant in high-stakes sectors like healthcare, where AI-driven diagnostic tools might make life-altering recommendations.

Transparency, or "explainability", is another ethical concern. AI systems, particularly deep learning models, are often seen as "black boxes" because their decision-making processes can be hard to interpret, even by experts. The lack of transparency hinders users' ability to fully understand and consent to the decisions AI makes on their behalf.

Secondly, the capacity of AI to process vast quantities of data generates ethical considerations about privacy and consent. Without stringent regulations and ethical guidelines, AI's data requirements could lead to invasive data collection practices or the misuse of personal information.

Finally, the pervasiveness of AI amplifies existing societal inequalities. Without careful attention, the proliferation of AI systems could unintentionally exacerbate discrimination. Biased data can lead to biased AI outputs, thereby potentially reinforcing existing stereotypes and discriminatory practices.

B. Regulatory Frameworks

Regulatory frameworks provide a formalized approach to managing the disruptive potential of AI and addressing the ethical issues it raises. However, developing these frameworks is a complex task, owing to the fast-paced nature of AI advancement and the diversity of its applications.

Frameworks should be flexible enough to accommodate future advancements, yet rigorous enough to protect societal interests. They should be built on broad principles that can apply to diverse AI applications, while allowing for sector-specific considerations. They should encourage innovation and competition, but also ensure fairness, safety, and respect for human rights.

Efforts to develop such regulatory frameworks are underway in several jurisdictions. For instance, the European Union's proposed Artificial Intelligence Act outlines provisions for transparency, accountability, and data governance among others.

C. Balancing Disruption, Ethics, and Regulation

Striking a balance between AI's disruptive potential, ethical considerations, and regulatory frameworks requires a collaborative, multidisciplinary approach. Stakeholders including technologists, ethicists, policymakers, and the public must engage in dialogue to shape AI's trajectory.

The aim should not be to hinder AI's disruptive potential but to harness it in a way that respects ethical principles and societal values. Ethical considerations should not be an afterthought but should be integrated into AI's design and deployment.

Similarly, regulatory frameworks should not be seen as a hindrance to innovation but as a means to ensure that AI's benefits are widely and fairly distributed.

By carefully balancing these dimensions, we can guide AI's disruption towards sustainable, inclusive, and beneficial outcomes for society.

## Conclusion: Navigating the Disruptive Force of AI

As we conclude this discourse, it is manifestly evident that artificial intelligence is a profoundly disruptive force with significant implications across the economic, social, and technological spectrum. Its promise and perils are intertwined, presenting a complex narrative of unparalleled opportunities and formidable challenges. Navigating this confluence requires a nuanced understanding of AI's capabilities, a profound appreciation of its ethical considerations, and the development of comprehensive regulatory frameworks.

The disruptive potential of AI is poised to catalyze a radical transformation across diverse sectors. E-commerce, healthcare, and finance, among others, have experienced substantial gains from AI-driven solutions, spanning efficiency improvements, novel insights, and the creation of new products and services. The introduction of AI into these sectors has disrupted traditional modes of operation, catalyzing business growth, enhancing competitiveness, and promoting sustainability.

However, the integration of AI into business operations is not an exercise that can be undertaken without strategic planning and foresight. Successful adoption requires businesses to be informed about the technology, strategically align AI adoption with their business goals, invest in necessary infrastructure and skills development, and ensure their organization is prepared to adapt to the transformative power of AI.

On the other side of the coin, AI also presents a multitude of challenges. The threat of job displacement, heightened privacy concerns, and security risks are notable issues that necessitate urgent attention. The integration of AI into complex, high-stakes decision-making processes engenders unique issues related to accountability, transparency, and bias, raising profound ethical considerations.

Balancing the disruptive potential of AI with these concerns necessitates comprehensive regulatory frameworks. These frameworks should facilitate innovation while safeguarding societal interests. They need to be flexible, future-oriented, and applicable across the myriad of AI applications, fostering an environment that encourages growth, competitiveness, and sustainability while ensuring fairness, safety, and respect for human rights.

In conclusion, as we stand on the precipice of this transformative era, the imperative to navigate the disruptive force of AI with prudence and foresight becomes ever more pressing. This is not a journey that can be taken by technologists alone. It requires the collective efforts of policymakers, ethicists, businesses, and the public. The road ahead is fraught with complexity, but with careful deliberation, strategic planning, and adherence to ethical principles, we can guide AI's disruption towards outcomes that enhance societal well-being, promote economic growth, and uphold the values we hold dear.

This is our challenge and our opportunity. Let us embrace it with thoughtfulness, creativity, and an unwavering commitment to building a future where AI serves as a tool of empowerment, a catalyst for innovation, and a testament to the limitless potential of human ingenuity.

**Synthesis of the key points discussed in the chapter.**

Artificial Intelligence (AI) represents a disruptive force that brings unprecedented opportunities for innovation and growth across a multitude of sectors, including e-commerce, healthcare, banking, and finance. Its implementation can enhance efficiency, provide novel insights, and lead to the creation of innovative products and services. However, successful adoption and integration of AI in businesses demand a strategic alignment with the organization's objectives, an investment in infrastructure, the development of requisite skills, and an organizational culture adaptive to change.

While AI's potential for disruption is clear, it concurrently presents substantial challenges. Notably, it poses threats relating to job displacement, privacy concerns, and security risks. Additionally, the use of AI in decision-making processes brings forth ethical considerations related to accountability, transparency, and bias.

In order to balance AI's disruptive potential with these ethical and societal concerns, the development and implementation of comprehensive and future-oriented regulatory frameworks are vital. These frameworks should encourage innovation, while safeguarding societal interests, fairness, safety, and respect for human rights.

In conclusion, to navigate this transformative era successfully, a collective and prudent effort is required from various stakeholders, including technologists, policymakers, ethicists, businesses, and the public. Guiding AI's disruption through careful deliberation, strategic planning, and ethical principles can lead to societal well-being, economic growth, and uphold human values. Thus, AI represents a dual-faced challenge and opportunity, requiring careful navigation and management to harness its full potential.

Embracing AI as a Catalyst for Innovation and Growth

Artificial Intelligence (AI) can appear to be a double-edged sword. On one side, it brings forth an array of opportunities for increased efficiency, productivity, and innovation. On the other side, it presents a set of challenges that include job displacement, privacy concerns, and ethical dilemmas. However, when viewed from the correct perspective, AI emerges not as a threat, but as an opportunity for organizations to innovate and achieve unprecedented growth.

The first step towards harnessing the power of AI lies in shifting the organizational mindset. Many companies view AI as a mere tool for automating tasks and reducing costs. Although automation is indeed a significant benefit of AI, such a narrow perspective could limit its potential. AI can enable businesses to create innovative products, services, and business models that could be the basis of competitive advantage and revenue growth.

The realm of E-commerce provides a prime example. Amazon, a pioneer in the industry, has effectively used AI to transform shopping experiences. Their recommendation systems, built on AI algorithms, have been instrumental in driving sales by personalizing customer experience. Similarly, chatbots have enabled businesses to offer round-the-clock customer service, improving customer satisfaction and retention.

In the healthcare sector, AI has transformed diagnostic procedures, drug discovery, and patient care. For instance, Google's DeepMind Health is helping medical professionals to spot early signs of diseases like Age-related Macular Degeneration (AMD) and Diabetic Retinopathy, which are amongst the leading causes of blindness in the world.

In banking and finance, companies like Ant Financial are using AI to extend micro-credit facilities, enabling financial inclusion for those who lack access to traditional banking services. Such examples demonstrate how AI can be a catalyst for innovation and growth.

It is essential for businesses to embrace AI strategically. The adoption of AI should align with the organization's overall objectives and involve careful planning. A dedicated AI strategy should consider which aspects of the business can most benefit from AI, identify the resources needed, and outline a timeline for implementation.

This approach helps to ensure that the adoption of AI contributes to the business's long-term growth.

Another crucial aspect is the cultivation of an AI-ready workforce. Upskilling employees to work effectively with AI technologies is crucial to maximizing the value derived from AI. Simultaneously, fostering an organizational culture that is adaptive to change can facilitate the integration of AI into business operations.

Equally vital is the need to approach AI ethically. Businesses should understand and address the potential risks and implications of AI. They should develop ethical guidelines to ensure fairness, accountability, and transparency in AI operations.

In conclusion, businesses need to approach AI as an opportunity rather than a threat. By shifting their perspective, organizations can unlock AI's potential as a catalyst for innovation and growth. AI does not spell doom for businesses; on the contrary, it can be their ticket to thrive in the digital economy.

**Closing thoughts on the future of AI and its continuing disruptive potential.**

As we stand at the intersection of the present and the future, it is abundantly clear that artificial intelligence (AI) is not merely a fleeting trend, but a powerful disruptive force that will continue to reshape our world in ways beyond our current comprehension. In contemplating the future of AI, it is crucial to balance optimism about its potential benefits with realism about the challenges it presents, and the need for proactive measures to navigate its implications.

The potential of AI to enhance efficiency and effectiveness across myriad sectors continues to expand. In the sphere of e-commerce, AI is set to drive more personalised and seamless customer experiences, while in healthcare, it promises to improve diagnoses and treatments and extend the reach of medical services to underserved populations. In banking and finance, AI is expected to transform risk management, fraud detection, and customer service, while also driving financial inclusivity.

Moreover, the burgeoning field of AI holds promise for addressing some of our most pressing global challenges, from climate change to food security. AI-powered predictive models could aid in forecasting environmental changes, while AI in agriculture could enhance productivity, optimise resources, and reduce waste, contributing significantly towards global food security.

Nevertheless, as we envisage this future, it is imperative to be cognisant of the disruptive challenges that AI presents. One of the primary concerns is the displacement of jobs. As AI and automation technologies continue to evolve, their capacity to perform tasks traditionally done by humans will undoubtedly increase, leading to significant shifts in the labour market. However, history informs us that technology-driven disruption often leads to the creation of new types of jobs, and AI is likely to follow a similar path.

Another profound concern pertains to privacy and security. The proliferation of AI-powered systems necessitates the collection and processing of vast amounts of data, which, if not managed properly, can pose significant risks to individual privacy and data security.

Moreover, the opaque nature of many AI algorithms presents challenges in terms of accountability and bias. In light of these concerns, there is an exigent need for robust regulatory frameworks to ensure that AI systems are developed and deployed in ways that are ethical, transparent, and accountable.

In parallel, an ongoing discourse on the ethical implications of AI is essential. As we delegate more decisions to AI, we must grapple with questions of fairness, justice, and human dignity. We must ensure that as we advance in our technological capabilities, we do not regress in our commitment to upholding our most cherished human values.

In conclusion, the future of AI is a journey into a landscape of immense potential and significant challenges. It is a journey that requires not just technological acumen, but also ethical wisdom and regulatory foresight. As we navigate this journey, let us approach AI not as a threat to be feared, but as a tool to be harnessed, a challenge to be met, and an opportunity to drive progress towards a future where technology serves humanity, and not the other way around. The continuous disruptive potential of AI is, indeed, a testament to its power – a power that lies in our hands to shape and direct for the benefit of all.

# Chapter 5: Ethical Considerations and Challenges

Artificial Intelligence (AI), with its manifold capabilities, has emerged as a double-edged sword. On one hand, it offers unparalleled potential for innovation, efficiency, and productivity. On the other hand, it introduces a labyrinth of ethical quandaries that, if left unaddressed, can have far-reaching consequences. The ethical issues engendered by AI are not only diverse but also deeply interconnected, often dealing with fundamental questions of fairness, privacy, and human agency.

At the core of these ethical considerations is the fact that AI systems, despite being machine constructs, make decisions that directly impact human lives. Whether it is determining creditworthiness in banking, driving treatment recommendations in healthcare, or personalizing online shopping experiences, the reach of AI is extensive and pervasive. However, the algorithms that drive these decisions are vulnerable to biases and inaccuracies that can lead to unjust outcomes and amplify social inequities.

Moreover, AI systems, especially those that employ machine learning, can be inscrutable "black boxes," leading to issues of transparency and explainability. Stakeholders, including those whose lives are affected by these algorithms, often lack the means to understand or question their workings and the decisions derived therefrom. This opacity in decision-making processes challenges principles of accountability and responsibility, both essential elements of ethical systems.

Furthermore, as AI systems increasingly process personal and sensitive data, the issue of privacy comes to the fore. The potential for misuse of such data, intentional or otherwise, and the risk of breaches that expose such data, raise ethical concerns about the duty of care incumbent on those who collect, store, and use such data.

Beyond these challenges, the development and deployment of AI also stimulate broader societal and philosophical debates. What happens when machines become capable of performing tasks traditionally reserved for humans, or when they start making decisions independently of human oversight? What are the ethical implications of potential job displacement, or the prospect of AI entities becoming sentient?

Acknowledging these ethical issues surrounding AI is not a mere academic exercise. It is integral to the responsible development and deployment of AI systems. Ethical considerations should not be an afterthought, but a guiding principle from the outset, embedded into the AI system's lifecycle, from conceptualization and design to testing, deployment, and auditing. This integration of ethics into AI underscores the need for a cross-disciplinary approach that incorporates perspectives from computer science, philosophy, social sciences, law, and other relevant fields.

Incorporating ethics into AI development is not only about preventing harm and mitigating risks, but it also offers a strategic advantage. Ethical AI systems are more likely to gain the trust of users and society at large, fostering acceptance and smooth integration into various sectors. Furthermore, proactively addressing ethical issues can help anticipate regulatory trends, enabling organizations to stay ahead of the curve and avoid punitive actions.

To navigate the complex ethical landscape of AI, it is essential to dissect and understand these issues in greater detail. This chapter aims to delve into three significant ethical considerations related to AI: Bias, data privacy and security, and the nature of the human-AI partnership. By dissecting these concerns, this discourse aims to provide a comprehensive overview of ethical challenges in AI and suggest strategies to address them effectively and responsibly.

# Addressing Bias in AI

One of the most prominent ethical issues engendered by the advancement of Artificial Intelligence is the phenomenon of bias. By bias, we refer to a systematic deviation from the truth or an unfair preference in decision-making. Bias in AI manifests in numerous ways, but it primarily arises from biased data and biased algorithms. These biases can exacerbate existing social inequalities, pose threats to fairness, and potentially harm individuals and groups that are unfairly treated by biased AI systems.

Origins of Bias in AI Systems

To grasp the problem of bias in AI, we must first understand its origins. AI systems, particularly those utilizing machine learning, are essentially sophisticated pattern recognition systems. They learn to make predictions or decisions by recognizing patterns in training data. If the training data reflect existing biases in society, the AI system is likely to learn and perpetuate these biases. For instance, a recruitment AI trained on historical hiring data may discriminate against certain demographics if those demographics were underrepresented or overlooked in past hires.

Bias can also stem from the design of the AI algorithm itself. While an algorithm might be mathematically impartial, its implementation might still lead to biased results if it disproportionately harms or benefits certain groups. The algorithm's design or the criteria for its success may inadvertently prioritize certain outcomes over others, leading to unintended bias. For example, a loan approval AI may prioritize minimizing risk to such an extent that it disproportionately denies loans to lower-income applicants.

Consequences of Bias in AI Systems

Bias in AI can have far-reaching and often deleterious consequences. In finance, a biased AI system may unduly deny loans to certain groups. In healthcare, it may recommend inappropriate treatments to certain demographics. In e-commerce, it may provide different levels of service to different customers. In essence, bias in AI can lead to unequal treatment, reinforce existing disparities, and create a vicious cycle of bias and discrimination.

Addressing Bias in AI Systems

Addressing bias in AI is a complex endeavor that requires a multi-faceted approach. First and foremost, it necessitates robust and diverse datasets for training AI systems. Such datasets should be representative of the environments in which the AI system will operate, and they should include meaningful input from all relevant groups. It is also critical to have clear guidelines on how these datasets are collected and used, to ensure that they do not violate privacy norms or ethical guidelines.

Second, there needs to be greater transparency and interpretability in AI algorithms. Black-box models, where the workings of the AI system are opaque, make it difficult to identify and correct biases. Approaches like explainable AI and algorithmic transparency can help expose underlying biases and ensure that AI systems are accountable.

Third, there should be an ongoing evaluation and monitoring of AI systems to detect and correct biases. Bias in AI is not a one-time issue that can be solved during the development stage; it is an ongoing challenge that requires continuous monitoring, evaluation, and refinement of AI systems.

Fourth, it is crucial to have diversity in AI development teams. Diverse teams bring diverse perspectives, helping to identify potential biases and blind spots that might otherwise be overlooked. This should be coupled with proper training for these teams on understanding and mitigating biases in AI.

Fifth, and perhaps most importantly, there needs to be a strong regulatory framework that sets out clear standards and guidelines on addressing bias in AI.

Regulations should not only deter and penalize biased AI practices but also encourage and incentivize fairness, transparency, and accountability in AI systems.

In conclusion, bias in AI is a profound ethical challenge that requires comprehensive solutions. By understanding and addressing bias in AI, we can harness the transformative potential of AI while ensuring that its benefits are equitably distributed. This is not merely an aspiration but an imperative, one that underscores the need for an ethically-guided approach in the era of AI.

## Defining and understanding bias in AI algorithms

Artificial Intelligence's potential to revolutionize various sectors is undeniable, yet it is essential to approach its implementation with cognizance of its possible pitfalls. A prominent ethical challenge in AI deployment is the occurrence of bias within AI algorithms. Such bias denotes the tendency of AI systems to display unfair preferences, leading to skewed outcomes. Algorithmic bias, as with any systemic bias, can have detrimental ramifications on societal equity and fairness, thereby necessitating an in-depth understanding and mitigation.

### Defining Algorithmic Bias

In the context of AI, bias represents a systematic deviation from the truth or fairness in the data or the algorithm itself. Algorithmic bias is a proclivity or predisposition in the functioning of an AI system, causing it to make decisions or predictions that unfairly favor one group or category over others. The bias may be inadvertent or unintentional but can have significant ramifications, leading to skewed outcomes that can perpetuate societal inequities or prejudices.

### Origins of Algorithmic Bias

Algorithmic bias typically originates from two primary sources: biased data and biased algorithm design.

Biased data is often the result of unrepresentative or incomplete datasets used in the training of AI models. If an AI model is trained on data that does not accurately represent the demographics of the population it will be serving, it can lead to biased outcomes. For instance, facial recognition technology trained primarily on images of light-skinned individuals has been found to have higher error rates when identifying individuals with darker skin tones.

Biased algorithm design, on the other hand, results from inherent biases in the design or structure of the algorithm itself. These biases can be inadvertent, arising from unconscious human biases in the algorithm's design or from the algorithm's optimization for certain objectives at the expense of others. For example, an algorithm may be optimized for accuracy, which might lead to biased results if the algorithm overgeneralizes or discriminates against less represented categories.

## Consequences of Algorithmic Bias

Algorithmic bias can have far-reaching and impactful consequences. It can lead to unfair treatment of individuals or groups, exacerbate existing social inequalities, and even violate legal norms of nondiscrimination. In the banking sector, for instance, a biased AI system might make unfair credit decisions, denying loans to individuals from certain socioeconomic backgrounds. In the realm of healthcare, algorithmic bias could lead to misdiagnosis or inadequate treatment recommendations for certain demographics.

## Understanding and Mitigating Algorithmic Bias

Algorithmic bias can be mitigated through various means. First, by diversifying the data used in the training of AI models. This involves ensuring that the training data is representative of all the groups that the model will be serving. Second, by increasing transparency in AI models to allow for the identification and rectification of biases. This might involve the use of explainable AI (XAI) methods that allow for better understanding and interpretability of the model's decisions.

Third, it requires regular auditing and testing of AI systems for bias and fairness. Bias in AI is not a one-time issue but a continuous challenge that requires ongoing scrutiny and refinement of the AI systems. Fourth, there should be a commitment to diversity in AI development teams. Diverse teams can bring varied perspectives, potentially identifying and rectifying biases that may be overlooked by a homogeneous group.

In summary, understanding and mitigating bias in AI algorithms is a crucial ethical consideration in AI implementation. The potential repercussions of algorithmic bias necessitate a comprehensive approach to ensure the equitable and fair deployment of AI systems.

## Illustrative examples of AI bias in various sectors

Understanding algorithmic bias through real-world instances elucidates the scale of its possible consequences and the necessity for its mitigation. This section explores various instances where AI bias has manifested in diverse sectors, namely, E-commerce, Healthcare, and Banking and Finance.

E-commerce: Personalized Recommendations

Artificial Intelligence plays an instrumental role in shaping the landscape of E-commerce by enabling personalized recommendations. Recommendation engines, powered by machine learning algorithms, sift through a vast amount of data, including user purchase history, browsing patterns, and even click responses, to tailor product recommendations. This personalization not only enhances user engagement but also boosts the probability of purchase, thereby driving business growth.

However, a shadow side of these AI-driven recommendation systems is the inadvertent propagation of societal biases. The algorithm is, in essence, a reflection of the data it is trained on, which might carry imprints of societal prejudices. As an illustration, consider a scenario where an E-commerce platform also serves as a job portal. The AI system, learning from historical data showing a trend of women occupying lower-wage positions, might, therefore, recommend lower-paying job postings to female users. This is a manifestation of algorithmic bias, where an AI system unwittingly echoes and perpetuates societal gender wage disparities.

Bias in personalized recommendations is not limited to gender alone. Other demographic characteristics such as age, location, and ethnicity, if misused, can lead to skewed recommendations. For example, if past data shows that people from a particular geographical area predominantly purchase low-cost products, the AI system might exclusively recommend budget items to users from this area, thereby restricting their exposure to a wider range of products. Similarly, a young user might be targeted with fashion-forward clothing or cutting-edge tech products, while an older user might only see comfort-oriented clothing or simpler gadgets. These examples illustrate how biases, when ingrained in AI systems, can lead to a limited and potentially unfair user experience.

Recognizing and mitigating such biases require a diligent and intentional approach. The process begins with scrutinizing the recommendation algorithm for any instances where it might unfairly associate certain characteristics with particular types of recommendations. It is equally important to assess the data the algorithm is learning from. Ensuring diversity in the training data can help the AI system make balanced recommendations. An audit of AI systems for fairness, regular monitoring, and algorithmic adjustments is essential to prevent the perpetuation of biases. While AI continues to revolutionize E-commerce, ethical considerations must remain a priority to ensure the benefits of personalization do not come at the cost of fairness and equality.

Healthcare: Predictive Healthcare Models

The healthcare sector has been profoundly transformed by Artificial Intelligence, with the implementation of predictive models offering promising advancements in disease diagnosis and treatment. By using machine learning algorithms, these predictive models analyse vast amounts of data to make forecasts about future outcomes. They can predict which patients might be at risk of developing a particular disease or suggest the most effective treatments based on a patient's unique medical history. The ultimate objective is to improve patient outcomes, increase efficiency, and streamline healthcare delivery.

Despite the enormous potential, biases embedded within these predictive models can lead to unequal healthcare outcomes. A telling example of this comes from a study conducted by Obermeyer et al., published in Science in 2019. The study analyzed an AI model used to predict which patients would most benefit from specific healthcare programs. The AI model's objective was to identify high-risk patients who would then be recommended for additional care programs. However, the model was less likely to refer Black patients than White patients, even if they were comparably ill.

This instance of algorithmic bias arose from an underlying flawed assumption. The AI model was trained to use healthcare costs as a proxy for health needs. The premise was that higher healthcare costs were indicative of higher health needs, and those incurring higher costs should be considered higher risk. However, this assumption overlooked critical systemic socio-economic factors. Black patients often incurred lower healthcare costs compared to White patients for the same level of illness, due to factors such as disparities in income and access to healthcare. As a result, Black patients were often categorized as lower risk, leading to fewer recommendations for beneficial care programs.

This case serves as a potent reminder of the impact of bias in AI applications within healthcare. It underscores the importance of carefully selecting unbiased and relevant proxies when training AI models. It is not sufficient to choose a seemingly straightforward and easily quantifiable parameter such as healthcare costs. Instead, one must consider a broader context, recognizing and adjusting for socio-economic factors that may affect healthcare access and cost.

Additionally, it highlights the critical role of continual monitoring and adjustment of AI models post-deployment. AI models in healthcare are not 'set and forget' tools; they require regular evaluation and refinement to ensure they deliver fair and optimal results. Addressing biases in AI models is not just about refining technology but also about ensuring equality and justice in healthcare. As we continue to leverage AI to revolutionize healthcare, we must strive to make this technology a tool for reducing disparities, not exacerbating them.

Banking and Finance: Credit Decisions

The Banking and Finance sector stands as a prime example of an industry that has profoundly integrated Artificial Intelligence into its decision-making processes. Particularly in the realm of credit and loan approvals, AI has substantially augmented the capacity for quick and efficient determinations. Despite these advantages, the potential for embedded biases in these systems to exacerbate economic inequalities is a critical concern.

Historically, decisions around credit and loan approvals were primarily driven by human judgment, which in turn relied on a set of established criteria. Today, machine learning algorithms have largely taken over this role, processing an enormous array of data to make determinations. These algorithms are generally trained on historical lending data to draw patterns that influence future decisions.

However, a pitfall lurks in this seemingly efficient process: the risk of perpetuating past discriminatory practices. If the data on which these algorithms train incorporate biases — consciously or unconsciously — the AI system may echo these prejudices in its own determinations. For instance, if historically, people of certain races or individuals residing in specific neighborhoods were disproportionately denied loans, an AI model trained on this data might continue this trend. Consequently, this perpetuates a vicious cycle of economic inequality, systematically disadvantaging certain groups.

This potential for algorithmic bias is not merely a technological challenge but represents a fundamental issue of fairness and economic justice. It reiterates the imperative of scrutinizing historical data for potential biases before deploying it in the training of AI models. Financial institutions and AI developers must work together to evaluate the representativeness and fairness of the data used.

Moreover, the deployment of AI in decision-making should not entirely replace human judgment. Instead, a combination of both AI and human oversight can provide an optimal solution, where AI models offer efficiency and scalability, and humans ensure fairness and ethical considerations. Such a balance can prevent the automation of bias and help institutions maintain trust with their customers.

This example, alongside those from E-commerce and Healthcare, illustrates the pervasive risk of algorithmic bias across various sectors. Despite the transformative potential of AI, unaddressed biases can perpetuate and even exacerbate existing social disparities. As we continue to harness AI's capabilities, a proactive approach in identifying and mitigating biases is crucial to unlocking its full potential and ensuring equitable outcomes.

## The potential consequences of AI bias on decision-making and social inequality

As the capabilities of Artificial Intelligence mature, its role in decision-making processes across various sectors is becoming increasingly pivotal. Whether in healthcare, banking, or e-commerce, AI systems hold tremendous promise in driving efficiency, scalability, and accuracy. Yet, it is precisely within this context that the implications of AI bias can become profoundly consequential, with the potential to entrench social inequalities further and distort decision-making.

Economic and Social Consequences

Expanding upon the economic and social consequences of AI bias necessitates a more granulated look into how systemic biases can be unwittingly encoded into algorithms, leading to disproportionate effects on various demographics.

In the E-commerce realm, personalized recommendation algorithms can inadvertently perpetuate gender wage disparities. These algorithms draw inferences based on past user behavior and historical data. Suppose the historical data holds societal biases, such as women being associated with lower-paying job roles. In that case, the recommendation system could potentially replicate this disparity, continuously suggesting lower-paying positions to female users. This insidious form of bias does not merely affect an individual's employment prospects. It entrenches gender wage disparity at a systemic level, contributing to broader economic inequality based on gender.

In the realm of healthcare, AI systems are increasingly employed for diagnosing diseases and personalizing treatments. Yet, biases embedded within these systems could inadvertently result in racial inequalities in healthcare access. For instance, a machine learning model trained on past data, which includes systemic biases against racial minorities, might not flag these individuals for necessary medical interventions, resulting in worse health outcomes. This biased decision-making could not only lead to reduced access to healthcare for racial minorities but also perpetuate health disparities at a societal level.

The Banking and Finance sector is another domain where AI systems' decisions have profound implications on individuals and society. Lending algorithms could deny financial resources to individuals based on their zip code, replicating historical biases against marginalized communities. This systemic financial exclusion could lead to a vicious cycle of poverty and economic stagnation within these communities, thereby exacerbating social inequality.

Unchecked, the implications of these biases are grave. Individual instances of biased decision-making coalesce into a pattern of systemic injustice that not only denies individuals equitable opportunities but also risks leading to a deeply polarized society. In such a society, opportunities and resources are not allocated based on merit but are determined by the algorithmic replication of existing societal biases. The socioeconomic fabric of our societies could become increasingly stratified, with existing disparities magnified by the unimpeded action of biased AI algorithms. Thus, it is imperative that these biases be actively recognized and mitigated to prevent such dystopian outcomes.

Legal Consequences

In an evolving legal landscape, where lawmakers worldwide are grappling with the implications of AI, failure to address and mitigate AI bias can engender significant legal repercussions. Legislation regarding fair and accountable AI is beginning to emerge, shifting the responsibility onto organizations to ensure the AI systems they deploy are free from discriminatory biases.

To illustrate this, consider the EU's proposed Artificial Intelligence Act. This draft legislation sets out a legal framework for AI, categorizing AI systems according to their potential to cause harm and imposing obligations accordingly. High-risk AI systems, including those used in recruitment, law enforcement, and credit scoring, require strict compliance to ensure transparency, robustness, and non-discrimination. Thus, any biases in such AI systems could trigger legal consequences, including fines of up to 6% of the global annual turnover for the most serious infringements.

Similarly, in the United States, the Algorithmic Accountability Act has been proposed to compel companies to scrutinize their automated decision systems, including AI, for potential biases and discrimination. Any violation could potentially lead to enforcement actions by the Federal Trade Commission, imposing severe penalties.

Such legal repercussions, coupled with public backlash, can significantly impact an organization's reputation and financial stability. Notably, it's not just the immediate financial penalty that could harm the organization. The associated reputational damage may erode customer trust, a valuable commodity in the digital age where data privacy and fairness are of increasing concern. This loss of trust could translate into decreased customer loyalty, leading to a potential loss of market share and profitability in the long term.

Beyond the economic consequences, these legal and reputational repercussions emphasize the need for ethical considerations in the development and deployment of AI systems. They serve as a potent reminder that the fight against AI bias isn't merely a matter of achieving technical accuracy, but a broader issue encompassing legal compliance, public trust, and ultimately, corporate survival. It behooves organizations to actively recognize and mitigate AI bias to harness AI's potential responsibly and equitably.

Decision-making Consequences

The allure of AI-based decision-making largely rests on its promise of objectivity and rationality, unencumbered by the prejudices that often plague human judgements. Yet, when confronted with the reality of algorithmic bias, this promise starts to falter. The realization that the pristine objectivity of AI can be compromised calls for a closer look at the decision-making process that AI upholds.

To comprehend this issue, we must delve deeper into the mechanics of AI, specifically machine learning algorithms, which learn patterns from historical data to make predictions or decisions. If the training data is tainted with bias, the algorithm inherits these biases, and any decision derived from such algorithms will essentially be a mirage of objectivity. Instead, it becomes a mere reflection of past prejudices embedded in the data.

A noteworthy dimension of this problem lies in the cyclical nature of this bias. Biased decisions feed back into the system as data, leading to an even greater reinforcement of the original bias. For example, if an AI recruiting tool, trained on historical hiring data skewed towards male employees, consistently selects male candidates over equally qualified female ones, it perpetuates the gender bias in hiring. As more men are hired, the disparity in the data grows, and the cycle continues, amplifying the initial bias.

This cyclical bias is not just an anomaly—it undermines the fundamental premise of AI's rational decision-making. It creates a self-perpetuating cycle of inequality, where the past prejudices dictate the future decisions, thus denying equal opportunities to those who were marginalized in the past. Such systemic bias can permeate all levels of decision-making, from individual to societal, leading to a profound and long-lasting impact.

Therefore, to realize the full potential of AI in making rational decisions, it is crucial to break this cycle of bias. This necessitates both a robust technical approach to de-biasing data and algorithms, and a broader socio-ethical dialogue to question and rectify the historical and systemic biases that seep into our data. Only by addressing these challenges can we hope to fulfill the promise of AI as a tool for fair and equitable decision-making.

Mitigating Consequences: A Way Forward

The potential consequences of AI bias are grave, yet they are not an inevitability. To ensure that the decision-making power of AI can be harnessed for the greater good, it is vital that all stakeholders—AI researchers, data scientists, policymakers, and society at large—commit to a rigorous and ongoing process of bias detection and mitigation.

This includes developing standards for transparent and interpretable AI, so it becomes possible to understand and challenge the decisions made by AI systems. It also involves implementing robust testing of AI systems for bias and fairness and establishing legal and ethical frameworks for AI.

Addressing AI bias is not a mere technical challenge to overcome; it is a moral imperative. If the promise of AI as a force for societal good is to be realized, the rectification of AI bias must be at the heart of all AI innovation. As we navigate the path ahead, we must ensure that our AI systems reflect our highest principles of fairness and justice, providing opportunities for all.

# Measures for detecting and mitigating AI bias

The far-reaching implications of AI bias, as discussed in the preceding sections, underscore the urgency of establishing robust measures to detect and mitigate it. This endeavor is not straightforward, given the complexity and often elusive nature of bias. Nevertheless, several techniques and best practices are emerging across multiple fields that aim to make AI systems more fair, accountable, and transparent. These range from proactive strategies focused on data collection and algorithm development to reactive approaches for post-hoc analysis and bias correction.

Proactive Measures: Focus on Data and Algorithm Design
A proactive approach to mitigating AI bias involves careful consideration at the data collection and algorithm design stages. It is important to remember that algorithms learn from the data they are provided with. If the training data is skewed or biased, it is likely that the trained model will be as well.

For instance, in an E-commerce recommendation system, if women are disproportionately shown and subsequently click on low-paying jobs, the model may learn this pattern and propagate it. This can be mitigated by ensuring a balanced representation of different demographic groups and outcomes in the training data.

Algorithm design also plays a crucial role. Machine learning algorithms need to be developed with a fairness objective in mind. Various statistical definitions of fairness have been proposed, such as demographic parity, equal opportunity, and equal odds, each suitable for different contexts. Researchers and developers should consider incorporating these fairness objectives into the learning process.

Reactive Measures: Post-hoc Analysis and Bias Correction
Reactive measures focus on detecting and correcting bias after a model has been trained. These include post-hoc analysis techniques, which assess a trained model for fairness metrics and bias patterns, and bias correction methods, which adjust a model's decisions to ensure fairness.

A key tool for post-hoc analysis is disparate impact analysis, a statistical test to identify discrimination in decision-making. For instance, in the healthcare scenario discussed earlier, a disparate impact analysis would assess whether the healthcare AI model unfairly disadvantages Black patients. If the analysis detects discrimination, the next step is bias correction.

Bias correction methods typically adjust a model's decisions or outputs to mitigate detected bias. These methods can be quite complex, involving mathematical programming and optimization techniques to satisfy a chosen fairness criterion.

Regulatory Measures: Legislation and Standards
In addition to technical measures, regulatory interventions play an increasingly important role in detecting and mitigating AI bias. Governments worldwide are starting to enact legislation aimed at ensuring the fairness and accountability of AI systems. An example is the Algorithmic Accountability Act in the United States, which requires companies to conduct impact assessments of their high-risk AI systems for fairness, accuracy, and bias.

Similarly, standards organizations such as the Institute of Electrical and Electronics Engineers (IEEE) are developing guidelines for ethically aligned design of autonomous and intelligent systems, which include recommendations for addressing bias.

Multidisciplinary Measures: The Importance of Diversity and Inclusion
Finally, addressing AI bias requires more than just technical or regulatory solutions. It necessitates a multidisciplinary approach, combining insights from computer science, social sciences, ethics, and law. A diverse and inclusive team can

better anticipate potential sources of bias and devise more effective strategies to combat them.

For instance, collaboration between computer scientists and sociologists can illuminate how social biases seep into data and algorithms, while legal experts can provide guidance on compliance with evolving regulations. By fostering a culture of diversity and inclusion, organizations can harness a broader range of perspectives to tackle AI bias.

In conclusion, addressing AI bias is a complex yet critical task. The measures described herein provide a starting point, but it is important to remember that mitigating AI bias is not a one-time activity but a continuous process that requires diligence, commitment, and a willingness to learn from mistakes. By adopting these measures, we can strive towards AI systems that are fair, transparent, and beneficial for all.

## The role of diversity in AI development to counteract bias

In recent years, there has been a growing recognition that diversity in AI development teams is not merely a nicety but a necessity. The heart of this realization lies in the acknowledgment of the fact that algorithms, contrary to their often-perceived objectivity, bear the imprints of their creators. The experiences, worldviews, and even implicit biases of the developers can, intentionally or unintentionally, shape the assumptions embedded in AI models and the decisions they make. Consequently, the lack of diversity in AI development can lead to biased systems that overlook or even exacerbate societal inequalities.

Why is Diversity Important?

Diverse teams bring a plurality of perspectives to the table, which can help uncover blind spots and challenge biased assumptions. For instance, consider a healthcare AI system designed to diagnose skin cancer. If the team training this model consists of individuals from predominantly lighter-skinned racial backgrounds, the data they collect may reflect their unconscious biases, leading to a model that performs well on lighter skin tones but poorly on darker ones. However, with a racially diverse development team, the chances of such an oversight occurring can be significantly reduced.

Additionally, diversity goes beyond demographic characteristics such as race and gender. It also encompasses different disciplines and fields of study. Technologists may be experts in algorithm development and data analysis, but they might lack the necessary understanding of the social, cultural, or ethical implications of the technology they build. A multidisciplinary team, with members from the social sciences, humanities, law, ethics, and other fields, can help bridge this gap.

Diversity and Fairness in AI Design

The link between diversity and fairness in AI systems can be elucidated through two primary pathways. The first pathway relates to the understanding of fairness itself. Fairness is not a monolithic concept but rather a multi-dimensional one, with different interpretations and implications across different contexts. A diverse team can help ensure a more comprehensive understanding of fairness and a more nuanced incorporation of it into AI systems.

The second pathway concerns the operationalization of fairness in AI systems. As discussed in the previous chapter, several technical measures can be employed to detect and mitigate bias in AI. However, implementing these measures requires making critical decisions about which fairness definition to prioritize, how to balance fairness with other objectives, and how to interpret and act upon the results. These decisions are inherently normative and contextual, requiring the kind of nuanced understanding that a diverse team can provide.

Diversity and Accountability in AI Development

Diversity in AI development can also foster greater accountability. When AI systems are designed and deployed by a homogeneous group, it can be challenging for those outside the group to scrutinize or challenge the systems. This can lead to a lack of transparency and accountability, with the potential for biased and harmful outcomes. However, a diverse team can help ensure that the development process is more open to external scrutiny and thus more accountable.

Challenges and Recommendations

Despite the evident importance of diversity in countering AI bias, achieving it in practice is fraught with challenges. These include systemic barriers to entry in the tech field, biases in recruitment and promotion, and a lack of supportive policies and practices. To overcome these challenges, concerted efforts are needed at multiple levels.

At the organizational level, policies and practices that promote diversity and inclusion should be prioritized. This can include targeted recruitment efforts, mentorship and development programs, and inclusive work culture initiatives. Organizations should also foster collaborations with external partners from diverse fields and backgrounds to ensure a wider range of perspectives.

At the societal level, interventions to enhance diversity in STEM education and careers are crucial. This can involve scholarships and support programs for underrepresented groups, initiatives to challenge stereotypes and biases, and efforts to create inclusive educational and professional environments.

In conclusion, diversity plays an indispensable role in countering AI bias. By fostering diverse and inclusive AI development teams, we can work towards AI systems that are not only technically robust but also ethically sound and socially equitable.

# Data Privacy and Security

The dawn of the Artificial Intelligence (AI) era has heralded unparalleled opportunities for growth, innovation, and enhancement in various fields, ranging from E-commerce and Healthcare to Banking and Finance. AI's capacity to digest and process enormous volumes of data, discern patterns, and make informed decisions or predictions surpasses human capabilities and offers a pathway to unprecedented advancements.

However, the inherent reliance of AI systems on data - often personal and sensitive - brings to the forefront critical issues concerning data privacy and security. This chapter aims to provide an in-depth introduction to these concepts, elucidating their significance, challenges, and potential solutions in the context of AI.

The Intersection of AI, Data Privacy, and Security
AI algorithms, particularly machine learning models, require vast amounts of data for training and functioning effectively. This data often includes personal information about individuals, such as their shopping habits, health records, financial history, or social interactions. Therefore, the use and handling of this data must be done with a high degree of caution to respect individuals' privacy rights and ensure the security of their personal information.

Data privacy refers to individuals' rights to control how their personal information is collected, used, stored, and shared. It encompasses principles such as consent (individuals should agree to their data being collected and used), purpose limitation (data should only be used for the purpose for which it was collected), and data minimization (only the necessary amount of data should be collected).

On the other hand, data security involves protecting data from unauthorized access, use, disclosure, disruption, modification, or destruction. It includes measures to ensure data confidentiality (information is accessible only to authorized individuals), integrity (information is accurate and complete), and availability (information is accessible when needed).

The Significance of Data Privacy and Security in AI
Respecting data privacy and ensuring data security are not only legal and ethical imperatives but also critical for the effective functioning and societal acceptance of AI systems.

First, a failure to protect data privacy and security can lead to severe consequences, such as identity theft, financial fraud, and damage to individuals' reputations. It can also expose sensitive information, leading to potential discrimination or stigmatization.

Second, data privacy breaches and security incidents can erode public trust in AI systems. If individuals fear their personal information could be misused or compromised, they may be reluctant to engage with AI systems, thereby hindering their deployment and effectiveness.

Lastly, legal frameworks around the world are increasingly emphasizing data protection, with stringent penalties for non-compliance. Therefore, upholding data privacy and security is essential for organizations to avoid legal repercussions and potential damage to their reputation.

Challenges to Data Privacy and Security in AI
The task of ensuring data privacy and security in AI, however, is fraught with challenges. AI's insatiable appetite for data, coupled with the increasing digitization and connectivity of our lives, create numerous potential avenues for privacy breaches and security threats. Further, advanced AI techniques like deep learning are often opaque, making it difficult to ascertain how they use and process data.

Additionally, traditional privacy-preserving and security measures might not be suitable or sufficient in the context of AI. For instance, anonymizing data – a commonly used technique to protect privacy – can be rendered ineffective by AI algorithms that can re-identify individuals from anonymized data.

Towards Privacy-Preserving and Secure AI
In response to these challenges, researchers, policymakers, and practitioners are exploring various approaches to ensure data privacy and security in AI.

These include technical measures such as differential privacy (a mathematical framework to add noise to data to prevent the identification of individuals), federated learning (an approach where AI models are trained on decentralized data, reducing the need to transfer and store data centrally), and encryption techniques (to protect data during storage and transmission).

Alongside technical measures, legal and policy measures play a crucial role. Regulations such as the General Data Protection Regulation (GDPR) in the European Union set forth comprehensive rules for data protection, including principles, rights, and obligations that organizations must follow when handling personal data.

Moreover, efforts are underway to make AI systems more transparent and explainable, enabling better understanding and control over how they use and process data. This field, known as Explainable AI or XAI, has become a vibrant area of research and development.

In conclusion, data privacy and security are critical considerations in the development and deployment of AI systems. Ensuring these aspects necessitates a multi-faceted approach, combining robust technical measures, stringent legal and policy frameworks, and a culture of respect for individuals' privacy rights and data protection.

## A. The ethical dimensions of data privacy and security in AI

As artificial intelligence (AI) permeates every facet of modern life, it brings forth profound questions about the ethical dimensions of data privacy and security. The proliferation of AI systems across E-commerce, Healthcare, Banking, and Finance, among other sectors, relies heavily on extensive data. This data often encompasses personal, sensitive, and critical information. Consequently, the management and protection of this data raise pressing ethical considerations that warrant rigorous exploration.

The Right to Privacy and Autonomy

The notion of individual autonomy, particularly the principle of self-determination, is the cornerstone of data privacy and one of the most cherished tenets in the broader sphere of human rights. Rooted in various international legal instruments such as the Universal Declaration of Human Rights and the European Convention on Human Rights, this fundamental right is interpreted as an individual's ability to govern their personal information.

In the context of artificial intelligence, individual autonomy and the right to privacy translate into an obligation for organizations to obtain informed consent from data subjects before gathering and processing their data. This duty is particularly vital, considering the wealth of personal data that AI systems can process and the myriad ways in which it can be utilized. Indeed, informed consent becomes the bulwark against unauthorized data usage and helps to uphold the sanctity of personal boundaries in the digital age.

Nonetheless, the opacity of AI systems, combined with the large-scale and complex nature of data they manage, can make it challenging for individuals to fully understand the scope and implications of their consent. This opacity may compromise an individual's ability to exercise their right to privacy and control over their personal data.

For instance, an AI system in an E-commerce platform could harness an individual's purchasing history, browsing habits, and even interactions on social media to formulate tailored product recommendations or targeted advertisements. In essence, the individual's behavior and preferences are codified into data points, processed by complex algorithms, and then utilized to drive certain business outcomes.

While such personalization may ostensibly enhance user experience by providing more relevant content, it can concurrently infringe upon the individual's privacy. If these data processing activities are conducted without transparent disclosure and without the individual's explicit consent, they violate the basic principles of autonomy and the right to privacy. The individual may find themselves unknowingly tracked and analyzed, and their data used in ways they have not authorized, highlighting a stark power asymmetry between data subjects and entities that manage AI systems.

Therefore, preserving individual autonomy in the age of AI necessitates rigorous efforts to foster transparency about data processing activities. Organizations should clearly communicate to users how their data is gathered, stored, processed, and protected. They must also provide mechanisms for individuals to express their consent explicitly, and opportunities to revoke it, promoting a more equitable and respectful interaction between AI systems and their human users.

Fairness and Non-Discrimination

The ethical dimensions of data privacy and security extend beyond the safeguarding of personal information, encompassing principles of fairness and non-discrimination. As artificial intelligence becomes increasingly integrated into societal fabric, its influence on decision-making processes and outcomes is ever-expanding. If AI systems are trained on biased data or developed without an inclusive consideration of diversity, they risk reinforcing, and even amplifying, societal disparities and prejudices that exist in the real world.

Consider, for example, the deployment of an AI system in healthcare that makes predictive analyses about patients' health risks. Such systems typically rely on large amounts of historical data to make accurate predictions. However, if this training data predominantly consists of records from certain demographic groups, there is a substantial risk that the AI system will learn, and subsequently internalize, the biases inherent in the data.

This scenario might manifest when an AI system is predominantly trained on data from a particular race or gender, thereby failing to adequately represent other demographics. Consequently, the system may deliver inaccurate or discriminatory health risk predictions for individuals outside of this group. Such misrepresentation can lead to uneven quality of care, exacerbating existing health disparities and disproportionately affecting underrepresented communities.

This illustration underscores the pivotal need for privacy-conscious and bias-free data collection and utilization practices within the field of AI. Data used to train AI models should be representative of the diversity of the population, and steps should be taken to ensure that the data does not unintentionally favor any particular group over others.

Further, to enhance fairness, organizations should implement robust privacy policies and data management practices. This includes providing transparency about how personal data is used, enabling individuals to control the processing of their data, and implementing measures to secure data and protect it from unauthorized access or use.

Fairness, in this context, also requires a constant vigilance to ensure that AI does not lead to discriminatory outcomes, even unintentionally. Regular audits of AI systems can help detect and mitigate any bias that may have inadvertently been introduced. Ultimately, the objective is to build and use AI systems in a way that respects individuals' rights to data privacy and security, and upholds the principles of fairness and non-discrimination.

Trust and Transparency

One cannot overstate the significance of trust and transparency in discussions on the ethical implications of data privacy and security in AI. As AI systems permeate various sectors, their ability to garner trust from users largely dictates their efficacy. This trust, in turn, depends on several key factors, among which robust privacy and security measures, as well as transparency regarding their implementation, hold prominent positions.

Let us take the domain of banking and finance as an illustrative example. As customers interact with their bank's AI-powered systems, they generate a trove of sensitive financial data. These customers need to trust that their data is not only secure from potential threats such as hackers and data breaches but also that it will be used responsibly and transparently by the bank's AI systems.

Achieving this level of trust requires the bank to be forthright about its data handling practices. The bank must clearly communicate how it collects, stores, and processes customer data. Importantly, it should also elucidate how its AI algorithms leverage this data to make decisions or predictions. If, for instance, an AI algorithm determines the creditworthiness of a customer, the bank should transparently disclose the factors the algorithm considers and how it weighs them.

Transparency in AI extends beyond merely the disclosure of data handling practices. It encompasses explainability of the AI systems themselves. Explainable AI, often termed XAI, refers to the ability of AI models to provide understandable and interpretable explanations for their predictions and decisions. As AI systems continue to grow in complexity, achieving such explainability becomes increasingly challenging, yet no less critical.

Thus, fostering trust in AI systems entails a two-fold challenge. Firstly, organizations must implement and uphold robust privacy and security measures to protect user data. Secondly, they must strive for transparency, not just in their data practices, but in the workings of their AI systems as well. By meeting these challenges, organizations can ensure that their AI systems are not only trusted but also ethically sound.

The Ethical Stewardship of Data

The ethical dimension of data privacy and security in AI development is anchored by the concept of stewardship. This notion firmly posits that organizations are not mere collectors or users of personal data, but are also its stewards. As stewards, they bear a profound moral responsibility to protect, manage, and use this personal data ethically. The ethical stewardship of data is underpinned by a number of distinct yet interconnected duties.

Firstly, organizations must ensure data accuracy. Accurate data is indispensable for AI systems, as it drives their effectiveness and credibility. Given that AI systems base their decision-making processes on the data at their disposal, any inaccuracies can lead to flawed outcomes, thereby compromising the trustworthiness of these systems. As such, organizations are ethically obliged to implement mechanisms that promote and maintain data accuracy.

Secondly, data stewardship mandates organizations to practice data minimization. This principle asserts that organizations should limit data collection to only what is necessary for their stated purposes. Excessive data collection not only increases the risk of privacy violations but also engenders an unnecessary risk of data breaches.

Along similar lines, data retention policies also fall within the purview of data stewardship. Organizations should retain personal data only for as long as it is needed to fulfill the purposes for which it was collected. Any retention beyond this period raises ethical questions and potentially infringes on individuals' privacy rights.

Furthermore, as data stewards, organizations must enact robust security measures to safeguard personal data from breaches, unauthorized access, or misuse. Security measures might include encryption, pseudonymization, and stringent access control mechanisms, among others.

However, despite the best measures, data breaches can and do occur. In such unfortunate events, ethical stewardship dictates a clear course of action for organizations. They must promptly notify the individuals affected by the breach and swiftly implement measures to mitigate the impact. Transparency, honesty, and swift action during such crises can help organizations maintain trust and demonstrate their commitment to ethical data stewardship.

In sum, the ethical stewardship of personal data is a multifaceted responsibility that transcends mere compliance with privacy laws and regulations. It is an ethical commitment that requires organizations to actively protect, manage, and use personal data with respect and care, ensuring that their AI systems are not only effective but also fair, transparent, and trustworthy.

Conclusion

In conclusion, the ethical dimensions of data privacy and security in AI are multifaceted and complex. Upholding these ethical principles requires a comprehensive approach that integrates technical measures, such as robust data protection and security protocols, with ethical considerations, like respect for privacy, fairness, and transparency. It requires a commitment from all stakeholders in the AI ecosystem, from data scientists and developers to policymakers and end-users, to ensure that AI serves the common good while respecting individual rights. To this end, fostering an ethical culture in AI development and usage, informed by ongoing dialogue and learning, is an imperative of our time.

## B. Risks and challenges associated with data breaches and misuse

As we progress through the age of digital innovation, where data has become the cornerstone of countless industries and Artificial Intelligence (AI) applications, we are also entering a period fraught with new risks and challenges. Of these, data breaches and the misuse of personal information have emerged as particularly critical issues. This chapter will explore these risks and challenges, focusing on the domains of E-commerce, healthcare, and banking and finance, and will elucidate potential countermeasures.

### Risks and Challenges in E-commerce

E-commerce platforms, acting as digital marketplaces for myriad goods and services, must handle and process an extensive array of sensitive customer data. This data takes many forms, encompassing everything from basic identifiers like names and addresses, to financial details such as credit card information, to nuanced personal insights derived from purchase history, browsing patterns, and product preferences. The management of such extensive and varied data brings with it inherent risks, among which data breaches stand prominent. The repercussions of these breaches can be wide-ranging and severe.

When a data breach occurs, the most immediate and tangible risk is that of financial fraud. Sophisticated cybercriminals can employ stolen credit card information or bank details to execute unauthorized transactions, draining victims' accounts or piling up charges that can lead to financial havoc. In certain instances, cybercriminals might not utilize the information directly but sell it on the dark web, where such personal data commands high prices.

However, the implications of data breaches extend beyond these immediate financial losses. The intangible, yet equally damaging, consequence lies in the erosion of customer trust. A breach signifies a failure in the organization's data security measures, and customers may view it as a violation of their privacy and an indication that the platform is incapable of protecting their personal data. The reputational harm that ensues can deter existing customers and dissuade potential ones, resulting in lost business that can be challenging to recoup.

The issue of trust becomes even more pronounced when we consider the misuse of personal data by third-party advertisers or data brokers. Personal data harvested by e-commerce platforms is a valuable commodity, often leveraged to personalize marketing efforts or to inform business strategies. Even if data brokers or advertisers obtain this information through ostensibly legitimate channels, the ways they utilize this data can have significant implications for customer privacy.

For instance, third parties could use the data to create highly targeted advertising campaigns that leverage intimate knowledge of customers' buying habits, preferences, and even their daily routines. While this might be intended to enhance the shopping experience, it can feel invasive and unwelcome to many consumers, and thus further erode trust. Moreover, if these third parties are not stringent about their data security, they could become targets for data breaches, adding another layer of risk.

In worst-case scenarios, third parties might use this data for purposes that were never disclosed to or approved by the customers, a practice that grossly infringes on their privacy rights. For example, selling personal data to employers, insurance companies, or lenders without consent can have serious ramifications for individuals, such as job loss, higher insurance rates, or loan denial.

In conclusion, data breaches and misuse of personal data in the realm of e-commerce are multi-faceted problems that pose substantial financial and reputational risks. Organizations must prioritize robust data protection strategies and uphold the principles of data ethics, not just to comply with regulations, but to maintain the trust and loyalty of their customer base.

Risks and Challenges in Healthcare

The healthcare sector, with its vast repository of sensitive data, represents a high-stakes battleground in the struggle for data security. The diverse array of healthcare data includes patient medical records, genetic profiles, prescription history, insurance information, and even notes from physician-patient interactions. This data holds immense value, both for its potential to improve patient outcomes and, unfortunately, for nefarious exploitation if it falls into the wrong hands. Consequently, the risks and consequences associated with data breaches in healthcare are particularly severe.

Foremost among the risks is the potential for identity theft. Patient medical records typically contain a wealth of personally identifiable information, including names, addresses, social security numbers, and dates of birth. Cybercriminals armed with these details can wreak substantial havoc, applying for credit, filing false insurance claims, or even obtaining medical care under the victim's identity. The personal and financial toll for the victims can be considerable, and untangling the ensuing web of fraudulent activity often proves daunting.

Beyond identity theft, healthcare data breaches can also pave the way for insurance fraud. Unauthorized individuals or entities possessing detailed medical records can fabricate or inflate claims, leading to substantial losses for insurance companies. These fraudulent activities eventually translate into higher premiums for consumers, turning data breaches into a widespread financial burden.

In a more sinister vein, healthcare data breaches can enable discriminatory practices. Genetic information or detailed medical histories could be exploited by unscrupulous employers, insurers, or lenders to discriminate against individuals based on their health status. Although laws exist to prevent such discrimination, enforcement can be challenging, especially when data breaches occur without immediate detection.

Moreover, the misuse of healthcare data represents a significant concern. For instance, the use of personal health information for targeted advertising without the explicit consent of the patient raises substantial ethical and legal questions. A pharmaceutical company, upon obtaining detailed patient medical histories, might target individuals with advertisements for their products. This practice not only violates the privacy rights of individuals but also potentially undermines the doctor-patient relationship, introducing commercial interests where clinical judgement should hold sway.

Additionally, in instances where healthcare data is used for research, there are potential ethical issues around informed consent, particularly if the data is used in ways that the patient did not initially agree to. The misuse of data in such cases may compromise the privacy of individuals and erode trust in healthcare providers and the wider medical research community.

In conclusion, the healthcare sector, due to the sensitivity and personal nature of the data it handles, faces significant challenges and risks related to data breaches and misuse. Unauthorized access and the illicit use of healthcare data can lead to a multitude of harmful consequences, ranging from identity theft and insurance fraud to potential discrimination and ethical breaches. Ensuring the privacy and security of health data is not only a legal requirement but also an ethical imperative for healthcare organizations, necessitating stringent measures for data protection and the responsible use of patient information.

Risks and Challenges in Banking and Finance

In an era of digital transactions and online banking, the finance sector holds a treasure trove of data that includes not only account details and transaction records but also a wealth of personally identifiable information. This vast and sensitive data pool, while enabling innovative financial services and better customer experience, concurrently attracts cybercriminals and presents a fertile ground for data breaches. As a result, the industry faces substantial risks and challenges associated with data security and misuse.

One of the most immediate threats in the aftermath of a data breach in the finance industry is the potential for identity theft. Cybercriminals armed with stolen personal information can create fake identities or take over existing accounts, executing unauthorized transactions, applying for loans, or setting up new accounts. Victims may find their credit ratings tarnished, their accounts drained, and their financial security jeopardized.

Beyond identity theft, financial data breaches can enable illicit transfers of funds. A hacker who gains access to a user's banking details can siphon funds to their own accounts, potentially bypassing standard security measures if they have enough information about the victim. This can lead to substantial financial loss for the targeted individuals and their banks.

Data misuse within the banking and finance industry carries its own set of perils. For instance, unauthorized trading — using a customer's data to perform transactions without their knowledge or consent — represents a serious breach of trust. Similarly, predatory lending practices, where lenders use data to identify vulnerable individuals and then exploit them with high-interest loans, highlight the potential for data misuse.

Importantly, the banking and finance industry's rapid adoption of artificial intelligence adds a fresh layer of complexity to data security concerns. AI systems, particularly those operating in the realm of machine learning, are typically trained on large datasets. However, if these systems are not properly managed, they can inadvertently leak sensitive information. A phenomenon known as 'model inversion' provides a stark illustration. In essence, an AI model trained on customer data could, when subjected to particular queries, reveal private information about individual customers. For instance, an attacker with some knowledge of a person's financial transactions could potentially use the model to infer other sensitive information about that person.

Moreover, the extensive use of AI in the industry introduces new challenges around algorithmic transparency and interpretability. Customers must trust not only that their data is secure but also that it is being used responsibly and fairly. Banks and financial institutions must, therefore, ensure a high degree of transparency in how their AI systems operate and make decisions. This includes providing clear and accessible explanations of how AI uses data, the rationale behind specific decisions, and what measures are in place to check for bias or errors.

In conclusion, the risks and challenges associated with data breaches and misuse in the banking and finance sector are substantial. They extend beyond immediate financial losses to encompass issues of customer trust, institutional reputation, ethical use of AI, and the necessity for robust transparency and accountability mechanisms. The industry, in its stride towards digital innovation, must therefore prioritize stringent data security measures and responsible data practices to safeguard individuals' financial data and uphold the sector's integrity.

Counteracting the Risks

The multiplicity of risks associated with data breaches and misuse underlines the urgent necessity for potent data security measures and ethical data management practices. Notably, the line between the two is often blurred. Effective data security inherently requires an ethical approach to data handling, and vice versa.

At the technological level, encryption serves as a powerful tool in securing data. Through the transformation of readable data into coded form, it ensures that only those with the correct decryption key can access the original information. Additionally, anonymization techniques, such as data masking or pseudonymization, can be employed to protect individuals' identities by making it difficult to link data back to them. These techniques are critical in contexts where data needs to be used but personal identifiers are not necessary, such as in research or AI model training.

In tandem with these technological solutions, rigorous access controls are indispensable. These include measures such as multi-factor authentication, stringent password policies, and role-based access controls that limit who can access what data based on their role within the organization. Coupled with regular audits to detect and rectify potential security gaps, these measures form a formidable defense against unauthorized access to data.

Beyond these security precautions, transparency and informed consent form the ethical bedrock of data usage. Particularly when AI systems or third-party advertisers are involved, individuals must be given a clear, jargon-free explanation of what data is being collected, why it is needed, how it will be used, and who will have access to it.

Further, they should have the right to withdraw consent and request the deletion of their data.

Regulatory frameworks also play an instrumental role in the landscape of data privacy and security. Legislative efforts such as the European Union's General Data Protection Regulation (GDPR) and California's Consumer Privacy Act (CCPA) have set rigorous standards for consumer data protection, ushering in obligations for transparency, data minimization, and robust security measures.

However, it is paramount to acknowledge that the battle against data breaches and misuse is an unending journey. As AI systems evolve and become more complex, and as cybercriminals devise increasingly sophisticated tactics, strategies to protect data must be continually reassessed and fortified. Therefore, building resilient and trustworthy AI systems necessitates not only technical acumen but also an unwavering commitment to ethical stewardship, customer privacy, and transparency in business practices.

The promise of AI and data-driven innovation hinges on our ability to navigate this intricate and dynamic landscape. Organizations that embrace these challenges, investing in robust security measures and ethical data handling practices, will not only safeguard their customers' data but also earn their trust and pave the way for a more secure and ethical digital future.

## C. Examples of data privacy and security issues in various fields

In the digital age, data has ascended to a pivotal role, acting as the backbone of myriad sectors such as E-commerce, Healthcare, and Banking and Finance. However, this data-centricity brings forth an array of privacy and security concerns that vary in nature and magnitude across different industries. In this chapter, we delve into examples of such issues in the aforementioned sectors, elucidating the complexities they present and their implications.

### I. E-commerce

E-commerce, an industry intrinsically rooted in digital technologies, has experienced exponential growth over the past few years. As businesses transition online and consumers embrace digital shopping, vast quantities of personal and financial data exchange hands daily.

One prominent concern in this sector is the tracking and profiling of consumers. Many E-commerce platforms employ sophisticated algorithms to collect and analyze consumers' browsing behavior, purchasing history, and other online interactions. This practice allows for highly personalized marketing and product recommendations. However, when executed without transparent disclosure or without explicit customer consent, it raises significant privacy issues, straying into the territory of surveillance capitalism.

Another pressing issue is data breaches. With platforms handling vast amounts of sensitive customer information, they present lucrative targets for cybercriminals. Breaches can lead to exposure of personal data and financial details, opening avenues for fraud and identity theft. A notable example is the 2014 eBay breach, where cyberattackers accessed approximately 145 million users' contact information and encrypted passwords, highlighting the scale and severity of such threats.

II. Healthcare

Healthcare is another sector where data privacy and security issues are exceptionally prominent, owing to the sensitivity of the data involved. Electronic health records, genomic data, and insurance details, among others, constitute a veritable treasure trove of information, valuable not only for care provision but also for research and public health tracking.

One major concern in this sector is unauthorized access or disclosure of health information, which can lead to serious violations of patient privacy. A breach could potentially expose a patient's most intimate health details, resulting in personal embarrassment, social stigma, or even discriminatory treatment.

Moreover, as AI and data analytics increasingly penetrate healthcare, the potential for misuse amplifies. For instance, AI algorithms used for disease prediction or health risk assessment, if not carefully managed, can inadvertently disclose sensitive information, a phenomenon known as 'model leakage'.

III. Banking and Finance

The Banking and Finance sector, often seen as the lifeblood of the economy, is another field where data privacy and security hold center stage. Financial institutions collect a wealth of sensitive data, ranging from account details and transaction histories to personal identification data.

Identity theft is one of the most dire consequences of data breaches in this sector, where criminals could potentially gain access to accounts, make unauthorized transactions, or even illicitly open new accounts in the victim's name. Notable instances include the 2017 Equifax data breach, which exposed the personal information of 147 million people, leading to a substantial surge in identity theft cases.

Moreover, the misuse of financial data, such as unauthorized trading or discriminatory lending practices, can lead to significant financial harm for customers. Additionally, as AI systems are increasingly deployed for tasks like credit scoring or fraud detection, it is imperative to ensure that these systems handle data responsibly and transparently, to prevent discriminatory or unfair practices.

In conclusion, as data continues to fuel innovation across sectors, the concurrent challenges of privacy and security grow increasingly intricate and pressing. Each sector, with its unique characteristics and contexts, presents its set of issues that must be addressed with a nuanced understanding and sector-specific strategies. The shared objective across sectors, however, remains unchanged: to respect and protect individual privacy, ensure robust data security, and ultimately build a digital ecosystem that is safe, fair, and trustworthy.

## D. The responsibilities of businesses and developers in protecting user data

In our increasingly interconnected digital landscape, businesses and developers find themselves entrusted with vast amounts of user data, ranging from demographic information to behavioral patterns to sensitive personal identifiers. With this profound responsibility comes an ethical and legal obligation to safeguard this data from breaches and misuse. In this chapter, we scrutinize these responsibilities, analyzing their implications for businesses, developers, and the broader data ecosystem.

Businesses and Data Stewardship

Businesses, especially those operating in the fields of E-commerce, Healthcare, and Banking and Finance, stand on the front line of data management and protection. Their responsibilities extend across various facets of data stewardship.

Primarily, businesses bear the responsibility of ensuring robust security measures to guard against data breaches. These measures include but are not limited to employing encryption techniques, implementing stringent access controls, and conducting regular security audits. It is critical that these measures are not static, but rather continuously updated to counter evolving threats in the cyber landscape.

Moreover, businesses must ensure data accuracy and integrity, as inaccurate or outdated data can lead to misguided business decisions and unfair or harmful outcomes for users. This involves validating and cleaning data regularly and providing avenues for users to update their information.

Ethical data collection is another crucial responsibility. Businesses should minimize data collection to what is necessary for providing their service, avoiding gratuitous or invasive collection practices. The principle of data minimization, as enshrined in the European Union's General Data Protection Regulation (GDPR), embodies this responsibility.

Lastly, businesses must uphold the principle of transparency, clearly communicating their data practices to users. This includes providing comprehensive and comprehensible privacy policies, disclosing third-party data sharing, and obtaining informed consent for data collection and usage.

Developers and Ethical Design

On the other side of the equation stand developers, the architects of the digital systems that collect, store, analyze, and utilize data. They too bear considerable responsibility in ensuring data privacy and security.

Developers need to imbibe the principle of 'privacy by design', which posits that privacy should not be an afterthought, but rather a fundamental component integrated into the system from the initial stages of design. This entails implementing robust security measures, such as secure coding practices and vulnerability testing, as part of the software development lifecycle.

Furthermore, developers must also ensure their systems incorporate mechanisms to facilitate user control over data. This includes providing users with the ability to view, update, and delete their data, in line with the rights of access, rectification, and erasure stipulated by the GDPR.

Another significant responsibility for developers lies in the design of AI systems. As AI systems rely on data to learn and make decisions, developers must ensure these systems handle data responsibly and ethically. This involves not only securing the data but also ensuring the AI systems do not propagate bias, discrimination, or any form of unfairness.

In conclusion, the responsibility of protecting user data is a collective endeavor, requiring concerted efforts from both businesses and developers. Through adherence to principles of ethical data stewardship and privacy-centered design, these stakeholders can foster a data ecosystem that respects user privacy, safeguards data, and ultimately builds user trust. By doing so, they contribute to a digital landscape where innovation and user protection go hand in hand, fueling progress without compromising on ethics and integrity.

## E. Strategies and tools to enhance data privacy and security in AI systems

In our increasingly interconnected digital landscape, businesses and developers find themselves entrusted with vast amounts of user data, ranging from demographic information to behavioral patterns to sensitive personal identifiers. With this profound responsibility comes an ethical and legal obligation to safeguard this data from breaches and misuse. In this chapter, we scrutinize these responsibilities, analyzing their implications for businesses, developers, and the broader data ecosystem.

### Businesses and Data Stewardship

Businesses, especially those operating in the fields of E-commerce, Healthcare, and Banking and Finance, stand on the front line of data management and protection. Their responsibilities extend across various facets of data stewardship.

Primarily, businesses bear the responsibility of ensuring robust security measures to guard against data breaches. These measures include but are not limited to employing encryption techniques, implementing stringent access controls, and conducting regular security audits. It is critical that these measures are not static, but rather continuously updated to counter evolving threats in the cyber landscape.

Moreover, businesses must ensure data accuracy and integrity, as inaccurate or outdated data can lead to misguided business decisions and unfair or harmful outcomes for users. This involves validating and cleaning data regularly and providing avenues for users to update their information.

Ethical data collection is another crucial responsibility. Businesses should minimize data collection to what is necessary for providing their service, avoiding gratuitous or invasive collection practices. The principle of data minimization, as enshrined in the European Union's General Data Protection Regulation (GDPR), embodies this responsibility.

Lastly, businesses must uphold the principle of transparency, clearly communicating their data practices to users. This includes providing comprehensive and comprehensible privacy policies, disclosing third-party data sharing, and obtaining informed consent for data collection and usage.

Developers and Ethical Design

On the other side of the equation stand developers, the architects of the digital systems that collect, store, analyze, and utilize data. They too bear considerable responsibility in ensuring data privacy and security.

Developers need to imbibe the principle of 'privacy by design', which posits that privacy should not be an afterthought, but rather a fundamental component integrated into the system from the initial stages of design. This entails implementing robust security measures, such as secure coding practices and vulnerability testing, as part of the software development lifecycle.

Furthermore, developers must also ensure their systems incorporate mechanisms to facilitate user control over data. This includes providing users with the ability to view, update, and delete their data, in line with the rights of access, rectification, and erasure stipulated by the GDPR.

Another significant responsibility for developers lies in the design of AI systems. As AI systems rely on data to learn and make decisions, developers must ensure these systems handle data responsibly and ethically. This involves not only securing the data but also ensuring the AI systems do not propagate bias, discrimination, or any form of unfairness.

In conclusion, the responsibility of protecting user data is a collective endeavor, requiring concerted efforts from both businesses and developers. Through adherence to principles of ethical data stewardship and privacy-centered design, these stakeholders can foster a data ecosystem that respects user privacy, safeguards data, and ultimately builds user trust. By doing so, they contribute to a digital landscape where innovation and user protection go hand in hand, fueling progress without compromising on ethics and integrity.

# The Human-AI Partnership

A. Debunking the myth: AI as a job augmenter rather than a job replacer
B. The role of AI in enhancing human capabilities
C. Real-world examples of AI augmenting human jobs in different fields
i. E-commerce
ii. Healthcare
iii. Banking and Finance
D. The importance of re-skilling and up-skilling in the AI era

## E. Shaping policy and workplace culture to foster a harmonious human-AI partnership

In our rapidly advancing digital epoch, Artificial Intelligence (AI) is steadily becoming integral to many aspects of our lives. The workplace, one of the pivotal spheres of human activity, is no exception. AI has the potential to enhance productivity, augment human capabilities, and revolutionize business processes. However, to fully realize these benefits, it is critical to foster a harmonious human-AI partnership. This requires shaping workplace policies and culture in a manner conducive to the ethical and effective integration of AI. This chapter presents an analysis of the strategies and considerations involved in cultivating such a partnership.

Policy-Making for Human-AI Collaboration

The first prong of approach involves policy-making. Policies function as the scaffolding for human-AI interaction, providing guidelines that govern the usage, implementation, and ethical considerations of AI in the workplace.

At the organizational level, businesses should formulate policies that foster responsible AI usage. These policies could encompass principles for ethical AI development and usage, guidelines for data handling, and procedures for AI auditing. They should promote transparency, accountability, and fairness, thereby reducing the risk of unethical AI practices, such as algorithmic bias or invasive data collection.

The banking and finance sector, for instance, can establish policies that regulate the use of AI in decision-making processes to ensure fairness and prevent discriminatory practices. In healthcare, organizations can outline guidelines for the secure and confidential handling of patient data by AI systems.

At the macro level, policy-makers should strive to enact legislation that promotes the ethical use of AI. Such legislation could stipulate the protection of user data, the right to explanation for AI decisions, and guidelines for AI safety and robustness. Examples include the European Union's General Data Protection Regulation (GDPR) and proposals for AI regulation.

Cultivating an AI-Savvy Workplace Culture

Workplace culture, the second pillar of human-AI partnership, plays a crucial role in how AI is adopted and used. A positive AI culture can enhance acceptance, drive responsible usage, and spur innovation.

Creating an AI-savvy culture necessitates education. Employers should invest in training employees about AI, its capabilities, its limitations, and its ethical implications. This will empower employees to use AI tools effectively and responsibly, and equip them to participate in discussions and decisions about AI in the workplace.

In the context of E-commerce, for example, training can help employees understand how AI can be used to personalize user experiences, while also understanding the privacy implications and the need for user consent. In healthcare, professionals can be trained to leverage AI for diagnostics while also understanding the importance of validating AI recommendations with their expertise.

Furthermore, an inclusive culture that invites dialogue and feedback about AI can help alleviate fears and dispel misconceptions. It can also foster an environment where ethical considerations are openly discussed, fostering collective responsibility for ethical AI usage.

In summary, fostering a harmonious human-AI partnership in the workplace is a multi-faceted endeavor. It requires concerted efforts to shape responsible policies and cultivate an AI-savvy culture. These steps, while challenging, are crucial to unlocking the full potential of AI, making it a trusted ally in the workplace, and navigating the course of our digital future with ethical integrity.

# Unit Conclusion: Embracing the AI Revolution

As we traverse the landscape of the 21st century, a salient feature of our journey is the epochal transformation being wrought by Artificial Intelligence (AI). From our daily interactions on e-commerce platforms, to complex healthcare diagnostics, to intricate financial transactions, AI systems are now an integral part of the tapestry of our lives. As we stand on the threshold of this revolution, it is incumbent upon us to look back and draw lessons from our collective experience, to understand and embrace the transformative potential of AI.

AI has indeed transformed the way we interact with the world. In the realm of e-commerce, it has personalized user experiences, optimized business processes, and enabled access to a global marketplace. It has revolutionized the healthcare sector, making diagnostics more precise, enhancing patient care, and expediting medical research. In banking and finance, AI has improved risk assessment, streamlined operations, and introduced a new level of sophistication in financial analysis.

Yet, while celebrating these advancements, we must also acknowledge the challenges and risks that have emerged. The potency of AI, and its capacity to shape our lives, brings with it significant ethical considerations. Data privacy and security have emerged as paramount concerns. AI systems have the potential to compromise user data, leading to breaches that can result in financial fraud, loss of trust, and even discriminatory practices. The issue of AI fairness and the specter of algorithmic bias have also come into sharp relief, necessitating a renewed focus on the ethics of AI.

In our journey with AI, businesses and developers have a crucial role to play. They are tasked with the responsibility of not only harnessing the power of AI but also protecting user data and ensuring the ethical usage of AI. This entails employing robust data security measures, upholding transparency, and obtaining informed consent for data usage. Moreover, it involves making conscientious decisions about the development and deployment of AI systems, to avoid the pitfalls of algorithmic bias and other ethical concerns.

Policy-making and workplace culture are pivotal in shaping our relationship with AI. Policies provide the framework for governing the use of AI, advocating for ethical practices, and promoting transparency and accountability. Workplace culture, on the other hand, shapes the human-AI interaction. An AI-savvy culture enhances acceptance, encourages responsible usage, and spurs innovation.

As we continue to embrace the AI revolution, we must strive for a balance. The balance between harnessing the transformative potential of AI and addressing the ethical and practical challenges that it presents. The balance between driving innovation and protecting user privacy. The balance between automating processes and respecting human autonomy. And, above all, the balance between viewing AI as a tool at our disposal and recognizing it as a partner in our shared journey into the future.

In conclusion, the AI revolution, like any great societal shift, brings with it profound opportunities and challenges. It is our responsibility to navigate these complexities, to ensure that we embrace AI in a manner that respects our ethical values, protects our rights, and enhances our capabilities. This journey is not one that we undertake in isolation, but collectively—as developers, as users, as policy-makers, as societies. And as we embark on this journey, let us remember that it is not about being swept away by the AI revolution, but about thoughtfully embracing it. Together, we can shape the future of AI, and in doing so, shape our collective future.

Summarizing the transformative power of AI in various industries.

Artificial Intelligence (AI) has indisputably reshaped the landscape of various industries. Its transformative power has brought about unprecedented changes in the operational, strategic, and competitive dynamics within these sectors. Let us now summarize these transformations, drawing on our detailed exploration of AI's impact in e-commerce, healthcare, and banking and finance.

E-commerce

E-commerce, the frontrunner in digital transformations, has witnessed considerable advancements fueled by AI. AI systems have enabled the creation of sophisticated recommendation engines, capable of analyzing vast datasets to offer personalized shopping experiences. These predictive models, equipped with machine learning algorithms, assess individual behavior, discern patterns, and predict consumer preferences with remarkable accuracy.

Furthermore, AI has revolutionized inventory management and supply chain processes in e-commerce. Through the utilization of predictive analytics, AI systems can forecast demand, optimize stock levels, and improve efficiency. Similarly, AI-driven chatbots have significantly enhanced customer service, providing immediate responses to customer inquiries, thereby increasing customer satisfaction and retention rates.

However, along with these transformative powers comes the need for careful navigation of data privacy and security concerns. It is incumbent upon e-commerce platforms to ensure the safe and ethical use of AI, thereby upholding consumer trust and regulatory compliance.

Healthcare

In healthcare, AI has paved the way for a new era of medical diagnosis, treatment, and patient care. Machine learning algorithms are being employed to identify disease patterns, predict patient outcomes, and even generate treatment plans. These technologies have the potential to augment the capabilities of medical professionals, enhance diagnostic accuracy, and expedite the process of medical research.

AI's role in healthcare extends beyond direct patient care. It plays a vital role in administrative tasks, such as scheduling appointments, managing patient records, and handling insurance claims. Furthermore, AI's potential in drug discovery and telemedicine marks a significant advancement in healthcare delivery.

Again, the same power that makes AI a potent tool in healthcare also poses ethical challenges. The secure handling of sensitive medical data and the question of algorithmic bias in healthcare predictions are among the key concerns that need to be addressed.

Banking and Finance

In banking and finance, AI has introduced a new level of sophistication. AI's potential to analyze large datasets enables it to accurately assess credit risk, detect fraudulent activities, automate customer service, and provide personalized financial advice. This has not only improved the efficiency and accuracy of financial services but also provided customers with a seamless banking experience.

AI has also influenced investment and trading activities. Through algorithmic trading, AI systems can analyze market trends, make predictions, and execute trades at speeds unattainable by human traders. In addition, AI models can help identify opportunities for mergers, acquisitions, and investments, leading to informed decision-making.

However, as in the other industries, the adoption of AI in banking and finance comes with its own set of ethical considerations. Issues related to data privacy, security, and the potential for algorithmic bias in financial decision-making are crucial and need to be managed with the utmost diligence.

In conclusion, AI has undoubtedly ushered in a new era of innovation across industries, redefining processes, enhancing efficiency, and paving the way for future advancements. However, the journey with AI is not devoid of challenges. As we further embed AI into our societal fabric, a keen awareness of its transformative power must be complemented by a thoughtful understanding of the accompanying ethical implications. As stewards of AI, it is our collective responsibility to navigate these transformative powers judiciously and ethically, thereby ensuring that AI serves as a tool for progress and prosperity.


**Encouraging businesses and individuals to seize the opportunities presented by AI.**

The rise of Artificial Intelligence (AI) represents a transformative epoch in human history. It ushers in an era where machines learn from experience, adapt to novel inputs, and execute tasks with a proficiency and speed that challenge or even surpass human capability. This chapter aims to encourage businesses and individuals alike to understand and seize the manifold opportunities that AI presents.

## A Call to Businesses

Businesses, irrespective of their domain or scale, stand to gain significantly from the prudent integration of AI technologies into their operational frameworks. However, the implementation of AI should not be perceived as a mere addition of technology, but a strategic alignment of business models, practices, and objectives with the capabilities of AI.

In e-commerce, AI can enhance customer experiences, refine marketing strategies, and optimize supply chain logistics. The usage of AI in healthcare can revolutionize diagnostic procedures, enable personalized treatment, streamline administrative tasks, and pave the way for innovative practices such as telemedicine. In banking and finance, AI can fortify security, automate customer service, improve risk assessment, and facilitate precision trading.

To harness the potential of AI, businesses should foster an environment that values innovation, encourages learning, and promotes digital literacy. AI, however, should not be treated as a panacea for all challenges. Businesses must understand that the adoption of AI requires careful planning, ethical considerations, and meticulous management of data privacy and security.

Moreover, businesses must remember that AI does not eliminate the need for human intervention. Instead, it necessitates an evolved role for human employees who can effectively collaborate with AI, harness its strengths, and address its limitations. Consequently, businesses should consider investing in continuous learning and skill development programs that prepare their human resources for a symbiotic partnership with AI.

## A Message to Individuals

For individuals, the AI revolution brings a plethora of opportunities, albeit with its own set of challenges. It is not just the tech-savvy or those engaged in STEM disciplines who stand to gain from this technological wave. AI's influence permeates various fields, rendering it an asset for artists, writers, educators, and many others.

The integration of AI into day-to-day lives is not a distant reality but a present circumstance. From smartphone assistants to recommendation engines, AI subtly yet significantly influences our lifestyle and decision-making processes. Understanding how AI works, and more importantly, how it can be utilized effectively and ethically, is thus a necessity, not a luxury.

Individuals must actively seek to understand AI, its implications, and its potential uses. Online courses, seminars, and workshops on AI offer ample opportunities for individuals to familiarize themselves with the subject matter. Learning programming languages such as Python, often used in AI development, or studying data analysis could provide individuals with valuable skills in an increasingly data-driven world.

In addition, individuals must understand the importance of data privacy and the ethical use of AI. By fostering a culture of digital literacy and ethical digital practices, individuals can ensure that they reap the benefits of AI without compromising their privacy or ethical standards.

In conclusion, the era of AI is a horizon of opportunities waiting to be seized. Businesses and individuals, by actively embracing the AI revolution, can gain a competitive edge, improve efficiency, and discover innovative solutions to age-old problems. However, while doing so, they must also ensure that they navigate this path with responsibility, thoughtfulness, and respect for ethical considerations. The AI revolution is not just about reaping benefits, but also about fostering a future where technology and humanity coexist harmoniously, enriching each other's capabilities.

## Acknowledging the challenges ahead and advocating for responsible AI development.

As we chart the course into an era marked by the ascendancy of Artificial Intelligence (AI), it is crucial to confront and cogitate upon the challenges this technological revolution presents. This concluding chapter aims to foster an understanding of these complexities, while simultaneously advocating for responsible AI development that synergizes technological advancement with ethical, social, and legal prudentials.

### A Panorama of Challenges

The AI revolution, while promising unprecedented benefits, ushers in a myriad of challenges that are technical, ethical, legal, and societal in nature. These must be acknowledged, examined, and addressed with the same fervour with which we embrace the benefits of AI.

### Technical Challenges

On the technical front, the problem of explainability, or the so-called "black box" issue, is significant. AI's capacity to reach conclusions or make predictions without offering a clear explanation of its reasoning process poses a challenge for its adoption and trustworthiness. How can we rely on a system whose workings we do not understand, particularly when critical areas such as healthcare or autonomous vehicles are involved?

Furthermore, the creation of robust, safe, and secure AI systems is a formidable challenge. AI models must be resilient against adversarial attacks, where small, intentional changes in input data can lead to dramatically incorrect outputs. These issues are of utmost importance as AI begins to permeate areas with serious consequences for failure, such as defence and cyber-security.

Ethical and Legal Challenges

The rise of AI ushers in profound ethical and legal concerns. Questions regarding data privacy and surveillance have taken centre stage. As our reliance on AI systems grows, so does the amount of data we produce and share, leading to privacy risks and potential misuse.

On the legal front, issues such as liability and accountability in the event of AI-related harm or damage remain largely unresolved. If an autonomous vehicle is involved in an accident, who is at fault: the manufacturer, the software developer, or the owner of the vehicle? The law must evolve to accommodate such complex scenarios.

Societal Challenges

AI also poses significant societal challenges. The issue of job displacement due to automation is a pressing concern, with certain industries and job roles at a higher risk of becoming obsolete. This could lead to economic disparities and social unrest if not handled carefully.

Advocating Responsible AI Development

Given these challenges, it becomes imperative to advocate for a responsible and inclusive approach to AI development.

First and foremost, the process of developing AI should be guided by a set of ethical standards that prioritizes respect for human rights, fairness, transparency, and accountability. AI technologies must be designed and used in a manner that respects user privacy and promotes trust.

Moreover, ensuring diversity and inclusivity in the field of AI is crucial. A lack of diversity could result in biased AI systems that perpetuate existing inequities, or overlook certain societal needs.

Additionally, there should be a clear legal framework governing AI use, including guidelines for data protection, privacy, and liability. It is also essential to establish norms and regulations for AI safety, to ensure robustness against adversarial attacks and prevent catastrophic mishaps.

From a societal perspective, there must be proactive measures to manage job displacement, including policies for job transition, education, and upskilling. A balance must be struck between automation and job creation to ensure an equitable economic transition.

In conclusion, while the journey towards AI dominance promises to be transformative, it is strewn with challenges that demand our attention. By acknowledging these challenges and advocating for responsible AI development, we can ensure a future where AI is not only technologically advanced, but also ethically sound, legally compliant, and socially beneficial.

# Part 2: Understanding AI: A Primer for Readers

In this second part of our exploration into the realms of Artificial Intelligence (AI), we delve into the fundamentals of AI technology, providing readers with a comprehensive primer to understand its underlying principles, capabilities, and limitations. By equipping ourselves with a solid understanding of AI, we can navigate the intricate landscape of this transformative field with clarity and confidence.

Unveiling the Essence of AI
Artificial Intelligence is a multidimensional concept, encompassing a spectrum of technologies and methodologies designed to enable machines to perform tasks that would typically require human intelligence. At its core, AI seeks to emulate human cognitive functions, such as perception, reasoning, learning, and problem-solving, and replicate them in machines.

The Building Blocks of AI
To comprehend the intricacies of AI, we must acquaint ourselves with its foundational building blocks. Machine Learning (ML), one of the most prominent branches of AI, enables machines to learn from data and improve their performance over time. ML algorithms are fed large datasets, from which they extract patterns and insights, allowing them to make predictions or take actions based on that knowledge.

Deep Learning (DL), a subset of ML, mimics the neural structure of the human brain through artificial neural networks. These networks consist of interconnected nodes called neurons, which process and transmit information. By utilizing multiple layers of neurons, deep neural networks can tackle complex tasks, such as image recognition, natural language processing, and speech synthesis.

The Role of Data in AI
Data is the lifeblood of AI. It fuels the learning process and empowers AI systems to extract meaningful information. The quality, diversity, and volume of data are crucial factors in the performance and effectiveness of AI algorithms. Consider an e-commerce platform that leverages customer data to personalize product recommendations. The more comprehensive and accurate the data, the better the system can understand individual preferences and deliver tailored suggestions.

## The Ethical Dimensions of AI

As we delve deeper into the realm of AI, we cannot overlook the ethical implications that arise from its deployment. AI brings with it a host of moral considerations, including the potential for bias, privacy invasion, job displacement, and the concentration of power. It is imperative that we address these ethical dimensions and strive to develop and utilize AI in a manner that aligns with societal values and respects human rights.

## The Promise and Limitations of AI

While AI holds immense promise in transforming industries, augmenting human capabilities, and solving complex problems, it is crucial to acknowledge its limitations. AI systems are susceptible to biases, can be brittle in unfamiliar situations, and lack true understanding or consciousness. It is important to manage expectations and approach AI as a powerful tool that complements human expertise, rather than a replacement for human intelligence.

## Conclusion

As we embark on this journey to unravel the intricacies of AI, we must cultivate a mindset of curiosity and critical thinking. Part 2 aims to equip readers with a solid foundation of AI principles, methodologies, and ethical considerations. Armed with this knowledge, we can explore the diverse applications of AI with discernment, navigate the complexities, and actively shape the future of this remarkable technology.

# Chapter 6: Overview of AI Technologies

In this chapter, we embark on a comprehensive exploration of the diverse technologies that underpin the fascinating world of Artificial Intelligence (AI). From classic techniques to cutting-edge advancements, we uncover the fundamental principles, methodologies, and applications that have propelled AI to the forefront of technological innovation. By delving into the depths of these AI technologies, we gain a deeper understanding of their capabilities, limitations, and the remarkable possibilities they offer across various fields.

The Spectrum of AI Technologies

AI technologies encompass a broad spectrum of methodologies and techniques, each serving unique purposes and addressing distinct challenges. Let us embark on a journey through this rich landscape to unveil the intricacies of the most prominent AI technologies.

1. Machine Learning (ML)

Machine Learning lies at the heart of AI, enabling systems to learn from data and improve their performance through experience. ML algorithms analyze large datasets, recognize patterns, and make predictions or take actions based on their learned knowledge. One notable example is the recommendation systems employed by e-commerce platforms, which leverage ML to provide personalized product suggestions based on user preferences and historical data.

2. Deep Learning (DL)

Deep Learning, a subset of ML, has revolutionized AI by emulating the human brain's neural structure. Deep Neural Networks (DNNs) consist of interconnected layers of artificial neurons that process and transmit information. DL has achieved remarkable success in complex tasks such as image and speech recognition, natural language processing, and autonomous driving. For instance, DL models power virtual voice assistants like Siri and Alexa, enabling natural language understanding and intelligent responses.

3. Natural Language Processing (NLP)

Natural Language Processing focuses on enabling machines to understand, interpret, and generate human language. NLP encompasses a wide range of techniques, including language understanding, sentiment analysis, machine translation, and text generation. NLP finds applications in various fields, from customer support chatbots in e-commerce to medical text analysis in healthcare.

4. Computer Vision

Computer Vision empowers machines to extract information from visual data, such as images and videos. By leveraging image recognition, object detection, and scene understanding, computer vision enables AI systems to perceive and interpret the visual world. Applications range from facial recognition for authentication and surveillance to autonomous vehicles that navigate and understand their surroundings.

5. Robotics and Autonomous Systems

Robotics and Autonomous Systems combine AI with physical embodiments to create intelligent machines capable of sensing, perceiving, and interacting with their environment. These systems range from autonomous drones and self-driving cars to advanced industrial robots that revolutionize manufacturing processes. AI-driven robotics promises to enhance efficiency, safety, and productivity across industries.

6. Reinforcement Learning

Reinforcement Learning focuses on training AI agents to make sequential decisions by interacting with an environment and receiving feedback in the form of rewards or penalties. This technology has demonstrated groundbreaking achievements in game-playing AI, such as AlphaGo, which defeated world champions in the ancient game of Go. Reinforcement Learning also finds applications in optimizing resource allocation, inventory management, and autonomous control systems.

Conclusion

As we conclude our journey through the overview of AI technologies, we gain a profound appreciation for the multifaceted nature of this field. Machine Learning, Deep Learning, Natural Language Processing, Computer Vision, Robotics, and Reinforcement Learning each contribute unique capabilities and drive advancements in AI. By leveraging these technologies, we unlock a world of possibilities in fields ranging from e-commerce and healthcare to banking and finance. With a solid understanding of these AI technologies, we are poised to explore their applications, tackle complex challenges, and shape the future of intelligent systems.

# What is Artificial Intelligence?: Defining AI and its objectives to mimic human intelligence and problem-solving capabilities.

Artificial Intelligence (AI) has become a buzzword that permeates various aspects of our lives, from virtual assistants to autonomous vehicles. But what exactly is AI, and what are its objectives? In this chapter, we delve into the essence of AI, exploring its definition and the goals it seeks to achieve—namely, the replication of human intelligence and problem-solving capabilities.

## Defining Artificial Intelligence

Artificial Intelligence, in its simplest form, refers to the development of intelligent machines that can perform tasks requiring human-like intelligence. It involves the creation of computer systems that can mimic cognitive functions such as learning, reasoning, problem-solving, perception, and decision-making. These systems aim to exhibit behaviors that imitate or surpass human capabilities, enabling them to tackle complex problems and adapt to new situations.

## Objectives of Artificial Intelligence

The primary objectives of AI revolve around emulating and augmenting human intelligence to enhance efficiency, productivity, and decision-making. Let's explore the key goals that AI seeks to achieve:

### 1. Problem Solving and Decision Making

One of the fundamental objectives of AI is to develop systems capable of solving complex problems and making informed decisions. By leveraging advanced algorithms, machine learning, and reasoning techniques, AI systems can analyze vast amounts of data, identify patterns, and derive meaningful insights. This ability to navigate complex problem spaces and provide intelligent solutions has far-reaching implications in diverse fields such as e-commerce, healthcare, banking, and finance.

### 2. Learning and Adaptation

AI aims to create systems that can learn from experience and adapt to changing circumstances. Machine Learning, a subset of AI, plays a crucial role in this objective. By training AI models on large datasets, they can acquire knowledge, recognize patterns, and improve their performance over time. This capability enables AI systems to continuously evolve and refine their understanding, making them invaluable in tasks like customer behavior analysis in e-commerce or medical diagnosis in healthcare.

3. Natural Language Processing and Communication

A significant objective of AI is to enable machines to understand and communicate with humans in natural language. Natural Language Processing (NLP) technologies, such as speech recognition and language generation, facilitate effective human-machine interaction. Virtual assistants like Amazon's Alexa and Apple's Siri exemplify the integration of NLP in everyday life, where users can engage in conversations, ask questions, and receive informative responses.

4. Perception and Sensing

AI aspires to endow machines with the ability to perceive and interpret the world around them. Computer Vision, a field within AI, focuses on enabling machines to understand and process visual information from images and videos. This capability finds applications in facial recognition, object detection, and autonomous driving. By replicating human-like perception, AI systems can interact with their environment in a more intuitive and intelligent manner.

Conclusion

As we conclude this chapter, we have gained a deeper understanding of Artificial Intelligence and its objectives. AI strives to replicate human intelligence and problem-solving capabilities, aiming to develop systems that can solve complex problems, learn from experience, communicate in natural language, and perceive and interact with the world. Through advancements in machine learning, deep learning, natural language processing, and computer vision, AI has made remarkable progress in various fields. By embracing AI and its objectives, we unlock the potential for transformative advancements and create a future where human and machine intelligence work in harmony to solve complex challenges and shape our world.

**Machine Learning (ML): Introducing the core concept of ML and its role in training AI models using data to make predictions and decisions.**

Machine Learning (ML) lies at the heart of the advancements we witness in Artificial Intelligence (AI) today. It empowers AI systems to learn from data, make predictions, and adapt to new information. In this chapter, we will delve into the core concept of Machine Learning and explore its pivotal role in training AI models to make intelligent decisions and predictions.

Understanding Machine Learning

Machine Learning is a subset of AI that focuses on creating algorithms and models capable of learning from data. Traditional programming involves explicitly defining rules and instructions for a system to follow. In contrast, Machine Learning enables systems to learn patterns and relationships directly from data, without explicitly being programmed.

At the core of Machine Learning is the concept of training. Instead of providing explicit instructions, ML algorithms are trained on large datasets, allowing them to extract patterns, discover insights, and make predictions. This process involves feeding the algorithm with labeled or unlabeled data, enabling it to iteratively adjust its internal parameters and improve its performance over time.

Training AI Models with Data

The training of AI models using data is a fundamental aspect of Machine Learning (ML). It involves collecting and preparing relevant data, extracting and representing meaningful features, selecting and training ML models, and evaluating their performance. In this chapter, we will delve into the key steps involved in training AI models, exploring the intricacies and significance of each stage.

1. Data Collection and Preparation

The first step in training AI models is to collect and prepare the data. The quality and relevance of the data play a crucial role in the performance of the trained model. In the field of e-commerce, for example, transaction records, customer behavior data, and product information are collected to build models that can predict customer preferences and recommend personalized products.

Once the data is collected, it needs to be processed and prepared for training. This involves cleaning the data by removing outliers, handling missing values, and addressing any inconsistencies. The data may also need to be transformed or normalized to ensure compatibility with the ML algorithms used for training.

2. Feature Extraction and Representation

Feature extraction is a critical step in training AI models. Features are the measurable properties or characteristics of the data that contribute to the learning process. In e-commerce, features could include customer demographics, purchase history, and product attributes. In healthcare, features could include patient medical records and vital signs.

Feature extraction involves selecting the most relevant features from the dataset and representing them in a way that captures their underlying patterns. This step requires domain expertise and an understanding of the problem at hand. Various techniques, such as dimensionality reduction and feature engineering, may be employed to enhance the discriminative power of the selected features.

3. Model Selection and Training

Choosing an appropriate ML model is crucial for achieving accurate predictions and optimal performance. The selection of the model depends on the nature of the problem, the available data, and the desired outcome. Linear regression models, decision trees, support vector machines, and neural networks are examples of ML models commonly used in different domains.

Once the model is selected, it is trained using the prepared data. During the training process, the ML algorithm adjusts its internal parameters to minimize errors and optimize its performance. The algorithm learns from the provided data and attempts to generalize patterns and relationships, enabling it to make predictions on new, unseen data.

4. Evaluation and Validation

After training the ML model, it is essential to evaluate its performance and validate its effectiveness. This is typically done using a separate dataset called the validation set or test set. The model's predictions on this dataset are compared to the ground truth values to measure its accuracy, precision, recall, and other performance metrics.

Evaluating the model helps identify any shortcomings or areas that require improvement. If the model meets the desired criteria, it can be deployed for real-world applications, such as predicting customer behavior, diagnosing diseases, or detecting fraudulent transactions.

Conclusion

The training of AI models is a multi-step process that involves collecting and preparing relevant data, extracting meaningful features, selecting and training ML models, and evaluating their performance. Each step is crucial in ensuring the accuracy and effectiveness of the trained models. By following these steps and leveraging the power of data, businesses and organizations can unlock the potential of AI and make informed decisions that drive innovation and progress in various fields.

Role of Machine Learning in AI

Machine Learning (ML) plays a pivotal role in the field of Artificial Intelligence (AI) by enabling systems to learn from data and make intelligent predictions and decisions. The ability of AI models to adapt and improve their performance based on real-world data makes them powerful tools in various fields. In this chapter, we will explore examples of how ML is transforming industries such as e-commerce, healthcare, and banking and finance.

E-commerce

In the realm of e-commerce, ML algorithms have revolutionized the way businesses understand and engage with their customers. By analyzing vast amounts of customer data, including behavior, purchase history, and browsing patterns, ML models can generate personalized product recommendations. These recommendations enhance the user experience, increase customer engagement, and drive sales. For example, popular e-commerce platforms like Amazon and Netflix leverage ML to provide tailored product suggestions and content recommendations to their users.

Healthcare

Machine Learning has brought significant advancements to the healthcare industry. ML models trained on extensive medical datasets can detect patterns and make accurate predictions, leading to improved diagnosis, personalized treatment plans, and early disease detection. By analyzing patient medical records, genetic information, and real-time health data, ML algorithms assist healthcare professionals in making informed decisions. This reduces medical errors, optimizes resource allocation, and improves patient outcomes. For instance, ML models have been used to identify early signs of diseases like cancer and predict patient responses to different treatments, enabling personalized and targeted care.

Banking and Finance

ML algorithms have transformed the banking and finance sector by enabling institutions to make data-driven decisions, mitigate risks, and provide personalized services to their customers. These algorithms analyze financial data, market trends, and customer profiles to detect fraudulent activities, predict market trends, and optimize investment strategies. By leveraging ML, banks can identify suspicious transactions, prevent fraudulent activities, and enhance security measures. ML-based trading systems can analyze vast amounts of market data to make real-time investment decisions, increasing efficiency and profitability. Moreover, ML models can help banks personalize financial services, such as offering customized loan rates or personalized investment portfolios, based on individual customer profiles.

Conclusion

Machine Learning is a powerful tool that underpins the development of Artificial Intelligence. By allowing systems to learn from data, ML algorithms enable AI models to make predictions, analyze complex patterns, and adapt to new information. Through data collection, feature extraction, model training, and evaluation, AI models become capable of making intelligent decisions and predictions in various fields, such as e-commerce, healthcare, and finance. As ML techniques continue to advance, we can expect AI to play an even more significant role in shaping the future.

**Natural Language Processing (NLP): Explaining NLP's significance in enabling AI systems to understand, interpret, and generate human language.**

Natural Language Processing (NLP) is a field of Artificial Intelligence (AI) that focuses on the interaction between computers and human language. It enables AI systems to understand, interpret, and generate human language, opening up a wide range of applications in various fields. In this chapter, we will explore the significance of NLP in enabling AI systems to process and communicate in human language.

The Complexity of Human Language

Human language is a complex and nuanced form of communication, encompassing grammar, semantics, syntax, and context. Understanding and processing human language is a formidable challenge for AI systems due to its inherent ambiguity, subtlety, and variations. NLP aims to bridge this gap by developing algorithms and models that can effectively process and analyze human language.

- Core Components of NLP

Natural Language Processing (NLP) is a branch of Artificial Intelligence (AI) that encompasses several core components working together to enable AI systems to interact with human language. In this chapter, we will delve into these components and explore their significance in enabling AI to understand, interpret, and generate human language.

- Text Preprocessing

     Text preprocessing is the initial step in NLP, where raw text data is cleaned and transformed to make it suitable for analysis. This process involves removing punctuation, converting text to lowercase, handling special characters or encoding issues, and other techniques. Proper text preprocessing sets the foundation for subsequent NLP tasks and ensures accurate and meaningful analysis.

- Tokenization

     Tokenization is the process of breaking down text into individual units called tokens. Tokens can be words, phrases, or even characters. By segmenting the text into tokens, AI systems can analyze and process the information at a granular level. For example, tokenization enables counting word frequencies, identifying key phrases, or building language models.

- Morphological Analysis

     Morphological analysis focuses on studying the internal structure and forms of words. It involves identifying prefixes, suffixes, and root words to understand word variations and derive their meaning. This knowledge is vital for tasks like word normalization, stemming, and lemmatization. For instance, understanding that "run," "running," and "ran" all have the same root helps in capturing the core meaning of the word.

- Part-of-Speech Tagging

     Part-of-speech (POS) tagging assigns grammatical tags to each word in a sentence, such as noun, verb, adjective, or adverb. This information helps AI systems understand the role and function of each word within the sentence. POS tagging forms the basis for syntactic analysis and aids in tasks like grammar checking, text summarization, and information extraction.

- Named Entity Recognition

     Named Entity Recognition (NER) aims to identify and classify named entities in text, such as person names, organizations, locations, dates, and more. NER plays a crucial role in understanding the context of the text and extracting relevant information. For example, in a news article, NER can identify the names of individuals involved, the organizations mentioned, and the locations discussed.

- Sentiment Analysis

Sentiment analysis focuses on determining the sentiment or opinion expressed in a piece of text. It helps AI systems understand the subjective tone of the text, whether it is positive, negative, or neutral. Sentiment analysis finds applications in customer feedback analysis, social media monitoring, and brand reputation management. For instance, e-commerce platforms can analyze customer reviews to gauge customer satisfaction and improve their products or services.

- Language Generation

Language generation involves the creation of coherent and contextually relevant text based on given input or prompts. This capability enables AI systems to generate automated responses in chatbots, write product descriptions, or even produce creative content like articles or stories. Language generation techniques range from rule-based approaches to more advanced methods like neural language models.

Conclusion

Natural Language Processing (NLP) is a vital component of AI that enables systems to understand, interpret, and generate human language. The core components of NLP, including text preprocessing, tokenization, morphological analysis, part-of-speech tagging, named entity recognition, sentiment analysis, and language generation, work together to unlock the power of language in AI applications. By harnessing the capabilities of NLP, AI systems can effectively analyze and derive insights from textual data, facilitate human-like interactions, and contribute to advancements in various fields, from customer service and healthcare to content generation and beyond.

Applications of NLP

Natural Language Processing (NLP) has emerged as a transformative technology that enables AI systems to interact with human language and perform tasks that were once exclusive to human intelligence. In this chapter, we will explore the profound impact of NLP on various industries and delve into the challenges it faces.

NLP Applications in Industries

E-commerce

In the realm of e-commerce, NLP plays a pivotal role in enhancing customer experience and improving business operations. NLP-powered chatbots and virtual assistants can understand and respond to customer inquiries, providing instant support and personalized recommendations. By analyzing customer interactions and preferences, e-commerce platforms can tailor their offerings, optimize product searches, and increase conversion rates.

Healthcare

NLP has revolutionized the healthcare industry by enabling the analysis of vast amounts of medical data, including electronic health records, research papers, and clinical notes. By extracting relevant information and identifying patterns, NLP assists in diagnosing diseases, predicting patient outcomes, and supporting clinical decision-making. Furthermore, NLP-powered systems automate data extraction, leading to improved efficiency and accuracy in healthcare workflows.

Banking and Finance

In the domain of banking and finance, NLP has become indispensable for extracting insights from financial news, market reports, and social media sentiment. By analyzing textual data, NLP can predict market trends, assess investment risks, and provide personalized financial advice to clients. Moreover, NLP-powered chatbots and virtual assistants enable customers to perform banking transactions, access account information, and receive tailored financial guidance.

Challenges in NLP

While NLP has made significant strides in enabling AI systems to understand and interact with human language, it still grapples with several challenges that demand ongoing research and innovation. Let's delve deeper into these key hurdles:

- Understanding Context:

    Understanding context is a fundamental challenge for NLP systems. Human language is rich with subtle nuances, cultural references, and context-dependent meanings that can vary across different situations. Ambiguities, idiomatic expressions, and sarcasm further complicate accurate comprehension. For example, the phrase "I'm in a pickle" may have a literal meaning of being in a difficult situation or a metaphorical meaning of being in a dilemma. NLP systems struggle to capture such nuances and accurately interpret the intended meaning within a specific context.

- Handling Ambiguity:

    Language is inherently ambiguous, and NLP models face challenges in disambiguating and interpreting meaning accurately. Ambiguity can arise from multiple sources, including homonyms (words with different meanings but the same spelling), polysemous words (words with multiple related meanings), and syntactic ambiguities (multiple interpretations of sentence structures). Resolving lexical and syntactic ambiguity remains a complex task that requires advanced language understanding and reasoning capabilities.

- Capturing Nuances:

    Language carries nuanced meanings that can be challenging to capture accurately with NLP systems. Subtle variations in tone, sentiment, and cultural references contribute to the intricate fabric of human communication. For instance, a simple statement like "That's interesting" can convey positive or negative sentiment depending on the context and intonation. Capturing these nuances requires sophisticated NLP techniques, including sentiment analysis, tone detection, and understanding cultural references, to ensure precise interpretation and response generation.

- Language Variations and Cultural Differences:

    Language varies across regions, dialects, and cultural contexts, posing a significant challenge for NLP systems. Different linguistic variations and cultural perspectives result in diverse ways of expressing ideas and nuances. For instance, the word "biscuit" refers to a different food item in the United Kingdom compared to the United States. NLP systems need to account for these variations and adapt to different dialects, regional expressions, and cultural nuances to ensure accurate understanding and generation of language. Developing universal NLP models that can handle such language variations and cultural differences is a complex and ongoing endeavor.

    Addressing these challenges requires continuous research and development efforts to enhance the performance and capabilities of NLP systems. Researchers are exploring advanced techniques, including deep learning, neural language models, and large-scale pre-training, to tackle these hurdles. Additionally, incorporating world knowledge, improving context understanding, and considering cultural and linguistic diversity are crucial for advancing NLP capabilities.

    By overcoming these challenges, NLP can reach new heights, enabling AI systems to understand and respond to human language with ever-increasing accuracy and sophistication. This paves the way for transformative applications in fields such as customer service, healthcare, education, and many more, where human-like interactions with AI systems can drive innovation and improve the human experience.

Conclusion

Natural Language Processing (NLP) has significantly impacted various industries, empowering AI systems to understand and interact with human language. From e-commerce to healthcare and banking and finance, NLP-driven applications have improved customer experiences, optimized operations, and enabled personalized services. However, challenges such as understanding context, handling ambiguity, capturing nuances, and addressing language variations and cultural differences persist. Advancements in NLP research and technology will be crucial in overcoming these challenges and unlocking the full potential of NLP in shaping the future of AI and human-machine interaction.Conclusion

**Computer Vision: Describing the importance of computer vision in AI systems' ability to interpret and process visual information.**

Computer Vision, a subfield of Artificial Intelligence (AI), equips AI systems with the ability to understand, interpret, and process visual information, much like humans do. It enables machines to perceive and analyze images and videos, opening up a world of possibilities across various industries. Let's delve into the significance of computer vision and its transformative impact.

The Power of Visual Perception

Human beings rely heavily on their visual perception to understand the world around them. Our brains effortlessly process the intricate details of our surroundings, recognize objects, and extract meaningful information from visual stimuli. Computer vision aims to replicate this remarkable human ability by enabling AI systems to perform similar tasks.

The Role of Computer Vision in AI Systems

Computer vision empowers AI systems to "see" and comprehend visual information, allowing them to analyze and interpret images and videos. By integrating computer vision capabilities, AI systems can extract valuable insights, make informed decisions, and automate tasks that require visual understanding. Here are some key applications of computer vision across various industries:

- Autonomous Vehicles: Revolutionizing Transportation with Computer Vision

One of the most significant advancements in the field of computer vision is its role in enabling self-driving or autonomous vehicles. By leveraging computer vision algorithms, AI systems can navigate and perceive their environment, making informed decisions and ensuring safe transportation. Let's explore how computer vision transforms the landscape of autonomous vehicles.

Perception and Decision-Making

Autonomous vehicles rely on computer vision to perceive and understand the world around them. By analyzing visual data captured by cameras and other sensors, AI systems equipped with computer vision algorithms can detect and recognize various objects, including pedestrians, vehicles, traffic signs, and road markings. This visual perception forms the foundation for decision-making in autonomous vehicles.

- Object Detection and Recognition

Computer vision algorithms excel at object detection and recognition, enabling autonomous vehicles to accurately identify and classify objects in real-time. This capability is crucial for ensuring the safety of passengers and pedestrians, as well as for making informed decisions on the road. For example, a computer vision system can detect a pedestrian crossing the road and adjust the vehicle's speed and trajectory accordingly.

- Lane Detection and Tracking

Computer vision also plays a key role in lane detection and tracking. By analyzing visual cues, such as road markings, AI systems can precisely determine the vehicle's position within the lanes. This information allows the autonomous vehicle to stay within its designated lane, accurately follow the road's curvature, and safely navigate through intersections and complex road scenarios.

- Traffic Sign Recognition

Recognizing and interpreting traffic signs is essential for autonomous vehicles to comply with traffic regulations. Computer vision algorithms can detect and interpret traffic signs, such as speed limits, stop signs, and traffic signals. By understanding and responding to these signs, autonomous vehicles can make informed decisions about speed adjustments, lane changes, and traffic signal adherence.

- Enhanced Safety and Efficiency

The integration of computer vision in autonomous vehicles significantly enhances safety and efficiency on the roads. Computer vision algorithms enable real-time monitoring of the vehicle's surroundings, detecting potential hazards and avoiding collisions. This technology reduces the risk of accidents caused by human errors, distractions, or fatigue. Moreover, autonomous vehicles can optimize their driving patterns based on traffic conditions, reducing congestion and improving overall road efficiency.

Conclusion

Computer vision's role in autonomous vehicles is transformative, revolutionizing transportation as we know it. By enabling vehicles to perceive and interpret their surroundings, computer vision algorithms empower autonomous vehicles to navigate safely, make informed decisions, and comply with traffic regulations. The use of computer vision in autonomous vehicles has the potential to enhance road safety, increase efficiency, and revolutionize the way we travel. As computer vision technology continues to advance, we can expect even greater advancements in autonomous vehicles, paving the way for a future where transportation is safer, more efficient, and more accessible for all.

The Challenges of Computer Vision

The Challenges of Computer Vision: Pushing the Boundaries of Perception

While computer vision has made remarkable progress, it faces several challenges that demand ongoing research and innovation. Overcoming these hurdles is crucial for unlocking the full potential of computer vision in various applications. Let's delve into some of the key challenges:

- Object Recognition in Complex Environments

Accurate object recognition in complex and dynamic environments poses a significant challenge for computer vision algorithms. Factors such as occlusions, varying lighting conditions, and object deformations can impede the recognition process. Robust algorithms that can handle these complexities and extract meaningful features from visual data are necessary to improve object recognition performance.

- Handling Large-Scale Data

The availability of vast amounts of visual data presents challenges in terms of processing, storage, and retrieval. Computer vision algorithms need to be scalable and efficient to handle large-scale datasets. Innovations in data management, parallel computing, and distributed processing are essential to address these challenges effectively.

- Interpretation of Visual Context

Extracting higher-level semantics and understanding the visual context from images and videos is a complex task. While computer vision algorithms can recognize objects, understanding their relationships, interactions, and the overall scene context is still an ongoing research challenge. Advancements in semantic understanding, scene understanding, and context reasoning are necessary to unlock the full potential of computer vision.

- Ethical Considerations

As computer vision technology becomes more pervasive, ethical considerations come to the forefront. Privacy concerns, surveillance implications, and potential biases in automated decision-making need careful consideration. It is vital to develop computer vision systems that respect privacy, protect individual rights, address bias issues, and ensure fair and unbiased outcomes.

- Data Bias and Generalization

Computer vision algorithms heavily rely on training data to learn and make predictions. However, biases present in the training data can propagate into the algorithm's behavior, leading to biased outcomes. Achieving robust and generalizable computer vision systems requires addressing issues of bias in data collection, data representation, and algorithmic decision-making.

Conclusion

The challenges faced by computer vision highlight the complexity of perception and understanding visual information. Overcoming these hurdles requires interdisciplinary research, innovative algorithm design, and ethical considerations. By addressing challenges such as object recognition in complex environments, handling large-scale data, interpreting visual context, and ensuring ethical considerations, computer vision can continue to advance and revolutionize various domains, including autonomous vehicles, healthcare, and surveillance. By pushing the boundaries of perception, computer vision has the potential to create profound impact and unlock new opportunities in the era of AI.

Computer vision is a cornerstone of AI, enabling machines to perceive, analyze, and understand visual information. It finds applications in various industries, including autonomous vehicles, healthcare, manufacturing, retail, and security. Despite the challenges it faces, ongoing research and development efforts continue to push the boundaries of computer vision. As computer vision technology advances, it holds the potential to revolutionize industries, enhance human capabilities, and pave the way for new and transformative applications.

**Robotics and Automation: Highlighting the integration of AI with robotics to create intelligent and autonomous systems.**

The integration of Artificial Intelligence (AI) with robotics has ushered in a new era of intelligent and autonomous systems. This fusion combines the power of AI algorithms with the physical capabilities of robots, enabling them to perceive, reason, and act in complex environments. Let us explore the realm of robotics and automation, where AI-driven robots are revolutionizing various industries.

The Rise of Intelligent Robots

Intelligent robots, driven by AI algorithms, have revolutionized the way they interact with the world around them. These robots possess the ability to sense their environment, interpret the sensory inputs, and make informed decisions based on the available information. Through the integration of AI with robotics, these systems can perform tasks with human-like intelligence and precision, opening up new possibilities across various industries.

One of the key components that enable intelligent robots to perceive and understand their environment is computer vision. By leveraging computer vision techniques, robots can analyze visual data, such as images or video streams, and extract meaningful information from them. This allows them to recognize objects, detect and track motion, and navigate through complex environments. For example, in industrial settings, robots with computer vision capabilities can identify and manipulate objects on a production line, ensuring efficient and accurate assembly processes.

Another essential aspect of intelligent robots is natural language processing (NLP), which enables them to understand and respond to human language. NLP algorithms allow robots to interpret and extract meaning from written or spoken language, enabling them to interact with humans in a more natural and intuitive manner. This is particularly valuable in scenarios where human-robot collaboration is required, such as customer service or healthcare assistance. By understanding and generating human language, robots can engage in meaningful conversations and carry out tasks based on verbal or written instructions.

Machine learning techniques also play a crucial role in empowering intelligent robots. Machine learning algorithms enable robots to learn from data and improve their performance over time. By analyzing patterns and extracting insights from large datasets, robots can acquire knowledge, adapt to different situations, and make informed decisions. This capability is particularly beneficial in dynamic and unpredictable environments where robots need to continuously update their understanding and behavior. For instance, autonomous drones equipped with machine learning algorithms can learn to navigate through complex environments, avoiding obstacles and adapting their flight paths based on real-time sensor data.

Integrating AI with robotics has unlocked tremendous potential for intelligent robots to perform tasks with precision, accuracy, and adaptability. Whether it's in manufacturing, healthcare, transportation, or other industries, these systems have the capacity to augment human capabilities, automate repetitive tasks, and operate in challenging or hazardous environments. As AI and robotics continue to advance, the synergy between the two fields will lead to even more sophisticated and capable intelligent robots, further transforming our world and pushing the boundaries of what was once thought possible.

Enhancing Efficiency and Productivity

In industries such as manufacturing and logistics, the integration of robotics and automation powered by AI has proven to be transformative. By leveraging AI algorithms, robots can handle repetitive and labor-intensive tasks with remarkable speed, accuracy, and consistency. This not only enhances productivity but also brings about significant cost reductions and improvements in safety conditions for workers.

A prominent example of the impact of AI-powered robots can be seen in automotive assembly lines. Robots equipped with advanced AI algorithms can efficiently perform intricate tasks such as welding, painting, and assembly. These tasks, which require precision and dexterity, can be executed by robots with unmatched consistency and quality. This not only reduces the risk of human errors but also increases the overall efficiency of the assembly process.

By automating these tasks, manufacturers can achieve higher production rates, reduce cycle times, and optimize resource allocation. This, in turn, leads to cost savings and improved competitiveness in the market. Moreover, robots do not experience fatigue or require breaks, enabling continuous and uninterrupted operation. This enhanced productivity contributes to the streamlining of manufacturing processes and the optimization of overall output.

Furthermore, the implementation of AI-powered robots in manufacturing environments significantly improves safety conditions for workers. Robots excel in performing tasks that are dangerous or physically demanding, mitigating the risk of injuries or accidents. By taking over these tasks, robots allow human workers to focus on more complex and value-added activities that require cognitive skills and decision-making abilities. This shift in roles creates a safer and more fulfilling work environment.

The impact of AI-powered robotics extends beyond the manufacturing sector. In logistics and warehousing, robots equipped with AI algorithms facilitate efficient material handling, inventory management, and order fulfillment. These robots can navigate through complex environments, autonomously transport goods, and optimize storage and retrieval operations. The integration of AI enables them to adapt to dynamic conditions, respond to changes in demand, and enhance overall supply chain efficiency.

While the integration of AI and robotics in manufacturing and logistics brings numerous benefits, it also presents challenges. For instance, the deployment and maintenance of robotic systems require specialized skills and expertise. Additionally, the ethical implications of automation, such as the potential displacement of human workers, need to be carefully considered and addressed.

Despite these challenges, the transformative potential of AI-powered robotics in industries such as manufacturing and logistics is undeniable. The combination of advanced algorithms, sensor technologies, and robotic systems opens up new possibilities for increased efficiency, productivity, and safety. As AI continues to advance and robotics become more intelligent and capable, we can expect further advancements that will reshape these industries and unlock new opportunities for growth and innovation.

Advancing Healthcare and Rehabilitation

AI-driven robots have indeed made significant contributions to the healthcare sector, particularly in the areas of surgical assistance and rehabilitation. These robots, equipped with advanced AI algorithms, have the potential to revolutionize medical procedures and improve patient outcomes.

In surgical settings, AI-driven robots assist surgeons in performing complex procedures with enhanced precision and accuracy. These robots are capable of carrying out intricate maneuvers that surpass human capabilities, such as performing microsurgeries or accessing hard-to-reach areas of the body. By leveraging AI algorithms, these robots can analyze real-time feedback from sensors and make adjustments to ensure precise movements. Surgeons can benefit from the robotic assistance by achieving greater accuracy, reducing the invasiveness of procedures, and minimizing the risk of complications. Ultimately, this can lead to improved patient outcomes, shorter recovery times, and reduced healthcare costs.

Another area where AI-driven robots have found significant applications is in rehabilitation. Rehabilitation robots help patients regain motor skills and facilitate their recovery from injuries or neurological conditions. These robots employ AI algorithms to adapt their actions based on the progress of individual patients. By analyzing patient responses and adjusting therapy parameters, these robots can provide personalized rehabilitation programs tailored to each patient's specific needs. This level of customization and adaptability allows for more effective and targeted rehabilitation, leading to improved patient outcomes and enhanced quality of life.

The integration of AI in robotic systems for surgical assistance and rehabilitation brings several advantages. First, it enhances the precision and accuracy of procedures, reducing the risk of human error. In surgical settings, this can result in improved outcomes and reduced complications. In rehabilitation, AI-driven robots can provide more precise and controlled therapy, optimizing the recovery process. Second, these robots can store and analyze large volumes of patient data, enabling clinicians to gain valuable insights and make data-driven decisions. This data-driven approach to healthcare can lead to more efficient and effective treatment plans. Lastly, AI-driven robots in healthcare can help alleviate the workload on medical professionals, allowing them to focus on complex decision-making tasks and providing higher levels of patient care.

However, the integration of AI-driven robots in healthcare also presents challenges. Safety and reliability are of utmost importance, as any malfunction or error in robotic systems could have serious consequences for patients. Moreover, ethical considerations regarding patient privacy and the responsibility of human oversight must be carefully addressed. Ensuring that these systems are transparent, explainable, and accountable is crucial to building trust in their deployment.

In conclusion, AI-driven robots have made significant advancements in the healthcare sector, particularly in surgical assistance and rehabilitation. By combining the precision and adaptability of AI algorithms with robotic systems, these technologies have the potential to revolutionize medical procedures and improve patient outcomes. As further research and development are carried out, we can expect to witness even more remarkable applications of AI-driven robots in healthcare, ultimately enhancing the quality of care and transforming the way medical treatments are delivered.

Enabling Autonomous Vehicles

The integration of AI and robotics has played a critical role in the development of autonomous vehicles, paving the way for a transformative revolution in transportation. Self-driving cars, powered by AI algorithms, computer vision, and sensor fusion, possess the ability to perceive their surrounding environment, make real-time decisions, and navigate safely without human intervention.

One of the key components of autonomous vehicles is their advanced perception system. Through computer vision techniques, these vehicles can analyze the visual input from cameras and other sensors to detect and recognize objects on the road, such as pedestrians, vehicles, traffic signs, and lane markings. AI algorithms process the data collected from these sensors, enabling the vehicle to build a comprehensive understanding of its surroundings. By leveraging sophisticated machine learning algorithms, autonomous vehicles can continuously improve their perception capabilities and adapt to changing road conditions.

However, perception alone is not sufficient for autonomous vehicles to operate safely. They also require robust decision-making capabilities. AI algorithms analyze the information gathered from sensors and make real-time decisions on acceleration, braking, and steering. These decisions are based on a combination of predefined rules, machine learning models, and probabilistic reasoning. By considering various factors such as traffic conditions, road markings, and the behavior of other vehicles, autonomous vehicles can navigate complex scenarios and make informed decisions that prioritize safety.

The potential benefits of autonomous vehicles are significant. Improved road safety is one of the key advantages, as human error is a major factor in a large number of traffic accidents. By removing the human element from the driving equation, autonomous vehicles can significantly reduce the occurrence of accidents caused by factors such as driver fatigue, distraction, or impaired driving. Furthermore, autonomous vehicles have the potential to optimize traffic flow, reducing congestion and improving overall transportation efficiency. Through advanced algorithms and coordination between vehicles, autonomous systems can achieve smoother and more predictable traffic patterns, leading to reduced travel times and fuel consumption.

Despite the promising potential of autonomous vehicles, there are still challenges that need to be addressed. Safety remains a top concern, as the reliability of AI algorithms and the robustness of the vehicle's decision-making systems must be ensured. Additionally, regulatory frameworks and legal considerations need to be established to govern the use and deployment of autonomous vehicles on public roads. Ethical dilemmas may also arise in situations where vehicles have to make split-second decisions that involve trade-offs and potential harm.

In conclusion, the synergy of AI and robotics has propelled the development of autonomous vehicles. These vehicles leverage AI algorithms, computer vision, and sensor fusion to perceive their environment, make real-time decisions, and navigate safely. By enhancing road safety, reducing congestion, and transforming the transportation landscape, autonomous vehicles hold immense potential to reshape our daily lives. However, addressing challenges related to safety, regulation, and ethics will be crucial for the successful and responsible integration of autonomous vehicles into our society.

### Ethical Considerations and Safety

As the autonomy of robots increases, it becomes imperative to address the ethical considerations and safety implications associated with their deployment. The rise of autonomous robots raises important questions regarding their ethical use, impact on employment, and potential for misuse. Careful consideration and the development of ethical frameworks are essential to ensure responsible integration of AI and robotics.

- Ethical Use of Autonomous Robots:

    The ethical use of autonomous robots involves examining their impact on society, human rights, and the potential consequences of their actions. For example, in the field of autonomous weapons, the development and deployment of lethal autonomous robots raise concerns about accountability, proportionality, and the potential for human rights violations. Clear guidelines and international agreements are necessary to govern the development and use of such technologies to prevent their misuse.

- Impact on Employment:

    The increasing automation of tasks through robots raises concerns about the impact on employment and the workforce. While robots can enhance productivity and efficiency, they also have the potential to replace human workers in certain job roles. This raises questions about job displacement, retraining, and the need for a social safety net to support individuals affected by automation. Society must proactively address these concerns to ensure a smooth transition and equitable distribution of the benefits brought about by AI and robotics.

✧   Safety of Humans Working Alongside Robots:

As robots become more autonomous and interact closely with humans, ensuring the safety of humans becomes paramount. Collaborative robots, also known as cobots, are designed to work alongside humans, sharing the same workspace and performing tasks collaboratively. However, safety protocols and standards must be in place to prevent accidents or harm to humans. Risk assessments, safety training, and the implementation of physical safeguards are crucial to protect human workers from potential hazards.

✧   Developing Robust Safety Protocols and Standards:

To ensure the safe integration of AI and robotics, robust safety protocols and standards need to be developed and implemented. This includes rigorous testing, certification processes, and continuous monitoring of robotic systems. Safety considerations should encompass not only the functionality of the robots but also the robustness of their AI algorithms and the reliability of their sensors and actuators. Regulatory bodies, industry collaborations, and research institutions play a vital role in establishing these protocols and standards.

✧   Ethical Frameworks for AI and Robotics:

Ethical frameworks are essential to guide the development, deployment, and use of AI and robotics. These frameworks should encompass principles such as transparency, accountability, fairness, and privacy. They should address issues such as algorithmic bias, the protection of personal data, and the potential for discriminatory outcomes. Establishing ethical guidelines ensures that AI and robotics technologies are aligned with societal values and do not compromise human rights or social well-being.

In conclusion, as robots become more autonomous, ethical considerations and safety become paramount. The ethical use of autonomous robots, their impact on employment, and the safety of humans working alongside them require careful consideration. Developing robust safety protocols and standards, along with ethical frameworks, is essential for the responsible integration of AI and robotics. By addressing these considerations proactively, we can ensure that AI and robotics technologies contribute positively to society while safeguarding human rights and well-being.

Conclusion: The Future of Robotics and Automation

The integration of AI with robotics is revolutionizing industries and ushering in a new era where intelligent machines work in collaboration with humans. By combining AI algorithms with physical robotics, these systems gain the ability to perceive their environment, reason about complex tasks, and make informed decisions. This fusion of AI and robotics has far-reaching implications and is reshaping various sectors, including healthcare, manufacturing, transportation, and beyond.

In healthcare, AI-driven robotics are transforming the way medical procedures are performed. Surgical robots equipped with AI algorithms assist surgeons in delicate procedures, enhancing precision and minimizing invasiveness. These systems can analyze vast amounts of medical data to support diagnosis, treatment planning, and personalized care. AI-driven robotics also enable the development of prosthetics and exoskeletons that restore mobility and improve the quality of life for individuals with disabilities.

In manufacturing, the synergy of AI and robotics has revolutionized production processes. Intelligent robots equipped with AI algorithms can handle repetitive and labor-intensive tasks with speed, accuracy, and consistency. This not only enhances productivity but also reduces costs and improves safety conditions for workers. For example, in automotive assembly lines, robots equipped with AI algorithms can efficiently perform intricate tasks like welding, painting, and assembly.

In transportation, the integration of AI and robotics is driving the development of autonomous vehicles. Self-driving cars leverage AI algorithms, computer vision, and sensor fusion to perceive the surrounding environment, make real-time decisions, and navigate safely. Through advanced perception and decision-making capabilities, autonomous vehicles have the potential to revolutionize transportation, improving road safety and reducing congestion.

While the integration of AI and robotics brings immense opportunities, it also raises ethical concerns and safety considerations. Ethical considerations involve ensuring the responsible development and deployment of AI-driven robotic systems, addressing issues of privacy, accountability, and potential biases. Safety considerations revolve around developing robust safety protocols, standards, and risk assessment frameworks to protect humans working alongside robots and prevent accidents.

To foster a harmonious coexistence between humans and intelligent machines, collaboration and cooperation are crucial. Humans and robots can complement each other's strengths, with humans providing creative problem-solving, empathy, and complex decision-making, while robots contribute efficiency, precision, and scalability. By fostering collaboration and establishing clear guidelines for ethical and safe practices, society can harness the full potential of AI-driven robotics while ensuring that the benefits are distributed equitably and the risks are mitigated.

In conclusion, the integration of AI with robotics is reshaping industries and paving the way for a future where intelligent machines work alongside humans. By combining AI algorithms and physical robotics, these systems gain advanced perception, reasoning, and decision-making capabilities. While this convergence offers tremendous opportunities for advancement, it is vital to address ethical concerns, prioritize safety, and foster collaboration between humans and intelligent machines. Through responsible development and deployment, we can create a harmonious coexistence that unlocks the full potential of AI-driven robotics and drives societal progress.

# Chapter 7: Unveiling the Essence of AI

As we delve deeper into the world of artificial intelligence (AI), it becomes increasingly crucial to uncover the essence of this transformative field. AI, with its ability to mimic human intelligence and perform complex tasks, has captured our imaginations and revolutionized industries across the globe. In this chapter, we embark on a journey to unravel the fundamental principles and components that underpin AI, shedding light on its capabilities, limitations, and potential implications for society.

The Essence and Core Principles of AI

At its core, AI aims to create intelligent systems that can perceive, reason, learn, and make decisions autonomously. These systems are designed to replicate and surpass human cognitive abilities, enabling them to perform tasks that were once exclusive to human intelligence. The essence of AI lies in its ability to process vast amounts of data, recognize patterns, and generate insights that drive informed decision-making.

Machine Learning: The Foundation of AI

To understand the essence of AI, we must first grasp the foundation upon which it stands: machine learning (ML). Machine learning algorithms empower AI systems to learn from data, adapt, and improve their performance over time. Through processes such as supervised learning, unsupervised learning, and reinforcement learning, ML algorithms extract meaningful information, identify patterns, and make accurate predictions.

Neural Networks: Unleashing the Power of Deep Learning

Deep learning, a subset of ML, has emerged as a powerful tool in AI. Central to deep learning are neural networks, computational models inspired by the structure and functioning of the human brain. Neural networks, with their interconnected layers of artificial neurons, excel at processing complex data, such as images and natural language. Through the advancement of deep learning techniques, AI systems can achieve remarkable accuracy and efficiency in tasks such as image recognition, speech synthesis, and language translation.

## Unleashing the Potential of Natural Language Processing (NLP)

One of the most captivating aspects of AI lies in its ability to understand and generate human language. Natural language processing (NLP) encompasses the techniques and algorithms that enable AI systems to interpret and respond to human language. By leveraging NLP, AI can comprehend text, extract meaningful information, and generate coherent responses. NLP finds applications in various fields, from virtual assistants and chatbots to language translation and sentiment analysis.

## Unveiling the Power of Computer Vision

The human ability to interpret and understand visual information is a fundamental aspect of intelligence. In AI, computer vision empowers systems to perceive and interpret visual data, just as humans do. By employing techniques such as image recognition, object detection, and image segmentation, AI systems can analyze visual information, enabling applications in fields like healthcare, manufacturing, and autonomous vehicles.

## The Synergy of AI and Robotics

The integration of AI with robotics represents a paradigm shift in automation and human-machine interaction. AI-driven robots possess the ability to sense their environment, understand it, and make decisions based on available information. These intelligent machines excel in handling repetitive, labor-intensive tasks, leading to increased productivity, cost reduction, and improved safety conditions in industries such as manufacturing, logistics, and healthcare.

## Ethical Considerations and Safety in AI

As AI becomes increasingly autonomous, ethical considerations and safety precautions take center stage. Questions surrounding the ethical use of autonomous robots, their impact on employment, and the potential for misuse necessitate careful deliberation. Ensuring the safety of humans working alongside robots is paramount, demanding the development of robust safety protocols, ethical frameworks, and guidelines for responsible AI practices.

## Looking Ahead: Balancing Progress and Societal Well-being

As we unveil the essence of AI, it is crucial to strike a delicate balance between technological progress and societal well-being. The transformative power of AI holds immense potential, but ethical considerations, privacy concerns, and the impact on human employment must be carefully addressed. By fostering collaboration between humans and intelligent machines, we can forge a path toward a future where AI and robotics coexist harmoniously, augmenting human capabilities and propelling society forward.

Conclusion

In this chapter, we have delved into the essence of AI, exploring its core principles, components, and applications across various domains. AI's ability to process data, learn from experience, and make intelligent decisions has the potential to reshape industries and enhance human lives. However, as AI continues to advance, ethical considerations, safety precautions, and the responsible development of AI systems remain imperative. By embracing the essence of AI with a steadfast commitment to ethical practices, we can unleash its full potential while ensuring a future that benefits all of humanity.

## The transformative power of AI

Artificial Intelligence (AI) has emerged as a transformative force, revolutionizing industries, shaping economies, and fundamentally altering the way we live and work. With its ability to mimic human intelligence and perform complex tasks, AI has become a catalyst for innovation and progress in fields such as e-commerce, healthcare, banking, and finance. In this section, we will explore the transformative power of AI and its impact on various sectors.

Enhancing Customer Experience in E-commerce

In the realm of e-commerce, AI has enabled personalized customer experiences, transforming the way businesses interact with their customers. Through AI-powered recommendation systems, e-commerce platforms can analyze customer preferences, browsing behavior, and purchase history to deliver tailored product suggestions. This enhances customer engagement, increases conversion rates, and drives sales. For instance, companies like Amazon and Netflix leverage AI algorithms to offer personalized product and content recommendations, ensuring a seamless shopping or streaming experience for their users.

Revolutionizing Healthcare with AI

AI has had a profound impact on healthcare, driving advancements in diagnosis, treatment, and patient care. Machine learning algorithms, trained on vast amounts of medical data, can assist healthcare professionals in detecting diseases, predicting patient outcomes, and recommending personalized treatment plans. AI-powered imaging technologies, such as computer-aided diagnosis systems, have improved the accuracy and efficiency of medical imaging interpretation, aiding in the early detection of diseases like cancer. Additionally, AI-enabled chatbots and virtual assistants are enhancing patient engagement and accessibility to healthcare services. Companies like IBM's Watson Health and Google's DeepMind are actively involved in developing AI solutions to tackle healthcare challenges and improve patient outcomes.

Optimizing Decision-making in Banking and Finance

AI has proven to be a game-changer in the banking and finance sector, empowering organizations to make data-driven decisions and optimize their operations. AI algorithms can analyze vast amounts of financial data, market trends, and customer behavior to detect patterns, predict market movements, and optimize investment strategies. In risk management, AI-powered systems can detect fraudulent activities, mitigate financial risks, and enhance security measures. For example, algorithmic trading, driven by AI, enables financial institutions to execute trades at lightning speed, leveraging complex algorithms that analyze market conditions and historical data to make investment decisions.

Addressing Challenges and Ethical Considerations

While the transformative power of AI is undeniable, it is essential to address challenges and ethical considerations that arise with its deployment. The potential for biases in AI algorithms, data privacy concerns, and the impact on employment are critical issues that demand careful attention. Biases in training data can lead to unfair and discriminatory outcomes, while the collection and use of personal data raise concerns about privacy and data security. Additionally, the automation of certain tasks through AI systems has raised questions about the potential displacement of jobs and the need for re-skilling the workforce.

In conclusion, the transformative power of AI is reshaping industries and societies, unlocking new possibilities and driving innovation. From personalized e-commerce experiences to revolutionizing healthcare and optimizing decision-making in banking and finance, AI is revolutionizing the way we interact with technology and transforming the landscape of various sectors. However, it is crucial to address challenges and ethical considerations to ensure that AI is deployed responsibly, with a focus on fairness, transparency, and inclusivity. By embracing the transformative power of AI while upholding ethical principles, we can harness its potential to drive positive change and create a better future for all.

# The essence and core principles of AI

Artificial Intelligence (AI) has become a buzzword in recent years, but what exactly is its essence? At its core, AI is a field of study and technology that aims to create intelligent machines capable of mimicking human cognitive abilities. AI systems are designed to perceive, reason, learn, and make decisions based on data and algorithms. In this section, we will delve into the essence of AI and explore its core principles.

## Machine Learning: The Foundation of AI

Machine learning is a fundamental principle underlying AI. It involves developing algorithms and models that enable machines to learn from data and improve their performance without being explicitly programmed. Through the process of training on vast amounts of data, AI systems can identify patterns, extract meaningful insights, and make predictions or decisions based on the acquired knowledge. Machine learning techniques include supervised learning, unsupervised learning, and reinforcement learning. Examples of machine learning applications can be found in various fields, such as image recognition, natural language processing, and recommendation systems in e-commerce.

## Neural Networks: The Brain of AI

Neural networks, inspired by the structure and functioning of the human brain, form the backbone of many AI systems. These interconnected layers of artificial neurons enable machines to process and analyze complex data, recognize patterns, and make inferences. Deep learning, a subfield of machine learning, harnesses the power of neural networks with multiple layers to solve intricate tasks. For instance, deep neural networks have achieved remarkable success in image and speech recognition, enabling AI systems to perceive and interpret the visual and auditory world with increasing accuracy.

## Natural Language Processing: Communicating with Machines

Natural language processing (NLP) focuses on the interaction between humans and machines through language. NLP enables machines to understand, interpret, and generate human language, opening up possibilities for voice assistants, chatbots, and language translation services. NLP algorithms employ techniques such as text analysis, sentiment analysis, and named entity recognition to extract meaning and context from written or spoken language. NLP finds applications in customer service, information retrieval, and sentiment analysis across industries.

Ethics and Bias in AI

As AI becomes more prevalent, it is essential to address ethical considerations and potential biases embedded in AI systems. AI algorithms are developed based on training data, and if the data used for training is biased, the algorithms can perpetuate or amplify those biases. For example, in facial recognition systems, biases in the training data have led to disproportionately higher error rates for certain ethnic groups. Recognizing these ethical challenges, researchers and policymakers are striving to develop AI systems that are fair, transparent, and accountable.

Human-AI Collaboration

Contrary to popular misconceptions, AI is not intended to replace humans but to augment human intelligence and capabilities. The concept of human-AI collaboration emphasizes the collaboration between humans and machines, leveraging their respective strengths to achieve better outcomes. AI can handle large-scale data analysis, automate repetitive tasks, and provide valuable insights, while humans bring creativity, critical thinking, and ethical judgment to the table. Successful applications of AI, such as in healthcare diagnosis or financial decision-making, often involve the harmonious integration of AI systems with human expertise.

In conclusion, the essence of AI lies in its ability to emulate human intelligence, enabling machines to perceive, reason, learn, and make decisions. Machine learning, neural networks, natural language processing, ethics, and human-AI collaboration form the core principles that underpin AI. As AI continues to evolve, it is crucial to address ethical considerations, strive for transparency, and foster collaboration between humans and machines to harness the full potential of AI and ensure its responsible and beneficial integration into society.

# Understanding the capabilities and limitations of AI

Artificial Intelligence (AI) has witnessed remarkable advancements in recent years, enabling machines to perform complex tasks with human-like intelligence. However, it is essential to grasp both the capabilities and limitations of AI to harness its potential effectively. In this section, we will explore the capabilities and limitations of AI, shedding light on what AI can accomplish and the challenges it still faces.

Capabilities of AI:

✧ Data Processing and Analysis:

AI systems possess an exceptional ability to process and analyze vast amounts of data rapidly and accurately. This capability enables them to uncover intricate patterns, correlations, and insights that may elude human perception. In the realm of e-commerce, AI algorithms play a pivotal role in understanding and predicting customer behavior, optimizing marketing strategies, and delivering personalized product recommendations.

The volume of data generated in the e-commerce industry is staggering, with each customer interaction, transaction, and browsing activity leaving digital footprints. AI algorithms can ingest and process this data, extracting valuable information and uncovering hidden patterns that human analysts may struggle to identify due to the sheer volume and complexity of the data.

By analyzing customer behavior, purchase history, and browsing patterns, AI algorithms can discern meaningful relationships and preferences. For instance, an AI-powered recommendation system can analyze a customer's previous purchases, search history, and interactions on the e-commerce platform to generate personalized product recommendations. These recommendations are based on the customer's unique preferences and buying patterns, increasing the likelihood of engagement and conversion.

Moreover, AI algorithms can optimize marketing strategies by leveraging data analytics to identify trends and patterns in customer behavior. By analyzing demographic information, browsing history, and purchase patterns, AI can segment customers into distinct groups and tailor marketing campaigns to specific target audiences. This targeted approach enhances the effectiveness of marketing efforts, leading to higher conversion rates and customer satisfaction.

An excellent example of AI-driven personalized recommendations can be observed in the success of major e-commerce platforms such as Amazon and Netflix. These platforms employ sophisticated AI algorithms that analyze user data to deliver personalized product suggestions and content recommendations. The AI algorithms continuously learn from user interactions, improving the accuracy and relevance of the recommendations over time.

In addition to personalized recommendations, AI can also optimize pricing strategies and inventory management in the e-commerce industry. By analyzing market trends, competitor pricing, and customer demand patterns, AI algorithms can identify optimal pricing points and adjust inventory levels to meet customer demands efficiently.

The ability of AI systems to process and analyze vast amounts of data with remarkable speed and accuracy gives e-commerce businesses a competitive edge. It allows them to understand customer preferences, make data-driven decisions, and deliver tailored experiences. As AI continues to advance, its impact on the e-commerce sector is expected to deepen, leading to improved customer satisfaction, increased sales, and enhanced operational efficiency.

- Pattern Recognition and Prediction:

AI's ability to recognize patterns and make predictions based on historical data is indeed a powerful tool across various industries. The healthcare and finance sectors are two prime examples where AI algorithms have made significant contributions.

In the healthcare industry, the analysis of medical records and patient data is crucial for accurate diagnosis, treatment planning, and disease management. AI algorithms can analyze vast amounts of medical data, including electronic health records, medical imaging, genetic information, and patient-reported outcomes. By recognizing patterns and correlations within this data, AI systems can predict disease outcomes, assist in early detection, and personalize treatment plans.

For instance, in cancer diagnosis, AI algorithms trained on large datasets of medical images can accurately identify cancerous cells or tumors, assisting radiologists in making timely and accurate diagnoses. AI-powered predictive models can analyze patient data and identify risk factors or early signs of diseases such as diabetes or cardiovascular conditions, enabling healthcare professionals to intervene proactively and provide personalized interventions.

Furthermore, AI can optimize treatment plans by considering individual patient characteristics, genetic information, and response data from similar patients. By analyzing historical patient data and treatment outcomes, AI algorithms can provide recommendations on the most effective treatment options for specific individuals, taking into account factors such as drug efficacy, potential side effects, and personalized dosage regimens.

In the finance industry, AI algorithms have proven to be valuable tools for analyzing market trends, historical data, and economic indicators to make predictions and optimize investment strategies. By recognizing patterns and correlations within financial data, AI systems can predict stock prices, identify market trends, and automate trading decisions.

For example, AI-powered trading systems can analyze vast amounts of historical financial data, news articles, social media sentiment, and other relevant information to identify patterns that indicate potential market movements. By continuously learning from historical data and adapting to changing market conditions, these systems can make predictions on stock price movements, enabling traders and investors to make more informed decisions.

AI algorithms can also optimize investment portfolios by considering risk profiles, financial goals, and market conditions. By analyzing historical market data and simulating various investment scenarios, AI systems can suggest optimal asset allocations and investment strategies that maximize returns while minimizing risks.

It is important to note that while AI algorithms have shown promise in these domains, they are not meant to replace human expertise and judgment. Rather, they act as powerful tools that augment human decision-making, providing valuable insights and predictions based on extensive data analysis.

In both healthcare and finance, the integration of AI algorithms has the potential to enhance decision-making, improve outcomes, and drive innovation. By leveraging the power of AI to analyze historical data and recognize patterns, these industries can unlock new possibilities for personalized medicine, more accurate diagnoses, optimized treatment plans, and informed financial decisions.

- Natural Language Processing and Understanding:

AI systems equipped with natural language processing (NLP) capabilities have revolutionized the way humans interact with machines. NLP enables AI systems to understand and interpret human language, bridging the communication gap between humans and machines. This capability has given rise to virtual assistants, chatbots, and voice-controlled devices that can understand and respond to user commands, queries, and conversations.

Virtual assistants, such as Amazon's Alexa, Apple's Siri, and Google Assistant, are prominent examples of AI applications that utilize NLP. These virtual assistants can process and understand spoken language, allowing users to interact with their devices using natural language. Users can ask questions, request information, set reminders, play music, and even control smart home devices through voice commands. NLP algorithms enable these virtual assistants to accurately interpret the user's intent and provide appropriate responses or perform requested tasks.

Chatbots are another application of NLP that has gained popularity across various industries. They are used in customer service, e-commerce, and other domains to provide automated and interactive customer support. Chatbots can understand natural language inputs, engage in conversations, and provide relevant information or assistance. By leveraging NLP techniques, chatbots can interpret user queries, analyze patterns, and generate contextually appropriate responses, delivering efficient and personalized customer experiences.

Voice-controlled devices, such as smart speakers, have become increasingly prevalent in households. These devices rely on NLP to understand spoken commands and perform various tasks, such as playing music, providing weather updates, or controlling smart home devices. By processing and interpreting human language, these devices enable seamless and intuitive interactions between users and their digital environments.

The advancements in NLP have also led to improvements in language translation and sentiment analysis. AI-powered translation systems can accurately translate text from one language to another, enabling cross-lingual communication and facilitating global interactions. Sentiment analysis algorithms can analyze text data, such as social media posts or customer reviews, to determine the sentiment expressed by users. This information is valuable for businesses in understanding customer feedback, monitoring brand reputation, and making data-driven decisions.

While NLP has made significant progress, challenges still exist in achieving human-like understanding and generating natural language responses. Ambiguity, context, and cultural nuances pose difficulties for NLP algorithms. Ongoing research is focused on developing more sophisticated models that can better comprehend and generate human language.

In summary, NLP plays a vital role in enabling AI systems to understand and interpret human language, facilitating conversational interactions between humans and machines. Virtual assistants, chatbots, and voice-controlled devices are tangible examples of how NLP empowers AI applications to deliver intuitive and personalized user experiences. As NLP continues to advance, we can expect even more sophisticated and natural interactions with AI systems, further blurring the lines between humans and machines.

Computer Vision and Image Recognition:

AI-powered computer vision systems have unlocked remarkable capabilities for machines to perceive and understand the visual world. Through the integration of sophisticated algorithms and deep learning techniques, these systems enable machines to analyze and interpret visual data, emulating the human ability to comprehend images and videos.

One notable application of AI-powered computer vision is in autonomous vehicles. These vehicles rely on a combination of sensors, including cameras, to capture real-time visual information about the surrounding environment. AI algorithms then process this visual data, extracting meaningful information and making informed decisions to navigate safely.

In the context of autonomous vehicles, computer vision algorithms play a critical role in recognizing and interpreting various objects and elements on the road. This includes identifying and classifying traffic signs, pedestrians, cyclists, other vehicles, and potential obstacles. By analyzing the visual cues captured by the cameras, AI algorithms can accurately determine the appropriate actions to be taken, such as stopping at a red light, yielding to pedestrians, or changing lanes to avoid collisions.

Computer vision systems can also contribute to advanced driver assistance systems (ADAS), which enhance the safety and convenience of human-driven vehicles. For example, computer vision algorithms can detect lane markings and monitor driver behavior to provide warnings or assistance in maintaining a safe driving trajectory. Additionally, computer vision can aid in recognizing and alerting drivers to potential hazards, such as sudden obstacles or drowsy driving.

Beyond the realm of transportation, computer vision has found applications in various industries. In e-commerce, computer vision algorithms can analyze product images to enable visual search capabilities. Customers can take a picture of an item and find similar products, enhancing the shopping experience. In manufacturing, computer vision is used for quality control, where AI algorithms can detect defects, measure dimensions, and ensure product conformity. In healthcare, computer vision assists in medical imaging analysis, aiding in the diagnosis of diseases and abnormalities.

Despite the impressive advancements in computer vision, challenges still persist. Variations in lighting conditions, occlusions, and complex scenes can pose difficulties for accurate object recognition and scene understanding. Continued research and development efforts are focused on improving the robustness and adaptability of computer vision algorithms to handle these challenges.

In summary, AI-powered computer vision systems have revolutionized the way machines perceive and understand visual information. In fields such as autonomous vehicles, computer vision enables machines to interpret camera data and respond to traffic signs, pedestrians, and obstacles, enhancing road safety. With continued progress and innovation, computer vision is poised to have a profound impact across various industries, enabling machines to interact with the visual world in a more intelligent and perceptive manner.

Limitations of AI:

- Contextual Understanding and Common Sense Reasoning:

While AI systems can excel at specific tasks, they often struggle with contextual understanding and common sense reasoning. Understanding the subtleties, nuances, and broader context of human language and real-world situations poses a challenge for AI. For instance, an AI language model may generate plausible-sounding sentences but lack true comprehension of the underlying concepts or implications.

- Ethical Considerations and Bias:

AI systems are only as unbiased and ethical as the data they are trained on. Biases present in the training data can lead to biased outcomes or reinforce existing societal inequalities. It is crucial to address ethical considerations, transparency, and fairness in AI algorithms to ensure they serve the best interests of society.

- Creativity and Abstract Thinking:

AI systems struggle with tasks that require creativity, imagination, and abstract thinking, which are inherent strengths of human intelligence. While AI can generate creative outputs within defined constraints, it often lacks the ability to think beyond predefined patterns or generate truly novel ideas.

- Limited Adaptability and Transfer Learning:

AI systems typically perform well within the domain they are trained on, but they may struggle when faced with unfamiliar situations or tasks outside their training data. Generalizing knowledge and skills learned from one domain to another, known as transfer learning, remains a challenge in AI research.

- Addressing the Limitations:

Researchers and practitioners in the field of AI are actively working to address these limitations and push the boundaries of what AI can achieve. Ongoing research in areas like explainable AI, ethical AI, and cognitive architectures aims to enhance AI systems' capabilities and address concerns related to bias, transparency, and interpretability.

Moreover, a multidisciplinary approach involving collaboration between AI experts, domain specialists, and ethicists can lead to more comprehensive solutions and better outcomes. By combining the strengths of AI with human intelligence, we can navigate the challenges and ensure that AI technology is developed and deployed in a responsible and beneficial manner.

In conclusion, AI possesses remarkable capabilities in data processing, pattern recognition, natural language understanding, and computer vision. However, it also has limitations in areas such as contextual understanding, common sense reasoning, ethical considerations, creativity, and adaptability. Understanding these capabilities and limitations is crucial for leveraging AI effectively and addressing the challenges to create a future where AI systems work in harmony with human intelligence for the benefit of society.

# The Foundation of AI: Machine Learning

Machine learning serves as the bedrock of artificial intelligence, providing the tools and techniques that enable systems to learn and make intelligent decisions without explicit programming. By leveraging vast amounts of data and powerful algorithms, machine learning unlocks the potential to solve complex problems and extract valuable insights. In this section, we will delve into the essence of machine learning, explore its key components, and examine its applications across various fields.

Understanding Machine Learning:

Machine learning is a subfield of artificial intelligence that focuses on the development of algorithms and models that allow computers to learn from data and improve their performance over time. At its core, machine learning enables systems to recognize patterns, make predictions, and uncover hidden relationships within data, thereby augmenting their decision-making capabilities.

The Components of Machine Learning:

a. Data Collection and Preparation:

To train machine learning models, high-quality and relevant data is essential. This data can be collected from various sources, such as e-commerce transactions, healthcare records, or financial market data. The collected data is then processed, cleaned, and transformed into a suitable format for analysis.

b. Feature Extraction and Representation:

Features are the measurable properties or characteristics of the data that contribute to the learning process. Feature extraction involves selecting relevant features from the dataset and representing them in a way that captures their underlying patterns. This step is crucial for the machine learning algorithm to learn meaningful relationships and make accurate predictions.

c. Model Selection and Training:

Choosing the appropriate machine learning model is crucial for the task at hand. There are various types of models, such as linear regression, decision trees, and neural networks. The selected model is then trained using the prepared data, where the algorithm adjusts its internal parameters to minimize errors and optimize its performance.

d. Evaluation and Validation:

After training the machine learning model, it needs to be evaluated and validated to ensure its effectiveness. This involves assessing its performance on a separate dataset, called the validation set or test set. Metrics such as accuracy, precision, recall, and F1 score are commonly used to measure the model's performance. If the model meets the desired criteria, it can be deployed for making predictions and decisions.

Applications of Machine Learning:

Machine learning finds applications across a wide range of fields, transforming industries and enhancing various processes. Let's explore some examples:

a. E-commerce:

Machine learning algorithms analyze customer behavior, purchase history, and browsing patterns to provide personalized product recommendations, optimize pricing strategies, and detect fraud.

b. Healthcare:

Machine learning models analyze medical data to predict disease outcomes, assist in diagnosis, and personalize treatment plans. This technology enhances patient care, reduces medical errors, and aids in medical research.

c. Banking and Finance:

Machine learning algorithms analyze financial data, market trends, and customer profiles to detect fraudulent activities, predict market trends, and optimize investment strategies. This enables banks and financial institutions to make data-driven decisions and provide tailored financial services.

Advancements and Challenges in Machine Learning:

Machine learning continues to evolve with advancements in algorithms, computational power, and data availability. However, challenges remain, such as handling high-dimensional data, ensuring model interpretability, and addressing bias and ethical considerations in the decision-making process. Ongoing research and innovation are crucial for overcoming these challenges and further unlocking the potential of machine learning.

Conclusion:

Machine learning forms the foundation of artificial intelligence, empowering systems to learn from data and make intelligent predictions and decisions. Through its key components, including data collection, feature extraction, model training, and evaluation, machine learning has found applications across various fields, revolutionizing industries and enhancing processes. As advancements and research progress, machine learning holds the promise of shaping a future where intelligent systems augment human capabilities and drive innovation.

## Exploring the fundamentals of machine learning

Machine learning is a fundamental pillar of artificial intelligence, enabling systems to learn from data and make intelligent predictions and decisions. By harnessing the power of algorithms and vast amounts of data, machine learning has the potential to transform industries, revolutionize processes, and unlock new insights. In this section, we will embark on a journey to explore the fundamental principles of machine learning, unravel its key components, and delve into its applications across diverse fields.

Understanding Machine Learning:

Machine learning is a branch of artificial intelligence that focuses on the development of algorithms and models capable of learning from data and improving their performance over time. At its core, machine learning empowers systems to recognize patterns, make predictions, and extract valuable knowledge from complex datasets. By learning from historical examples, machine learning models can generalize their understanding and apply it to new, unseen data.

The Building Blocks of Machine Learning:

a. Data Acquisition and Preparation:

High-quality and relevant data is the foundation of machine learning. It involves acquiring and preparing datasets that adequately represent the problem at hand. For example, in e-commerce, customer purchase history, browsing patterns, and demographic information can be collected and processed to train models for personalized product recommendations.

b. Feature Engineering:

Feature engineering is the process of transforming raw data into a suitable format that captures the underlying patterns and relationships. It involves selecting relevant features, creating new derived features, and encoding categorical variables. Effective feature engineering is crucial for enabling machine learning models to learn meaningful representations and make accurate predictions.

c. Model Selection and Training:

Machine learning offers a plethora of algorithms, each designed for specific tasks and data characteristics. Model selection involves choosing the most appropriate algorithm based on the problem's nature, available data, and desired outcomes. The selected model is then trained using the prepared data, where the algorithm adjusts its internal parameters to minimize errors and optimize its performance.

d. Evaluation and Validation:

Evaluating the performance of machine learning models is essential to ensure their effectiveness. This involves using separate datasets, called the validation set or test set, to assess the model's accuracy, precision, recall, and other performance metrics. Through rigorous evaluation and validation, models can be fine-tuned and refined to meet the desired criteria.

Applications of Machine Learning:

Machine learning has permeated numerous industries, driving innovation and delivering tangible value. Let's explore some examples:

a. E-commerce:

Machine learning algorithms analyze customer behavior, purchase history, and browsing patterns to provide personalized product recommendations, optimize pricing strategies, and detect fraudulent activities.

b. Healthcare:

Machine learning models analyze medical records, imaging data, and genetic information to aid in disease diagnosis, predict patient outcomes, and enable personalized treatment plans.

c. Banking and Finance:

Machine learning algorithms analyze financial market data, customer transactions, and credit histories to detect anomalies, predict market trends, and optimize investment portfolios.

Advancements and Challenges in Machine Learning:

Machine learning is a rapidly evolving field, driven by advancements in algorithms, computational power, and data availability. However, challenges persist. These include handling high-dimensional data, addressing issues of bias and fairness, and ensuring the interpretability and transparency of machine learning models. Ongoing research and innovation are essential to overcome these challenges and unlock the full potential of machine learning.

Conclusion:

Machine learning lies at the heart of artificial intelligence, offering powerful techniques to analyze data, make predictions, and uncover hidden insights. By understanding the principles of machine learning, harnessing the right algorithms, and leveraging high-quality data, industries can embrace the transformative potential of this technology. As machine learning continues to advance, it holds the promise of reshaping industries, driving innovation, and pushing the boundaries of what is possible in the realm of artificial intelligence.

Supervised learning:

Supervised learning is a cornerstone of machine learning, enabling systems to learn from labeled data and make accurate predictions or decisions. It encompasses a range of algorithms and techniques that have revolutionized various industries by leveraging the power of labeled examples. In this section, we will embark on an exploration of supervised learning, its underlying principles, and its wide-ranging applications across diverse fields.

Understanding Supervised Learning:

Supervised learning is a machine learning paradigm where algorithms learn from labeled training data to generalize patterns and make predictions on unseen data. In this paradigm, the training data consists of input-output pairs, where the input represents the features or attributes of the data, and the output is the corresponding label or target variable. By mapping the input features to the output labels, supervised learning algorithms aim to learn the underlying patterns and relationships to make accurate predictions on new, unseen data.

The Key Components of Supervised Learning:

a. Training Data:

In supervised learning, high-quality labeled training data is crucial. The training data should represent the problem domain adequately and provide a diverse range of examples for the algorithm to learn from. For instance, in e-commerce, the training data may include customer demographic information, purchase history, and corresponding labels indicating whether a customer made a purchase or not.

b. Feature Extraction and Selection:

Extracting relevant features from the input data is essential for effective supervised learning. Feature extraction involves transforming the raw data into a format that captures the relevant information. Feature selection helps to identify the most informative features and discard irrelevant ones. For example, in healthcare, features such as age, blood pressure, and cholesterol levels can be extracted from patient records to predict the likelihood of cardiovascular diseases.

c. Model Selection:

Supervised learning offers a wide range of algorithms, each with its own strengths and assumptions. Model selection involves choosing the most appropriate algorithm that suits the problem at hand. For example, linear regression can be used for predicting continuous variables, while decision trees or support vector machines (SVMs) may be suitable for classification tasks.

d. Model Training and Evaluation:

The selected model is trained using the labeled training data. The algorithm adjusts its internal parameters to minimize the difference between its predictions and the true labels in the training data. Model evaluation involves assessing the performance of the trained model on separate validation or test data. Various evaluation metrics, such as accuracy, precision, recall, and F1 score, can be used to gauge the model's effectiveness.

Applications of Supervised Learning:

Supervised learning has found extensive applications across numerous industries. Let's explore some examples:

a. E-commerce:

Supervised learning algorithms can predict customer churn, classify customer preferences, and optimize pricing strategies based on historical customer data.

b. Healthcare:

Supervised learning models can assist in medical diagnosis, predict patient outcomes, and identify high-risk patients based on medical records, imaging data, and genetic information.

c. Banking and Finance:

Supervised learning algorithms can be employed to detect fraudulent transactions, predict credit default risk, and optimize investment portfolios based on historical financial data.

Advancements and Limitations of Supervised Learning:

Supervised learning has witnessed significant advancements due to the availability of large labeled datasets and the development of powerful algorithms. However, it does have limitations. One of the key challenges is the reliance on labeled data, which may be expensive or time-consuming to acquire. Additionally, supervised learning algorithms may struggle with new, unseen data that deviates significantly from the training distribution. Exploring techniques such as transfer learning and active learning can help mitigate some of these limitations.

Conclusion:

Supervised learning is a powerful paradigm within machine learning that empowers systems to make accurate predictions and decisions based on labeled data. By understanding the principles of supervised learning, harnessing the right algorithms, and curating high-quality labeled datasets, industries can unlock the full potential of this approach. As supervised learning continues to advance, it promises to reshape industries, drive innovation, and unlock new frontiers in the realm of artificial intelligence.


Unsupervised learning:

While supervised learning has garnered significant attention, unsupervised learning offers a distinct avenue for uncovering hidden patterns and structures within data. Unlike supervised learning, which relies on labeled examples, unsupervised learning algorithms delve into unlabeled data to extract meaningful insights and discover inherent structures without explicit guidance. In this section, we will embark on a journey into the realm of unsupervised learning, exploring its fundamental principles, techniques, and applications across various domains.

Understanding Unsupervised Learning:

Unsupervised learning encompasses a collection of algorithms designed to discover patterns, relationships, and underlying structures within unlabeled data. Unlike in supervised learning, there are no explicit labels or targets provided during the training process. Instead, the algorithms autonomously learn to identify similarities, clusters, or associations within the data, without any predefined notions of what to look for. Unsupervised learning allows for the exploration and extraction of hidden patterns, enabling valuable insights and knowledge discovery.

Key Techniques in Unsupervised Learning:

a. Clustering:

Clustering algorithms play a crucial role in organizing and categorizing data points based on their intrinsic similarities. By grouping similar data points together, clustering algorithms enable us to gain insights into the underlying patterns and structures within the data. One popular clustering algorithm is K-means clustering, which partitions the data into K distinct clusters based on their proximity to cluster centroids. Hierarchical clustering, on the other hand, creates a hierarchical structure of clusters by recursively merging or splitting clusters. Density-based clustering algorithms, such as DBSCAN (Density-Based Spatial Clustering of Applications with Noise), identify clusters based on regions of high data density.

The applications of clustering are vast and diverse. In e-commerce, clustering can be used for customer segmentation, where customers with similar purchasing behaviors or preferences are grouped together. This allows businesses to tailor their marketing strategies and product offerings to specific customer segments, enhancing customer satisfaction and improving sales. In healthcare, clustering algorithms can assist in disease subtyping, where patients with similar clinical characteristics are grouped together. This can aid in the identification of personalized treatment plans or the discovery of subtypes within a disease, leading to more targeted and effective interventions.

b. Dimensionality Reduction:

In many real-world scenarios, datasets often have high dimensionality, meaning they contain a large number of features or variables. However, high-dimensional data can be challenging to analyze and visualize. Dimensionality reduction techniques address this challenge by reducing the number of features while preserving the essential structure and relationships within the data.

Principal Component Analysis (PCA) is a widely used dimensionality reduction technique that transforms high-dimensional data into a lower-dimensional representation. It achieves this by identifying the principal components, which are the directions of maximum variance in the data. By selecting a subset of these components, PCA can capture the most significant information while discarding the less important ones. This compression of the data into a lower-dimensional space not only simplifies visualization but also aids in tasks such as data analysis, anomaly detection, and feature selection.

Another dimensionality reduction technique, t-SNE (t-Distributed Stochastic Neighbor Embedding), focuses on preserving local similarities between data points in a lower-dimensional space. It is particularly effective in visualizing high-dimensional data clusters and revealing intricate relationships between data points. t-SNE has found applications in various fields, including image analysis, where it can help visualize and explore complex image datasets.

c. Association Rule Mining:

Association rule mining is a technique used to discover interesting relationships, co-occurrences, or patterns within datasets. It aims to identify items that frequently appear together in transactions or events. One prominent example of association rule mining is market basket analysis, often used in the retail industry to uncover item combinations frequently purchased together by customers.

The insights gained from association rule mining can have significant implications for businesses. By understanding the relationships between items, retailers can make informed decisions on product placement, pricing, and cross-selling strategies. For instance, if market basket analysis reveals that customers who purchase diapers are also likely to buy baby formula, retailers can strategically position these items in proximity to each other, leading to increased sales and customer satisfaction. Furthermore, association rule mining enables personalized recommendations, where customers are presented with relevant items based on their previous purchases or preferences, enhancing their overall shopping experience.

In conclusion, clustering, dimensionality reduction, and association rule mining are powerful techniques within the realm of AI that enable us to gain valuable insights from data. By grouping similar data points, reducing the complexity of high-dimensional data, and discovering meaningful relationships, these techniques provide the foundation for various applications in fields such as e-commerce, healthcare, and retail. The ability to uncover hidden patterns and structures within datasets fuels advancements in personalized marketing, disease subtyping, and data analysis, leading to more informed decision-making and improved outcomes.

Applications of Unsupervised Learning:

a. Anomaly Detection:

Anomaly detection is a critical application of unsupervised learning that involves identifying unusual patterns or outliers in data. By analyzing the inherent structure and distribution of the data, unsupervised learning algorithms can detect deviations from the norm. This capability finds application in various domains, such as cybersecurity, fraud detection, and equipment failure prediction.

In cybersecurity, unsupervised learning can analyze network traffic data to identify anomalous behavior that may indicate a cyber attack or intrusion. By learning the normal patterns of network activity, these algorithms can detect any deviations or suspicious activities that may pose a threat to the system's security.

Similarly, in fraud detection, unsupervised learning techniques can analyze transactional data to identify abnormal patterns that may indicate fraudulent activities. By identifying outliers or unusual behaviors, such as unusual spending patterns or suspicious transactions, financial institutions can take proactive measures to prevent fraud and protect their customers' assets.

In the realm of predictive maintenance, unsupervised learning can be used to detect anomalies in equipment sensor data. By monitoring the normal behavior of machinery, any deviations from the expected patterns can signal potential equipment failures. This allows businesses to take preventive actions, such as scheduling maintenance or replacing faulty components, before major breakdowns occur.

b. Text Mining and Topic Modeling:

Unsupervised learning algorithms are valuable tools for extracting meaningful information and topics from unstructured text data. Text mining and topic modeling techniques enable the organization, analysis, and understanding of large volumes of textual data.

One popular unsupervised learning technique in this domain is latent Dirichlet allocation (LDA), which is used for topic modeling. LDA can identify latent topics within a collection of documents by analyzing the co-occurrence patterns of words. By assigning words to different topics, LDA allows for the discovery of underlying themes or subjects present in the text data.

Text mining and topic modeling have diverse applications. Sentiment analysis, for example, leverages unsupervised learning to determine the sentiment expressed in text, such as social media posts or customer reviews. This information is valuable for understanding public opinion, monitoring brand reputation, or guiding marketing strategies.

Document clustering is another application where unsupervised learning algorithms group similar documents together based on their content, enabling efficient organization and retrieval of information. This technique finds use in various domains, such as news categorization, document recommendation, and search engines.

c. Image and Video Analysis:

Unsupervised learning plays a vital role in image and video analysis, enabling machines to automatically understand and interpret visual data. By applying unsupervised learning algorithms to image and video datasets, machines can categorize objects, detect features, and organize visual content.

In the field of autonomous vehicles, unsupervised learning algorithms can analyze camera data to recognize objects, such as pedestrians, vehicles, and traffic signs. This information is crucial for the decision-making processes of self-driving cars, allowing them to navigate safely and respond appropriately to the surrounding environment.

In surveillance systems, unsupervised learning can identify and track objects of interest, such as individuals or vehicles, in video streams. This capability enhances security measures and assists in identifying suspicious activities or potential threats.

Unsupervised learning algorithms are also employed in content organization and recommendation systems. By clustering similar images or videos based on their visual features, these algorithms can aid in content organization, facilitating efficient retrieval and browsing experiences for users. Additionally, by analyzing user behavior and preferences, unsupervised learning techniques can recommend relevant visual content, enhancing personalized recommendations in platforms such as streaming services or e-commerce websites.

In summary, unsupervised learning is a powerful approach that finds diverse applications across various domains. From anomaly detection in cybersecurity to text mining and topic modeling in natural language processing, and image and video analysis in autonomous vehicles and surveillance, unsupervised learning enables machines to uncover meaningful patterns and structures in data without the need for labeled examples. This capability expands the boundaries of AI and empowers industries to leverage untapped insights, improve decision-making processes, and enhance user experiences.

Advancements and Challenges in Unsupervised Learning:

Unsupervised learning is a rapidly evolving field, driven by advancements in algorithmic techniques and computing power. However, challenges remain. The absence of labeled data makes evaluation and validation complex, as there is no ground truth to compare against. The curse of dimensionality and the identification of meaningful structures within complex datasets pose additional hurdles. Researchers and practitioners are actively exploring innovative solutions, such as generative adversarial networks (GANs) and self-supervised learning, to overcome these challenges.

Conclusion:

Unsupervised learning provides a powerful means of extracting valuable insights and discovering hidden patterns within unlabeled data. By leveraging clustering, dimensionality reduction, and association rule mining, unsupervised learning algorithms unlock the potential for knowledge discovery and data exploration. While challenges exist, the ongoing advancements in unsupervised learning techniques hold promise for revolutionizing various domains and driving innovation. As we delve deeper into the realm of artificial intelligence, unsupervised learning stands as a vital pillar in our quest for understanding and harnessing the immense potential of data.

Reinforcement learning:

Reinforcement learning stands at the forefront of artificial intelligence, offering a unique approach to learning and decision-making through interaction with the environment. It embodies the principles of trial and error, as well as feedback-driven improvement. In this section, we embark on a journey into the realm of reinforcement learning, exploring its fundamental concepts, underlying mechanisms, and its remarkable applications across diverse fields.

Understanding Reinforcement Learning:

Reinforcement learning is a learning paradigm where an agent learns to make optimal decisions by interacting with an environment. It involves an agent, which takes actions in a given state, and an environment, which provides feedback in the form of rewards or penalties. The agent's objective is to maximize cumulative rewards by discovering the optimal policy—a mapping of states to actions. Reinforcement learning differs from supervised and unsupervised learning as it learns from experiences rather than labeled or unlabeled data.

Key Elements of Reinforcement Learning:

a. Agents and Environments:

In the realm of reinforcement learning, an agent interacts with an environment, forming a dynamic relationship where the agent learns to make decisions based on the environment's feedback. The agent perceives the current state of the environment, takes actions, and receives feedback or observations as a result of those actions. The environment, on the other hand, can take various forms, ranging from virtual simulations to physical robots.

Consider an example of an autonomous robot navigating a maze. The robot serves as the agent, while the maze represents the environment. The robot observes its current location within the maze, selects an action to move in a certain direction, and receives feedback in the form of rewards or penalties based on its progress towards the goal. The interaction between the agent (robot) and the environment (maze) allows the agent to learn and improve its decision-making process over time.

b. State, Action, and Reward:

At each time step, the agent perceives the current state of the environment, which encapsulates the relevant information needed to make decisions. The state can include factors such as the robot's position, the presence of obstacles, or any other relevant variables. Based on this perception, the agent selects an action to perform. Actions can be as simple as moving in a specific direction, or they can involve more complex behaviors and decisions.

After the agent takes an action, it receives a feedback signal known as a reward. The reward serves as a measure of the desirability or quality of the action taken by the agent in a given state. The reward signal can be positive, negative, or zero, reflecting the agent's success or failure in achieving its goals. The agent's objective is to learn a policy, which is a mapping from states to actions, that maximizes the cumulative rewards it receives over time.

For example, in a game-playing scenario, the state could represent the current game board, and the actions could be the possible moves the player can make. The rewards would reflect the player's success in the game, such as gaining points or winning.

c. Policy and Value Functions:

The policy in reinforcement learning determines the agent's behavior by mapping states to actions. It represents a strategy or a set of rules that guide the agent's decision-making process. The policy can be deterministic, meaning it always selects the same action given a particular state, or it can be stochastic, with probabilities assigned to different actions.

Value functions play a crucial role in reinforcement learning as they assess the expected rewards or quality associated with different states or state-action pairs. The state-value function, also known as the V-function, estimates the expected cumulative rewards starting from a given state under a specific policy. The action-value function, also called the Q-function, estimates the expected cumulative rewards starting from a given state-action pair under a specific policy.

These value functions guide the agent's decision-making process by evaluating the long-term consequences of different actions. By estimating the values of states or state-action pairs, the agent can assess the potential outcomes and select actions that lead to higher rewards.

In summary, the concepts of agents and environments, along with the notions of state, action, reward, policy, and value functions, form the foundation of reinforcement learning. Understanding these components is crucial for developing effective learning algorithms and enabling agents to learn optimal strategies in various domains, ranging from game playing to robotics.

Reinforcement Learning Algorithms:

Reinforcement learning algorithms employ various techniques to learn optimal policies and value functions. Some notable algorithms include Q-learning, SARSA (State-Action-Reward-State-Action), and deep reinforcement learning, which combines reinforcement learning with deep neural networks to handle complex and high-dimensional problems. These algorithms employ exploration-exploitation strategies to balance the exploration of unknown states with the exploitation of learned knowledge.

Applications of Reinforcement Learning:

a. Game Playing:

Reinforcement learning has demonstrated its prowess in the domain of game playing, showcasing the ability of AI agents to surpass human performance. Notable examples include DeepMind's AlphaGo, which defeated world champion Go player Lee Sedol, and OpenAI's Dota 2 AI, which achieved professional-level gameplay. In these instances, reinforcement learning algorithms learn strategies and tactics by playing against themselves or human players, leveraging vast amounts of gameplay data. Through continuous learning and optimization, these algorithms acquire an exceptional level of skill, surpassing even the most skilled human players.

b. Robotics and Control Systems:

Reinforcement learning plays a crucial role in enabling robots to learn complex tasks and acquire new skills. By interacting with the environment, robots can learn to perform various actions, such as grasping objects, walking, or manipulating objects. Reinforcement learning algorithms provide a framework for robots to learn from trial and error, where they receive feedback in the form of rewards or penalties based on their actions. Over time, the robots adapt their behavior and refine their motor skills to achieve desired objectives. This ability has significant implications in fields like manufacturing, where robots can autonomously learn to perform intricate tasks and adapt to new scenarios.

c. Finance and Trading:

Reinforcement learning algorithms have found applications in finance and trading, offering the potential to optimize trading strategies and improve financial outcomes. By learning from historical market data, these algorithms can identify patterns and correlations, enabling them to make profitable decisions. Reinforcement learning agents can adapt to changing market conditions and learn from the feedback provided by financial indicators, allowing them to adjust their trading strategies accordingly. This adaptive and data-driven approach has the potential to enhance investment decisions and generate more favorable trading results.

For instance, a reinforcement learning agent could learn to navigate complex financial markets and dynamically adjust its portfolio allocation based on market trends and risk factors. By continuously interacting with the market and optimizing its trading decisions, the agent aims to maximize financial returns while managing risks effectively.

These examples highlight the versatility and power of reinforcement learning across various domains. By leveraging the principles of reward-based learning and iterative improvement, reinforcement learning algorithms enable machines to acquire complex skills, make intelligent decisions, and achieve remarkable performance in challenging tasks.

Advancements and Challenges in Reinforcement Learning:

Reinforcement learning is an active area of research with ongoing advancements. Deep reinforcement learning, which combines deep neural networks with reinforcement learning, has shown remarkable success in complex tasks. However, challenges persist, such as sample inefficiency, long training times, and the need for extensive exploration. Researchers are addressing these challenges through techniques like transfer learning, meta-learning, and improved exploration-exploitation strategies.

Conclusion:

Reinforcement learning represents a paradigm that combines learning, decision-making, and interaction to achieve optimal outcomes in dynamic and uncertain environments. By leveraging the principles of trial and error, agents can learn to navigate complex tasks and find optimal solutions. As we delve deeper into the realm of artificial intelligence, reinforcement learning continues to push the boundaries of what machines can achieve. Its applications in game playing, robotics, finance, and beyond exemplify its potential to revolutionize various fields. Embracing the power of reinforcement learning will pave the way for intelligent systems that can learn and adapt in a dynamic world, enabling humanity to tackle complex challenges and unlock new frontiers of possibility.

The role of data in machine learning

In the realm of machine learning, data is the lifeblood that nourishes the algorithms and empowers them to unlock the mysteries hidden within. It serves as the foundation upon which intelligent models are built, providing the necessary raw material for learning, training, and generating valuable insights. In this section, we delve into the critical role of data in machine learning, exploring its significance, challenges, and the ways it shapes the future of AI.

The Significance of Data:

Data is the driving force behind machine learning, serving as the primary source of information that algorithms utilize to discover patterns, make predictions, and generate meaningful outputs. The quality, quantity, and diversity of data directly influence the performance and effectiveness of machine learning models. In essence, data fuels the learning process, allowing algorithms to extract valuable knowledge from vast and complex datasets.

a. Data Collection:

The data lifecycle begins with data collection, where various sources are tapped into to gather relevant information. In the context of AI applications, data can come from diverse sources such as E-commerce transactions, electronic health records, banking transactions, or user interactions. Careful selection of data sources is essential to ensure the availability of accurate and representative data for training and modeling purposes. During the data collection phase, data may need to be preprocessed, anonymized, or transformed to adhere to privacy regulations and ethical considerations.

b. Data Preparation:

Once the data is collected, it undergoes a series of preprocessing steps to transform it into a suitable format for machine learning algorithms. This includes cleaning the data by removing any irrelevant or redundant information, handling missing values, and addressing inconsistencies. Data normalization techniques may be applied to scale the data and bring it into a standardized range. Feature extraction methods are employed to identify and extract relevant features from the raw data, which can significantly influence the performance of machine learning models. This stage is crucial for ensuring data quality and removing noise or biases that could adversely impact the accuracy and reliability of the models.

c. Data Labeling and Annotation:

In certain scenarios, labeled or annotated data is required to train supervised learning models. This process involves human experts assigning class labels or annotations to data instances, providing ground truth information for the models to learn from. For example, in the healthcare field, medical images may be annotated to indicate regions of interest or specific diagnoses. Data labeling and annotation can be a time-consuming and labor-intensive task, often requiring domain expertise and careful consideration to ensure accurate and consistent labeling.

d. Training and Validation:

The prepared data is divided into training and validation sets. The training set is used to teach the machine learning model by exposing it to labeled data and allowing it to learn patterns and relationships. The model learns from the training data by adjusting its internal parameters based on the provided labels or annotations. The validation set serves as an independent measure to assess the model's performance. It helps evaluate how well the trained model generalizes to unseen data, providing insights into its accuracy, precision, recall, and other performance metrics. This step is crucial for fine-tuning the model and optimizing its performance.

e. Model Evaluation and Iteration:

After the model is trained, it needs to be evaluated using a separate test dataset to assess its performance in real-world scenarios. This evaluation provides an unbiased measure of how well the model can make accurate predictions on new and unseen data. The model's performance is analyzed based on various metrics, such as accuracy, F1 score, or area under the curve (AUC). Based on the evaluation results, iterative processes can be employed to improve the model's accuracy and robustness. This may involve fine-tuning the model, optimizing hyperparameters, or employing ensemble learning techniques to combine the predictions of multiple models. Iterative refinement and improvement are crucial for enhancing the model's performance and ensuring its effectiveness in real-world applications.

In summary, the data lifecycle encompasses various stages, from data collection and preparation to labeling, model training, and evaluation. Each stage plays a vital role in building accurate and reliable AI models. By following a systematic approach to the data lifecycle, organizations can harness the power of data to develop AI systems that deliver meaningful insights and predictive capabilities in a wide range of fields and industries.

Challenges in Data-driven Machine Learning:

a. Data Quality and Bias:

Data quality plays a crucial role in the accuracy and reliability of AI models. Flawed or biased data can lead to incorrect or unfair predictions, impacting decision-making processes. Biases can arise due to various factors, such as sampling bias, label bias, or societal biases embedded in the data. For example, if a dataset used for credit scoring predominantly includes data from a certain demographic group, the resulting model may exhibit bias and discrimination against other groups. Addressing data quality and bias requires careful attention to data collection methods, diverse representation in the data, and rigorous data preprocessing techniques. Data augmentation can be employed to balance the representation of different classes or groups within the data. Additionally, techniques like bias detection and fairness-aware models can help identify and mitigate biases in the training process, promoting fairness and equity in AI applications.

b. Data Privacy and Security:

With the increasing reliance on data in machine learning, ensuring data privacy and security is of paramount importance. Organizations must handle sensitive personal or confidential information with utmost care to protect individuals' privacy rights and comply with regulations such as the General Data Protection Regulation (GDPR). Techniques such as differential privacy can be applied to anonymize and aggregate data in a way that prevents individual reidentification. Secure multiparty computation allows for collaborative analysis of encrypted data without exposing the underlying sensitive information. Robust data governance practices, including data access controls, encryption, and secure storage, are essential to safeguard data throughout its lifecycle. By maintaining strong data privacy and security measures, organizations can build trust with users and stakeholders, fostering responsible and ethical AI practices.

c. Data Availability and Scalability:

Acquiring large-scale, high-quality datasets can be a challenge, particularly in domains where data is scarce or access is restricted due to privacy concerns. For example, in healthcare, obtaining labeled medical data for training machine learning models can be difficult due to privacy regulations and the limited availability of annotated data. Collaborative efforts, such as federated learning and data sharing frameworks, provide solutions for data availability and scalability. Federated learning allows models to be trained across multiple devices or organizations without the need for data sharing, preserving data privacy. Data sharing frameworks, with appropriate anonymization and consent mechanisms, enable the pooling of data resources to enhance the quality and diversity of datasets. These collaborative approaches promote advancements in AI while respecting privacy regulations and ethical considerations.

By addressing data quality, bias, privacy, and scalability challenges, organizations can build AI models that are not only accurate and reliable but also fair, privacy-preserving, and scalable. By prioritizing these aspects of data management, we can ensure that AI technologies are developed and deployed in a responsible and trustworthy manner, benefiting society as a whole.

The Future of Data in Machine Learning:

a. Domain-Specific Data Applications:

In various domains, data applications have revolutionized decision-making processes and improved outcomes. For instance, in e-commerce, companies analyze customer behavior data to gain insights into preferences, purchasing patterns, and browsing habits. These insights are then used to personalize product recommendations, optimize marketing strategies, and enhance customer experiences. By leveraging data-driven machine learning techniques, e-commerce platforms can tailor their offerings to individual customers, increasing customer satisfaction and driving sales.

Similarly, in healthcare, data applications have the potential to transform patient care. Electronic health records, medical imaging data, and genomic data can be analyzed to uncover patterns and insights that aid in disease diagnosis, treatment planning, and patient monitoring. Machine learning models can be trained on large-scale healthcare datasets to predict disease outcomes, identify high-risk patients, and optimize treatment protocols. This data-driven approach allows healthcare providers to deliver personalized and targeted care, ultimately improving patient outcomes and reducing costs.

In the banking and finance industry, data applications play a crucial role in risk assessment, fraud detection, and investment strategies. Financial institutions analyze vast amounts of transaction data, market data, and customer information to identify patterns of fraudulent activities, assess creditworthiness, and optimize investment portfolios. Machine learning algorithms can detect anomalies, predict market trends, and make data-driven decisions that improve risk management and financial performance.

b. Advanced Data Analytics:

Advancements in technology have brought forth advanced data analytics techniques that enable deeper insights from complex and interconnected datasets. Deep learning, a subset of machine learning, has proven to be highly effective in tasks such as image and speech recognition, natural language processing, and generative modeling. Deep neural networks with multiple layers can automatically learn hierarchical representations of data, capturing intricate patterns and relationships that were previously challenging to extract. This has led to significant advancements in fields like computer vision, speech processing, and autonomous systems.

Graph analytics is another powerful approach that leverages the inherent structure of interconnected data to uncover hidden relationships and perform network analysis. It finds applications in social network analysis, recommendation systems, fraud detection, and cybersecurity. By analyzing the complex relationships between entities, graph analytics provides insights into community detection, influence propagation, and anomaly detection.

Probabilistic programming combines probability theory with programming languages, allowing for the modeling and inference of complex probabilistic systems. This approach enables the modeling of uncertainty and facilitates reasoning under uncertainty. Probabilistic programming is particularly valuable in domains such as healthcare, where predictive modeling and decision-making often involve inherent uncertainties.

c. Ethical Considerations and Responsible Data Usage:

As data becomes a crucial component of decision-making processes, it is essential to address ethical considerations and ensure responsible data usage. Privacy concerns, transparency, and fairness are key areas of focus. Organizations must adopt ethical practices to protect individuals' privacy rights and comply with data protection regulations. Transparent data usage, where individuals have a clear understanding of how their data is collected and used, builds trust and fosters a positive relationship between organizations and users.

Explainable AI is an emerging field that aims to make machine learning models more interpretable and understandable. It enables users to understand how models arrive at their predictions or decisions, providing insights into the underlying factors and contributing variables. This transparency is vital for ensuring fairness, accountability, and avoiding biases in automated decision-making.

Algorithmic audits and AI governance frameworks are essential components of responsible data usage. Auditing algorithms helps identify and mitigate biases, discriminatory practices, and unintended consequences. AI governance frameworks provide guidelines and policies for organizations to follow when developing and deploying AI systems. These frameworks encompass aspects such as accountability, fairness, transparency, and compliance with ethical standards.

By incorporating these ethical considerations and responsible data usage practices, organizations can harness the power of data while upholding societal values and ensuring that AI technologies are deployed in a fair, unbiased, and transparent manner. This approach not only protects individuals' privacy and rights but also builds public trust and enables the continued advancement and adoption of AI applications.

Conclusion:

Data is the backbone of machine learning, serving as the catalyst for unlocking the potential of AI. Its quality, diversity, and accessibility shape the effectiveness and accuracy of machine learning models. By harnessing the power of data, industries such as e-commerce, healthcare, banking, and finance can gain valuable insights, improve decision-making, and enhance customer experiences. However, challenges related to data quality, privacy, and bias must be addressed through responsible practices and ethical considerations. As we navigate the intricacies of data-driven machine learning, a conscientious and thoughtful approach will pave the way for a future where data becomes the key to unraveling the mysteries of our complex world.

# Training and evaluating AI models

In the world of AI, training and evaluating models lie at the heart of developing intelligent systems. Training entails the process of imparting knowledge and patterns to AI algorithms, while evaluation serves as the litmus test to measure their performance and assess their capabilities. In this section, we delve into the intricacies of training and evaluating AI models, exploring the methodologies, challenges, and the importance of continuous learning and improvement.

Training AI Models:

a. Supervised Learning:

Supervised learning is a popular approach where AI models learn from labeled examples, mapping input data to corresponding output labels. Through an iterative process, the models adjust their parameters and internal representations to minimize prediction errors. In e-commerce, supervised learning models can be trained to predict customer preferences based on historical purchase data, enabling personalized recommendations.

b. Unsupervised Learning:

Unsupervised learning involves training models on unlabeled data to discover hidden patterns and structures. These models autonomously explore the data, identifying clusters, similarities, and anomalies. In healthcare, unsupervised learning can be employed to uncover patterns in patient data, leading to insights for personalized medicine or disease detection.

c. Reinforcement Learning:

Reinforcement learning focuses on training AI agents to interact with an environment and learn optimal actions through trial and error. The agents receive feedback in the form of rewards or penalties, guiding their decision-making process. In finance, reinforcement learning can be used to optimize investment strategies by training agents to make decisions based on market conditions and historical data.

d. Semi-Supervised Learning:

Semi-supervised learning combines elements of both supervised and unsupervised learning. In this approach, models are trained on a combination of labeled and unlabeled data, leveraging the advantages of both. Semi-supervised learning is particularly useful when labeled data is scarce or expensive to obtain. In the field of natural language processing, semi-supervised learning can be applied to improve language understanding models by leveraging large amounts of unlabeled text data alongside a smaller set of labeled examples.

e. Transfer Learning:

Transfer learning involves training a model on a source task and transferring the acquired knowledge to a related target task. By leveraging pre-trained models and their learned representations, transfer learning allows for efficient training on smaller datasets or domains with limited labeled data. In image classification, for example, a model pre-trained on a large dataset like ImageNet can be fine-tuned on a smaller dataset specific to a particular domain, such as medical imaging or satellite imagery.

Each of these learning paradigms offers unique capabilities and is suited to different problem domains and data availability. By understanding the characteristics and strengths of each approach, AI practitioners can select the most appropriate learning paradigm for their specific applications, thereby unlocking the full potential of machine learning algorithms.

Evaluating AI Models:

a. Metrics and Performance Evaluation:

Evaluating AI models requires carefully selecting appropriate metrics that align with the desired objectives. Accuracy, precision, recall, and F1-score are commonly used metrics for classification tasks, while mean squared error (MSE) or root mean squared error (RMSE) are used for regression tasks. These metrics provide quantitative measures of model performance and guide decision-making processes.

b. Cross-Validation and Holdout Testing:

Cross-validation techniques help estimate a model's performance by training and testing on different subsets of the available data. K-fold cross-validation and stratified sampling ensure robustness in performance estimation. Holdout testing, where a separate test dataset is reserved, provides an independent evaluation of the model's generalization capabilities.

c. Bias, Fairness, and Interpretability:

Evaluating AI models goes beyond mere performance metrics. It involves examining potential biases in the data or the model's decision-making process. Ensuring fairness and avoiding discrimination is crucial, especially in domains like banking and finance where biased models can lead to unequal treatment. Interpretability techniques, such as attention mechanisms or feature importance analysis, shed light on how the model arrives at its decisions.

d. Trade-offs and Performance-Complexity:

Evaluating AI models requires considering trade-offs between performance and complexity. Models with higher complexity, such as deep neural networks, may achieve higher accuracy but require more computational resources and longer training times. On the other hand, simpler models, such as linear regression or decision trees, are computationally efficient but may sacrifice some accuracy. Understanding these trade-offs helps in selecting the most suitable model for a given task.

e. Robustness and Generalization:

Assessing the robustness and generalization capabilities of AI models is essential. Robust models should perform consistently across different datasets or real-world scenarios, even when the data distribution shifts. Techniques like data augmentation, regularization, and adversarial testing can enhance model robustness and detect vulnerabilities to adversarial attacks.

f. Continuous Monitoring and Model Maintenance:

AI models require continuous monitoring and maintenance to ensure their performance remains optimal over time. As data distributions change or new patterns emerge, models may require updates or retraining. Ongoing evaluation and monitoring allow for identifying and addressing performance degradation or concept drift, ensuring the reliability and effectiveness of AI systems.

By considering these aspects of metrics, evaluation techniques, biases, interpretability, trade-offs, robustness, and monitoring, stakeholders can make informed decisions about the deployment and usage of AI models. Evaluating and understanding the performance of AI systems in real-world contexts is vital for building trust, mitigating risks, and maximizing their positive impact across various domains.

### Challenges in Training and Evaluation:

a. Data Quantity and Quality:

Adequate training data is essential for AI models to learn effectively. In domains with limited data availability, such as healthcare, acquiring labeled data can be challenging. Data quality, including noise, missing values, or biases, also affects the training process and model performance. Employing techniques like data augmentation, transfer learning, or active learning can mitigate these challenges.

b. Overfitting and Generalization:

Overfitting occurs when a model performs well on the training data but fails to generalize to unseen data. Regularization techniques, such as dropout or weight decay, help prevent overfitting by introducing constraints on model complexity. Ensuring proper validation and testing procedures minimize the risk of overfitting and improve the model's generalization capabilities.

c. Ethical Considerations and Responsible AI:

Training and evaluating AI models necessitate ethical considerations to address potential biases, discrimination, and fairness issues. Bias detection and mitigation techniques, algorithmic audits, and diverse representation in training data help foster responsible AI practices. Transparency, accountability, and inclusivity are pivotal in building trustworthy and ethically sound AI systems.

Conclusion:

Training and evaluating AI models form the bedrock of developing intelligent systems that make informed decisions and generate valuable insights. Whether through supervised, unsupervised, or reinforcement learning, training empowers AI algorithms to acquire knowledge and patterns from data. Evaluation, on the other hand, serves as the critical step to assess model performance, identify biases, and ensure fairness. Overcoming challenges related to data quality, overfitting, and ethical considerations paves the way for responsible AI development. As we navigate the complexities of training and evaluation, continuous learning, and improvement fuel the growth of AI, empowering us to unlock the true potential of intelligent machines.

## Case studies and examples from various industries

In this section, we will explore real-world case studies and examples of how artificial intelligence (AI) is transforming various industries, including e-commerce, healthcare, banking, and finance. Through these examples, we will witness the diverse applications and the profound impact that AI is having on different sectors. From optimizing customer experiences to enhancing medical diagnoses and revolutionizing financial decision-making, AI is reshaping the way industries operate.

E-commerce:

In the e-commerce industry, AI-driven technologies are driving personalized experiences and revolutionizing the customer journey. Companies like Amazon and Alibaba leverage AI algorithms to analyze vast amounts of customer data, including purchase history, browsing behavior, and demographic information. These insights enable them to provide personalized product recommendations, targeted advertisements, and optimized pricing strategies. For example, Amazon's recommendation system, based on collaborative filtering and machine learning, suggests products that align with a customer's interests, increasing sales and customer satisfaction.

Additionally, AI-powered chatbots and virtual assistants are transforming customer service by providing instant and personalized support. These intelligent agents can answer inquiries, assist with product searches, and even process transactions. E-commerce giant, Shopify, utilizes AI-powered chatbots to provide 24/7 customer support, resulting in improved response times and reduced customer wait times.

Healthcare:

In the healthcare industry, AI is revolutionizing patient care, diagnosis, and treatment. Medical imaging, such as X-rays, MRIs, and CT scans, can be analyzed by AI algorithms to assist radiologists in detecting abnormalities and making accurate diagnoses. For instance, Google's DeepMind has developed AI systems that can detect and classify various diseases, including diabetic retinopathy and breast cancer, with high accuracy.

AI-powered virtual assistants are also being utilized in healthcare settings. These assistants can process and understand natural language, allowing patients to ask questions about symptoms, medications, and treatment options. Ada, an AI-powered symptom checker, uses machine learning algorithms to ask relevant questions and provide initial assessments of a patient's condition, helping individuals make informed decisions about seeking medical attention.

Banking and Finance:

The banking and finance sector is embracing AI to improve decision-making, risk management, and fraud detection. AI algorithms can analyze vast amounts of financial data, market trends, and historical patterns to optimize investment strategies, predict stock prices, and manage portfolios. Hedge funds and investment banks are employing AI-powered trading algorithms to make split-second decisions based on real-time market data, resulting in improved financial outcomes.

Moreover, AI-driven technologies are enhancing fraud detection and security measures in the financial industry. By analyzing transactional data, customer behavior, and patterns of fraudulent activities, AI algorithms can identify and flag suspicious transactions, preventing potential financial losses and protecting customers' assets.

PayPal's AI-powered fraud detection system analyzes millions of transactions daily, identifying patterns and anomalies to minimize fraudulent activities.

Conclusion:

These case studies and examples demonstrate the profound impact of AI across diverse industries. From personalized e-commerce experiences to improved healthcare diagnostics and optimized financial decision-making, AI is transforming the way businesses operate and enhancing the lives of individuals. The integration of AI algorithms, machine learning, and big data analytics is enabling industries to gain valuable insights, make data-driven decisions, and provide exceptional customer experiences. However, as AI continues to advance, it is crucial to address ethical considerations, privacy concerns, and ensure transparent and responsible usage of AI technologies. Through ongoing research, innovation, and responsible implementation, the potential for AI to drive further advancements across industries is immense.

# The Role of Neural Networks in AI

Artificial intelligence (AI) has made remarkable advancements in recent years, largely due to the significant contributions of neural networks. Neural networks, inspired by the structure and functionality of the human brain, are at the core of many AI systems. These networks have the remarkable ability to learn and adapt from data, enabling machines to perform complex tasks and make intelligent decisions. In this chapter, we will delve into the fundamental concepts of neural networks and explore their pivotal role in AI.

The Neural Network Paradigm:

Neural networks are computational models consisting of interconnected nodes, or artificial neurons, that work collectively to process and interpret information. These nodes, also known as artificial neurons or perceptrons, receive inputs, apply mathematical operations, and produce output signals. Through the network's architecture and the interaction of these artificial neurons, complex computations can be performed, allowing the network to learn and make predictions.

Learning from Data:

The true power of neural networks lies in their ability to learn from data. By exposing the network to a large volume of labeled examples, known as training data, the network can adjust its internal parameters, known as weights and biases, to minimize prediction errors. This process, called training, involves iteratively refining the network's parameters until it can accurately generalize to new, unseen data.

Feedforward and Backpropagation:

Neural networks operate through two fundamental processes: feedforward and backpropagation. In the feedforward phase, data is propagated through the network, layer by layer, with each layer performing computations and passing the results to the next layer. This process allows the network to generate predictions or make classifications.

During the backpropagation phase, the network compares its predictions with the known labels from the training data and calculates the prediction errors. These errors are then propagated backward through the network, allowing the network to adjust its weights and biases in a way that minimizes the errors. This iterative process of feedforward and backpropagation enables the network to fine-tune its parameters and improve its performance.

Deep Neural Networks:

Deep neural networks, often referred to as deep learning models, are neural networks with multiple hidden layers. These deep architectures enable the network to learn increasingly complex representations of the data, leading to higher levels of abstraction and better performance in tasks such as image recognition, natural language processing, and speech recognition.

Deep learning has been a driving force behind many breakthroughs in AI. For instance, in the field of computer vision, deep convolutional neural networks (CNNs) have achieved remarkable accuracy in image classification tasks. Examples include the ImageNet Large Scale Visual Recognition Challenge, where deep learning models have outperformed traditional computer vision techniques.

Applications across Industries:

Neural networks have found applications across various industries, transforming fields such as e-commerce, healthcare, banking, and finance. In e-commerce, neural networks power personalized product recommendations and customer segmentation, enhancing the shopping experience. In healthcare, neural networks aid in medical image analysis, disease diagnosis, and drug discovery. In the financial sector, neural networks optimize investment strategies, predict market trends, and detect fraud.

Conclusion:

Neural networks play a pivotal role in the field of artificial intelligence. Their ability to learn from data, model complex relationships, and make accurate predictions has opened up new frontiers in AI research and applications. By mimicking the intricate workings of the human brain, neural networks have propelled advancements in image recognition, natural language processing, and decision-making tasks. As AI continues to evolve, further innovations in neural network architectures and learning algorithms hold the potential to drive even more transformative changes across industries.

## Understanding neural networks and their architecture

Neural networks have emerged as a powerful tool in the realm of artificial intelligence, revolutionizing various industries and pushing the boundaries of what machines can accomplish. In this chapter, we will embark on a journey to understand the intricacies of neural networks and explore their architecture, shedding light on how these remarkable systems learn, process information, and make predictions.

The Building Blocks of Neural Networks:

Neural networks are constructed using interconnected nodes, also known as artificial neurons, that aim to mimic the behavior of neurons in the human brain. These artificial neurons are the fundamental units responsible for processing information and making predictions in neural networks.

Each artificial neuron receives inputs from other neurons or external sources, performs computations on those inputs, and produces an output. The inputs to a neuron are multiplied by corresponding weights, which represent the significance or impact of each input on the neuron's activation. The weighted inputs are then combined and passed through an activation function, which introduces non-linearity and determines the output of the neuron.

Layers and Connections:

Neural networks are typically organized into layers, which are composed of groups of artificial neurons. The most common architecture is the feedforward network, which consists of an input layer, one or more hidden layers, and an output layer. In this architecture, information flows sequentially from the input layer through the hidden layers to the output layer, with each layer performing computations and transforming the data.

Connections between neurons in adjacent layers are established through weights. Each connection represents the strength and impact of the input neuron's activation on the subsequent neuron in the next layer. These weights are adjustable parameters that are learned during the training process of the neural network. By adjusting the weights, the network can adapt its behavior and improve its ability to make accurate predictions.

The process of training a neural network involves iteratively adjusting the weights based on the errors between the network's predictions and the true labels of the training data. This adjustment is achieved using optimization algorithms, such as gradient descent, which aim to minimize the prediction errors and optimize the network's performance.

The ability of neural networks to learn and generalize from data is due to the collective behavior of interconnected neurons and the adaptive nature of their weights. As the network is exposed to more training examples, it can refine its internal representations and capture complex relationships within the data. This process of learning and adaptation is what allows neural networks to excel in various fields, such as e-commerce, healthcare, banking, and finance, by providing valuable insights and making accurate predictions.

In summary, neural networks are composed of interconnected artificial neurons organized into layers. The connections between neurons are represented by adjustable weights, which determine the influence of input signals on subsequent neurons. This architecture allows neural networks to process information, learn from data, and make predictions, making them a powerful tool in the realm of artificial intelligence.

Activation Functions:

Activation functions are essential components of neural networks as they introduce non-linear transformations to the input data. These functions determine the output of an artificial neuron based on its weighted inputs, adding the necessary non-linearity to capture complex relationships in the data. By applying activation functions, neural networks can model more sophisticated patterns and improve their ability to learn and make accurate predictions.

There are several commonly used activation functions in neural networks. The sigmoid function, represented by the mathematical formula $1 / (1 + e^{-x})$, maps the weighted sum of inputs to a value between 0 and 1. It is particularly useful in binary classification problems where the output represents a probability. The hyperbolic tangent (tanh) function, defined as $(e^x - e^{-x}) / (e^x + e^{-x})$, maps the inputs to values between -1 and 1, providing a more balanced range of outputs compared to the sigmoid function.

Another popular activation function is the Rectified Linear Unit (ReLU), defined as max(0, x). It outputs the input value if it is positive and sets it to zero otherwise. ReLU has gained popularity due to its simplicity and ability to mitigate the vanishing gradient problem, which can impede learning in deep neural networks. ReLU and its variants, such as Leaky ReLU and Parametric ReLU, have demonstrated excellent performance in many deep learning applications.

Training Neural Networks:

The training of neural networks involves presenting the network with a set of labeled examples, known as the training data. The network learns to adjust its weights through an iterative process by comparing its predictions with the true labels. The difference between the predictions and the true labels is quantified using a loss function, which measures the error.

Optimization algorithms, such as gradient descent, are employed to minimize the loss and update the weights accordingly. During each iteration, gradients are computed to determine the direction and magnitude of weight adjustments. The learning rate, a hyperparameter, determines the step size in each update. By iteratively adjusting the weights based on the gradients, the network gradually improves its performance and learns to make more accurate predictions.

Deep Neural Networks:

Deep neural networks, also known as deep learning models, are neural networks with multiple hidden layers between the input and output layers. These deep architectures enable the network to learn hierarchical representations of the data, extracting increasingly abstract features as the information flows through the layers.

Deep learning has revolutionized various fields, such as computer vision and natural language processing. For example, in computer vision, deep neural networks can automatically learn to recognize complex patterns, enabling applications like object detection, image segmentation, and facial recognition. In natural language processing, deep learning models can comprehend and generate human-like text, leading to advancements in machine translation, sentiment analysis, and chatbots.

The depth of a neural network allows it to capture intricate patterns and relationships in the data, making it highly expressive and capable of achieving state-of-the-art performance in many AI tasks. However, training deep neural networks can be challenging due to the vanishing gradient problem and the increased computational requirements. Nevertheless, advancements in algorithms, hardware acceleration, and distributed computing have facilitated the training of deep neural networks, fueling their widespread adoption and success in various industries.

In summary, activation functions introduce non-linear transformations to neural networks, enabling them to capture complex relationships in the data. The training of neural networks involves adjusting the weights based on prediction errors using optimization algorithms. Deep neural networks leverage multiple hidden layers to learn hierarchical representations, allowing them to extract increasingly abstract features from the data. These advancements have revolutionized fields like computer vision and natural language processing, leading to remarkable progress in AI applications.

Convolutional Neural Networks (CNNs):

Convolutional Neural Networks (CNNs) are a specialized type of neural network architecture that has revolutionized computer vision tasks. CNNs are designed to process grid-like data, such as images, by incorporating convolutional layers. These layers efficiently extract features by applying filters or kernels across the input data. The convolution operation allows the network to detect local patterns and spatial relationships in the data.

CNNs consist of multiple layers, including convolutional layers, pooling layers, and fully connected layers. Convolutional layers apply filters to the input data, capturing different features at various scales. Pooling layers downsample the data, reducing the spatial dimensions while preserving the important features. Fully connected layers connect the extracted features to the final output layer for classification or regression.

The power of CNNs lies in their ability to automatically learn relevant features from raw image data, eliminating the need for manual feature engineering. By learning hierarchical representations, CNNs can detect low-level features like edges and textures in early layers and combine them to recognize complex objects or scenes in deeper layers. This hierarchical approach enables CNNs to achieve remarkable performance in image classification, object detection, and image generation tasks.

For example, in image classification, a CNN can learn to recognize different objects or categories in images. By training on a large dataset of labeled images, the network can learn discriminative features and generalize to new, unseen images. CNNs have outperformed traditional computer vision algorithms and have become the backbone of many image-based AI applications, including autonomous driving, medical image analysis, and facial recognition.

Recurrent Neural Networks (RNNs):

Recurrent Neural Networks (RNNs) are a type of neural network architecture specifically designed for processing sequential data. Unlike feedforward networks that process each input independently, RNNs maintain an internal memory state that allows them to capture temporal dependencies and make predictions based on context.

RNNs are characterized by their recurrent connections, where the output of a neuron is fed back as input to the same neuron in the next time step. This cyclic structure allows information to persist over time, enabling RNNs to model sequential patterns effectively. The internal memory state, often referred to as the hidden state, serves as a form of short-term memory, retaining information from previous time steps.

RNNs have found applications in various sequential data tasks, such as natural language processing, speech recognition, and time series analysis. For example, in language modeling, RNNs can generate coherent text by predicting the next word based on the context of the previous words. In machine translation, RNNs have been used to build models that can translate sentences from one language to another. The ability of RNNs to capture dependencies in sequential data makes them well-suited for tasks where context and temporal information are crucial.

Generative Adversarial Networks (GANs):

Generative Adversarial Networks (GANs) are a fascinating type of neural network architecture that involves two networks: a generator and a discriminator. GANs are primarily used for generating new, realistic data samples, such as images or text. The generator network aims to produce data that resembles real samples, while the discriminator network tries to distinguish between the generated samples and real data.

The generator network takes random input, often referred to as the latent space, and maps it to the space of the target data. Through a series of layers, the generator learns to transform the latent input into meaningful data samples that approximate the real distribution. The discriminator network, on the other hand, is trained to classify whether a given sample is real or generated. The two networks are trained simultaneously in an adversarial fashion, with the generator aiming to fool the discriminator, and the discriminator striving to accurately differentiate between real and generated data.

The iterative process of training GANs results in the generator improving its ability to produce more realistic data samples, while the discriminator becomes more discerning. GANs have achieved remarkable success in tasks such as image synthesis, where they can generate highly convincing images that resemble real photographs. They have also been applied to other domains, such as text generation and video synthesis.

In summary, convolutional neural networks (CNNs) are specialized architectures for computer vision tasks, recurrent neural networks (RNNs) excel at processing sequential data, and generative adversarial networks (GANs) are used for data generation tasks. These neural network architectures have significantly advanced the field of AI, enabling breakthroughs in image classification, natural language processing, and data synthesis.

Conclusion:

Neural networks are the backbone of modern artificial intelligence, enabling machines to process information, learn from data, and make intelligent predictions. By understanding the architecture of neural networks, we gain insight into how these complex systems operate and contribute to advancements in various fields, including e-commerce, healthcare, banking, and finance. As AI continues to evolve, the exploration of neural network architectures and their capabilities will pave the way for new breakthroughs, empowering machines to perform increasingly sophisticated tasks.

## Deep learning: Going beyond shallow networks

In the realm of artificial intelligence, deep learning has emerged as a powerful technique that surpasses the limitations of shallow neural networks. By leveraging deep architectures with multiple hidden layers, deep learning models can learn hierarchical representations of data, allowing them to capture intricate patterns and relationships. This chapter delves into the principles and applications of deep learning, showcasing its transformative impact on various industries.

The Power of Depth in Neural Networks:

Shallow neural networks, with just a few layers, have their limitations when it comes to learning complex representations. As the depth of a network increases, it gains the ability to extract progressively more abstract features from the input data. Each hidden layer of a deep network learns to capture different levels of abstraction, allowing the model to understand the data at multiple levels of granularity.

The rise of deep learning can be attributed to the availability of massive amounts of data and computational power. With an abundance of labeled data, deep learning models can effectively learn intricate representations without the need for manual feature engineering. This characteristic has transformed the landscape of artificial intelligence, enabling breakthroughs in computer vision, natural language processing, and many other domains.

Convolutional Neural Networks (CNNs) for Vision Tasks:

Deep learning has revolutionized computer vision through the development of Convolutional Neural Networks (CNNs). CNNs leverage the hierarchical structure of deep architectures to automatically learn and extract meaningful features from images. By using convolutional layers, these networks can effectively capture local patterns, edges, and textures. This ability allows CNNs to excel in tasks such as image classification, object detection, and image generation.

For instance, in e-commerce, CNNs can analyze product images and automatically categorize them into relevant classes. This enables accurate product recommendations, improving the user experience and driving sales. In healthcare, CNNs can assist in the diagnosis of diseases by analyzing medical images, aiding doctors in making more precise and timely decisions.

Recurrent Neural Networks (RNNs) for Sequential Data:

Deep learning extends its capabilities beyond static data with Recurrent Neural Networks (RNNs). RNNs are specifically designed to process sequential data, such as time series, speech, and text. Unlike traditional feedforward networks, RNNs maintain an internal memory state that enables them to capture temporal dependencies and contextual information.

RNNs have had a significant impact on natural language processing tasks. For example, in language translation, RNNs can process sequences of words and generate accurate translations by considering the context and order of words. In speech recognition, RNNs can analyze spoken language and convert it into text, improving voice assistants and transcription services.

Generative Models and Unsupervised Learning:

Deep learning has also revolutionized generative modeling and unsupervised learning. Generative models, such as Variational Autoencoders (VAEs) and Generative Adversarial Networks (GANs), can generate new data samples that resemble the training data distribution. These models have applications in image synthesis, text generation, and even music composition.

Unsupervised learning, where models learn from unlabeled data, has seen great advancements with deep learning. Autoencoders, for example, can learn compact representations of data and reconstruct the input, enabling tasks such as data compression and denoising. These unsupervised techniques have opened doors to new possibilities in fields like anomaly detection, where identifying rare events in large datasets is crucial.

The Limitations and Challenges of Deep Learning:

While deep learning has achieved remarkable success, it is not without limitations and challenges. One significant limitation is the need for large amounts of labeled data, which can be challenging and costly to obtain, especially in domains with limited data availability. Additionally, deep learning models can be computationally intensive and require substantial computational resources.

Another challenge lies in the interpretability of deep learning models. With their numerous layers and millions of parameters, understanding how these models arrive at their decisions can be difficult. This lack of interpretability raises concerns in critical domains where transparency and explainability are essential.

Conclusion:

Deep learning has propelled the field of artificial intelligence to new heights, enabling breakthroughs in computer vision, natural language processing, and generative modeling. Through the power of deep architectures and the ability to learn hierarchical representations, deep learning models have revolutionized numerous industries, including e-commerce, healthcare, and finance.

However, challenges remain in terms of data availability, computational requirements, and model interpretability. Addressing these challenges will pave the way for further advancements in deep learning and ensure its responsible and ethical integration into various domains.

As we move forward, understanding the strengths and limitations of deep learning is crucial. By harnessing its potential while being mindful of its constraints, we can unlock the transformative power of deep learning and continue pushing the boundaries of artificial intelligence.

# Convolutional neural networks for image processing

In the realm of artificial intelligence, Convolutional Neural Networks (CNNs) have emerged as a transformative tool for image processing. With their ability to automatically learn and extract meaningful features from images, CNNs have revolutionized tasks such as image classification, object detection, and image generation. In this chapter, we delve into the inner workings of CNNs, explore their architecture, and showcase their applications across various domains.

Understanding Convolutional Layers:

At the heart of CNNs lies the convolutional layer, a key component that enables these networks to capture local patterns and structures in images. The convolutional layer applies filters, also known as kernels, across the input image to produce feature maps. Each filter learns to detect specific features, such as edges, textures, or shapes, by convolving across the image using a sliding window approach.

The power of convolutional layers lies in their ability to preserve spatial relationships. By leveraging local connections and weight sharing, CNNs can effectively capture patterns regardless of their position in the image. This property makes CNNs robust to translations, rotations, and scale variations, enabling them to generalize well to unseen data.

MaxPooling and Strides for Spatial Subsampling:

In addition to convolutional layers, CNNs often employ max pooling and strides for spatial subsampling. Max pooling reduces the spatial dimensions of feature maps by retaining only the maximum values within a pooling window. This operation helps to reduce the computational burden and introduces a form of translation invariance, making the network more robust to small variations in the position of detected features.

Strides, on the other hand, determine the amount of pixel shift when sliding the convolutional kernel across the image. By increasing the stride, the network reduces the spatial dimensions of the feature maps even further, allowing for higher-level features to be learned while maintaining a broader receptive field. This process facilitates hierarchical feature extraction and contributes to the network's ability to capture complex visual representations.

Deep Architectures for Hierarchical Learning:

One of the key strengths of CNNs lies in their ability to learn hierarchical representations of data. Through the stacking of multiple convolutional and pooling layers, CNNs can extract increasingly abstract features as the information flows through the network. Lower layers learn basic features such as edges and textures, while higher layers learn more complex representations, such as object parts or semantic concepts.

This hierarchical learning enables CNNs to excel in image classification tasks. For instance, in e-commerce, CNNs can automatically categorize products based on their visual appearance, allowing for more accurate and efficient product search and recommendation systems. In healthcare, CNNs can aid in the analysis of medical images, assisting doctors in diagnosing diseases or detecting abnormalities.

Training and Optimization of CNNs:

Training CNNs involves two main steps: forward propagation, where the input data is fed through the network to generate predictions, and backpropagation, where the errors between the predictions and the ground truth labels are propagated backward to update the network's weights. This process iteratively adjusts the network's parameters, minimizing the prediction errors and improving its performance.

To optimize CNNs, various techniques and algorithms are employed. The choice of optimization algorithm, such as stochastic gradient descent (SGD) or Adam, plays a crucial role in finding the optimal set of weights. Regularization techniques, like dropout or weight decay, prevent overfitting and enhance the generalization capabilities of the network. Hyperparameter tuning, such as adjusting the learning rate or batch size, also influences the network's performance.

Real-World Applications of CNNs:

The impact of CNNs extends far beyond theoretical concepts. Across industries, CNNs have found remarkable applications in solving real-world problems. In autonomous vehicles, CNNs enable object detection and recognition, allowing vehicles to perceive their surroundings and make informed decisions. In security and surveillance, CNNs aid in face recognition and anomaly detection, enhancing public safety. In agriculture, CNNs can analyze satellite images to monitor crop health and optimize irrigation.

One prominent example of CNN's prowess is in the field of art. DeepArt, a popular mobile application, employs CNNs to transform ordinary photos into artistic masterpieces by applying the style of famous paintings. This showcases the versatility and creative potential of CNNs, pushing the boundaries of human imagination and machine collaboration.

Challenges and Future Directions:

While CNNs have achieved remarkable success, challenges and opportunities lie ahead. One of the challenges is the need for large annotated datasets, as CNNs heavily rely on labeled data for training. In some domains, such as medical imaging, acquiring large-scale labeled data can be time-consuming and costly. Techniques like transfer learning and data augmentation mitigate this challenge by leveraging pre-trained models and generating synthetic data.

Interpretability is another area of concern in CNNs. Despite their exceptional performance, understanding how CNNs arrive at their decisions can be challenging. This lack of interpretability raises questions about trust, accountability, and fairness. Researchers are actively exploring techniques such as attention mechanisms, visualization tools, and explainable AI to shed light on the decision-making process of CNNs.

Conclusion:

Convolutional Neural Networks (CNNs) have revolutionized image processing, enabling computers to understand and interpret visual information with exceptional accuracy. Through the power of convolutional layers, hierarchical learning, and deep architectures, CNNs have reshaped industries, ranging from e-commerce and healthcare to autonomous vehicles and agriculture.

As we continue to push the boundaries of artificial intelligence, it is crucial to address the challenges and ethical considerations associated with CNNs. By further advancing the interpretability of CNNs and ensuring responsible and unbiased deployment, we can unlock the full potential of this technology and pave the way for a future where AI-powered visual understanding enhances our lives in profound ways.

## Recurrent neural networks for sequential data

In the realm of artificial intelligence, Recurrent Neural Networks (RNNs) have emerged as a powerful tool for processing sequential data. Unlike feedforward neural networks, which process data in a one-time manner, RNNs possess a memory mechanism that allows them to capture dependencies and patterns across time. This unique ability makes RNNs well-suited for tasks such as speech recognition, language translation, and sentiment analysis. In this chapter, we delve into the intricacies of RNNs, explore their architecture, and showcase their applications across various domains.

Understanding the Temporal Nature of RNNs:

RNNs are specifically designed to handle data with a sequential or temporal structure. They can process input sequences of varying lengths and remember information from previous time steps. This is accomplished through the use of recurrent connections, which form a feedback loop, allowing information to flow from one time step to the next.

The core element of an RNN is the hidden state, which serves as the memory of the network. At each time step, the hidden state takes into account the current input and the previous hidden state, generating a new hidden state that encapsulates the information from both time steps. This sequential processing enables RNNs to capture long-term dependencies and context in the data.

Types of RNN Architectures:

RNNs come in several architectural variants, each with its own strengths and use cases. The most common type is the Vanilla RNN, also known as the Elman network, which uses a simple recurrent structure. However, Vanilla RNNs suffer from the vanishing gradient problem, where the gradients diminish exponentially, hindering the learning process for long sequences.

To address this issue, more advanced RNN architectures have been developed, such as the Long Short-Term Memory (LSTM) and Gated Recurrent Unit (GRU). These architectures incorporate specialized gating mechanisms that regulate the flow of information within the network, allowing for better gradient flow and alleviating the vanishing gradient problem. LSTMs and GRUs have become the de facto choice for many sequential tasks, as they can effectively capture long-term dependencies and mitigate the challenges associated with training deep RNNs.

Applications of RNNs in Sequential Data Analysis:

RNNs have found extensive applications across various domains that deal with sequential data. In natural language processing, RNNs excel at tasks such as language modeling, sentiment analysis, and machine translation. For example, in healthcare, RNNs can analyze electronic health records and predict patient outcomes based on their medical history. In finance, RNNs can predict stock market trends based on historical trading data. Moreover, in music composition, RNNs can generate new melodies that mimic the patterns observed in a given musical genre.

One notable application of RNNs is in the field of speech recognition. By processing audio signals as sequential data, RNNs can convert spoken language into written text, enabling voice assistants, transcription services, and other speech-based applications. For instance, virtual assistants like Amazon's Alexa and Apple's Siri employ RNN-based models to understand and respond to spoken commands.

Challenges and Future Directions:

While RNNs have demonstrated their efficacy in processing sequential data, they also come with challenges that researchers and practitioners are actively addressing. One key challenge is the issue of long-term dependencies. Although LSTM and GRU architectures alleviate the vanishing gradient problem, they can still struggle with capturing dependencies that span very long sequences. Techniques like attention mechanisms, which allow the network to focus on relevant parts of the input, have shown promising results in addressing this challenge.

Another challenge is the computational cost associated with training RNNs, particularly with large-scale datasets. The sequential nature of RNNs makes it difficult to parallelize the computations, leading to slower training times. To mitigate this, researchers have explored techniques such as mini-batch training, distributed computing, and hardware accelerators like graphical processing units (GPUs) to expedite the training process.

Conclusion:

Recurrent Neural Networks (RNNs) have emerged as a formidable tool for analyzing and understanding sequential data. Their ability to capture dependencies across time has paved the way for significant advancements in fields such as natural language processing, speech recognition, and healthcare. By incorporating memory and feedback mechanisms, RNNs have unlocked the potential to model and predict complex patterns in sequential data.

As research in RNNs continues to progress, addressing challenges related to long-term dependencies and computational efficiency will be paramount. By pushing the boundaries of RNN architectures, optimization techniques, and training methodologies, we can further harness the power of RNNs and unlock their potential in a wide range of applications. The future of sequential data analysis holds immense promise, and RNNs are at the forefront of this exciting journey.

# Case studies and breakthroughs in neural network applications

Neural networks, with their remarkable ability to learn from data and make complex predictions, have revolutionized various industries and brought forth significant breakthroughs. In this chapter, we delve into real-world case studies and highlight the transformative impact of neural network applications across diverse fields such as E-commerce, Healthcare, Banking, and Finance. These examples will underscore the potential of neural networks and provide insights into their practical implementation.

## E-commerce: Personalized Recommendations

In the realm of E-commerce, neural networks have played a pivotal role in enhancing customer experiences through personalized recommendations. Take, for instance, the giant online retailer Amazon. Their recommendation engine employs sophisticated deep learning models that analyze customer browsing and purchase history to suggest relevant products. By leveraging the power of neural networks, Amazon can tailor recommendations to individual preferences, leading to increased customer engagement and sales. Similar approaches have been adopted by other E-commerce platforms, such as Netflix and Spotify, to deliver personalized movie and music recommendations, respectively.

## Healthcare: Disease Diagnosis and Prognosis

Neural networks have made significant strides in the healthcare industry, particularly in disease diagnosis and prognosis. For instance, in medical imaging, convolutional neural networks (CNNs) have demonstrated remarkable accuracy in identifying abnormalities and diagnosing diseases. In a study published in Nature Medicine, researchers developed a deep learning model that outperformed human dermatologists in detecting skin cancer from images. This breakthrough paves the way for improved early detection and potentially life-saving interventions. Additionally, recurrent neural networks (RNNs) have been utilized for predicting patient outcomes based on electronic health records, aiding healthcare professionals in making informed decisions and providing personalized treatment plans.

## Banking and Finance: Fraud Detection

The banking and finance sector has harnessed the power of neural networks to combat fraud and enhance security. Machine learning models, including neural networks, are employed to analyze vast amounts of financial data and detect suspicious activities. For instance, banks use anomaly detection algorithms based on deep learning architectures to identify fraudulent transactions in real-time. By analyzing patterns, trends, and historical data, these models can flag potentially fraudulent transactions and reduce financial losses. This proactive approach to fraud detection is instrumental in maintaining the integrity of the financial system and safeguarding customers' assets.

Autonomous Vehicles: Vision-based Perception

The advent of autonomous vehicles has spurred advancements in computer vision, and neural networks are at the forefront of these developments. Deep learning models, particularly CNNs, are used to process visual data from cameras and sensors mounted on vehicles, enabling them to perceive and understand the surrounding environment. Companies like Tesla have leveraged neural networks to enhance their Autopilot feature, which enables autonomous driving capabilities. By analyzing real-time data and learning from millions of miles driven, neural networks can identify objects, detect lane boundaries, and make critical decisions for safe navigation. This breakthrough in vision-based perception brings us closer to a future where self-driving cars become a reality.

Counterarguments and Challenges:

While neural networks have achieved remarkable breakthroughs, it is important to acknowledge potential counterarguments and challenges. One common criticism is the "black box" nature of neural networks, where the inner workings of the model may be difficult to interpret or explain. This lack of interpretability raises concerns about bias, ethics, and accountability. Researchers and practitioners are actively working on developing techniques for explainable AI, allowing us to understand how neural networks arrive at their predictions.

Another challenge is the need for large amounts of labeled data to train neural networks effectively. In certain domains, such as healthcare, acquiring labeled data can be challenging due to privacy concerns and limited access to medical records. However, approaches like transfer learning and semi-supervised learning aim to mitigate this challenge by leveraging pre-trained models or making use of unlabeled data.

Conclusion:

Neural networks have led to groundbreaking advancements across a myriad of industries, demonstrating their potential to transform the way we live and work. Through personalized recommendations, disease diagnosis, fraud detection, autonomous vehicles, and more, neural networks have revolutionized the capabilities of AI systems. However, challenges related to interpretability, data availability, and ethical considerations remain important areas of research and development.

As we witness the continued evolution of neural network applications, it is crucial to strike a balance between innovation and responsible AI practices. By addressing challenges and pushing the boundaries of neural network technology, we can unlock even more transformative breakthroughs and shape a future where AI systems augment human capabilities in a beneficial and ethical manner.

# Natural Language Processing: Enabling AI to Understand Human Language

Natural Language Processing (NLP) is a subfield of artificial intelligence (AI) that focuses on the interaction between computers and human language. It encompasses a range of techniques and algorithms that enable machines to understand, interpret, and generate human language in a way that mimics human cognitive abilities. NLP has emerged as a fundamental technology for a variety of applications, including language translation, sentiment analysis, chatbots, and information retrieval.

Understanding human language is a complex and nuanced task. Humans possess a remarkable ability to comprehend and generate language effortlessly, but for machines, it remains a challenging endeavor. NLP aims to bridge this gap by developing algorithms and models that can process and make sense of textual data.

Key Concepts in Natural Language Processing:

To appreciate the intricacies of NLP, it is essential to grasp some key concepts that form the foundation of this field. Let's explore these concepts in detail:

Text Preprocessing:

Text preprocessing is a crucial step in NLP that involves transforming raw text into a clean and structured format that is easier for machines to understand. This process typically involves tasks such as tokenization (splitting text into individual words or tokens), stemming (reducing words to their base or root form), and removing stopwords (common words that do not contribute much meaning). By preparing the text data appropriately, NLP algorithms can operate more effectively.

Language Modeling:

Language modeling involves building statistical models that capture the underlying structure and patterns of a particular language. These models enable machines to predict the likelihood of certain words or phrases occurring in a given context. For example, language models have been used to generate coherent and contextually relevant text, such as in the case of language generation in chatbots or automated article writing.

Sentiment Analysis:

Sentiment analysis is a valuable application of NLP that aims to understand and interpret the subjective information conveyed in text, such as opinions, emotions, and attitudes. By analyzing the sentiment of customer reviews, social media posts, or product feedback, businesses can gain valuable insights into public perception and make data-driven decisions. Sentiment analysis has found widespread use in areas like brand monitoring, customer feedback analysis, and market research.

Named Entity Recognition:

Named Entity Recognition (NER) is the task of identifying and classifying named entities in text, such as names of people, organizations, locations, and dates. This task is vital for information extraction and knowledge representation. For example, in the healthcare industry, NER can be employed to automatically extract relevant information from medical records, such as patient names, diagnosis, and treatment details.

Machine Translation:

Machine translation is one of the most well-known and impactful applications of NLP. It involves the automatic translation of text from one language to another. Neural machine translation models, powered by deep learning techniques, have demonstrated remarkable improvements in translation quality, enabling more accurate and fluent translations. Companies like Google and Microsoft have leveraged NLP to develop advanced translation systems that make communication across languages more accessible and efficient.

Case Studies and Examples:

To further illustrate the practical applications of NLP, let's delve into a few case studies and examples:

E-commerce: Customer Support Chatbots

E-commerce platforms often employ NLP techniques to develop chatbots that can understand customer queries and provide relevant assistance. These chatbots use techniques such as natural language understanding (NLU) to extract user intent from text and generate appropriate responses. For example, Amazon's Alexa virtual assistant leverages NLP capabilities to understand and respond to user commands, enabling a more intuitive and interactive shopping experience.

Healthcare: Clinical Text Analysis

In the healthcare domain, NLP is used to analyze clinical text data, such as medical records and research articles. By applying NLP techniques, researchers can extract valuable insights from a vast amount of unstructured data, leading to advancements in disease diagnosis, treatment recommendations, and drug discovery. For instance, IBM's Watson for Oncology utilizes NLP to analyze medical literature and provide evidence-based treatment recommendations for cancer patients.

Banking and Finance: Fraud Detection

NLP techniques are leveraged in the banking and finance sector to detect fraudulent activities and enhance security measures. By analyzing textual data, such as transaction descriptions and customer communication, NLP algorithms can identify suspicious patterns, uncover hidden relationships, and flag potentially fraudulent transactions. This helps financial institutions mitigate risks and protect their customers from fraudulent activities.

Conclusion:

Natural Language Processing (NLP) is a fascinating field that enables machines to understand and interact with human language. Through techniques like text preprocessing, language modeling, sentiment analysis, named entity recognition, and machine translation, NLP has found applications in various domains, including E-commerce, Healthcare, and Banking and Finance. The ability of machines to comprehend and generate human language opens doors to enhanced customer experiences, improved healthcare outcomes, and more efficient business processes.

However, challenges remain in the form of language ambiguity, cultural nuances, and the need for large labeled datasets for training NLP models effectively. Continued research and advancements in NLP techniques, coupled with the ethical considerations surrounding language processing, will pave the way for further breakthroughs and ensure responsible and impactful use of AI in understanding human language.

## Exploring the challenges and complexities of natural language processing

Natural Language Processing (NLP) is a field of study that aims to bridge the gap between human language and machine understanding. While NLP has made significant advancements in recent years, it is not without its challenges and complexities. In this section, we will delve into some of these challenges and explore the complexities that researchers and practitioners face in the realm of NLP.

Language Ambiguity and Context:

In the realm of natural language processing (NLP), one of the most significant challenges is dealing with the inherent ambiguity of human language. Words and phrases can have multiple meanings depending on the context in which they are used. This ambiguity poses a complex task for NLP models, as they need to accurately interpret and understand the intended meaning of words or phrases in different contexts.

For example, consider the word "bank." Without additional context, it could refer to a financial institution where people deposit and withdraw money. However, in a different context, it could also refer to the edge of a river. Understanding the intended meaning of "bank" in a given sentence or text requires considering the surrounding words and the broader context.

To address this challenge, NLP models leverage techniques such as semantic analysis and contextual embeddings. These techniques aim to capture the context and meaning of words by considering their neighboring words and the overall structure of the sentence or text. By analyzing the surrounding context, NLP models can make more accurate interpretations and disambiguate between different meanings.

Cultural Nuances and Language Variations:

Human language is incredibly diverse, encompassing a wide range of cultural nuances, variations, and idiosyncrasies. Different languages, dialects, and cultural contexts introduce unique challenges in NLP. Idiomatic expressions, slang, and cultural references add layers of complexity to accurately interpreting and understanding text.

For instance, idiomatic expressions such as "raining cats and dogs" or "kick the bucket" cannot be understood by simply analyzing the individual words. They require knowledge of the specific cultural context and familiarity with the associated idioms. Slang terms, which can differ significantly across regions and social groups, further complicate language processing tasks.

To address cultural nuances and language variations, NLP models need to be trained on diverse datasets that capture the linguistic and cultural richness of different communities. By exposing models to a wide range of linguistic expressions, idioms, and cultural references, they can learn to understand and interpret text more accurately across different languages and cultural contexts. Additionally, leveraging techniques such as transfer learning, where models learn from one language or culture and transfer that knowledge to another, can also help bridge the gap between different linguistic variations.

Lack of Annotated Data:

Training NLP models, especially those based on supervised learning, relies heavily on annotated data, where human experts label data examples with their corresponding meanings or categories. However, creating annotated datasets is a labor-intensive and time-consuming process that requires domain expertise and meticulous attention to detail. Annotating data at scale becomes even more challenging in specialized domains or languages with limited resources.

The lack of annotated data poses a significant obstacle to building accurate NLP models in specific domains such as healthcare or legal texts. Without sufficient annotated examples, models may struggle to generalize well to real-world scenarios, leading to suboptimal performance and limited applicability.

To mitigate the scarcity of annotated data, researchers explore various techniques. Active learning approaches, for example, prioritize the annotation of data points that are uncertain or difficult for the model, allowing for more efficient use of annotation resources. Additionally, techniques like transfer learning can help leverage annotated data from related domains or languages to bootstrap the learning process and reduce the need for extensive annotation efforts.

By addressing the challenges of language ambiguity, cultural nuances, and data scarcity, NLP researchers and practitioners can pave the way for more robust and accurate language understanding models. These advancements will enable a wide range of applications across diverse fields, including e-commerce, healthcare, banking, and finance, where precise and context-aware language processing is paramount.

Ethical and Bias Concerns:

Ethics and bias in natural language processing (NLP) are critical considerations as AI models can inadvertently perpetuate biases present in the data they are trained on. If the training data contains biases or reflects societal prejudices, NLP models may learn and reproduce those biases, leading to biased or unfair outcomes in their predictions or language generation.

For instance, if an NLP model is trained on a dataset that contains imbalanced representations of different social groups, it may learn to associate certain attributes or stereotypes with specific groups. This can result in biased predictions or reinforce discriminatory behavior. These biases can manifest in various applications, such as automated hiring processes or language translation systems, where the output may exhibit unfair treatment or perpetuate stereotypes.

Addressing these ethical concerns requires a proactive and multidimensional approach. Firstly, it is essential to ensure that the training data used for NLP models is diverse, representative, and inclusive. This means collecting and curating datasets that capture a wide range of perspectives, avoiding skewed or imbalanced data that may reinforce biases.

Furthermore, developing fairness-aware and bias-detection techniques is crucial to identify and mitigate biases in NLP models. Researchers are actively exploring methods to quantify and measure bias in AI systems, such as analyzing the representation and treatment of different social groups. By integrating fairness metrics into the training process, developers can strive to minimize biases and promote equitable outcomes.

In addition to technical solutions, there is a need for ethical guidelines and frameworks to guide the responsible development and deployment of NLP models. Ethical considerations should include transparency, accountability, and explainability. Users and developers should have access to information about how NLP models make predictions, what data they were trained on, and the potential biases they may exhibit.

Language Understanding Beyond Syntax:

While NLP models have made remarkable progress in understanding syntactic structures of language, semantic understanding remains a challenge. While syntactic parsing enables the analysis of sentence structure and grammatical relationships, it falls short in capturing the deeper meanings, nuances, and context-dependent interpretations of language.

Semantic understanding requires going beyond mere syntactic analysis and incorporating knowledge from various domains. Integrating world knowledge, commonsense reasoning, and domain-specific knowledge can enhance the semantic understanding capabilities of NLP models. For example, understanding the meaning of idiomatic expressions, resolving ambiguities, and grasping the intent behind a text often require drawing upon external knowledge sources.

Advancements in semantic understanding have the potential to revolutionize NLP applications across industries. In e-commerce, it can enable more sophisticated product recommendations based on customer preferences and intents. In healthcare, it can support accurate interpretation of medical records and enable personalized treatment recommendations. Furthermore, in banking and finance, semantic understanding can facilitate intelligent sentiment analysis of financial news and aid in making informed investment decisions.

Researchers are exploring various techniques to enhance semantic understanding, including the integration of knowledge graphs, semantic role labeling, and pre-training models on vast amounts of text data. By furthering our understanding of semantic representations and reasoning mechanisms, we can unlock the full potential of NLP in understanding and processing human language.

Case Studies and Examples:

To further illustrate the challenges and complexities of NLP, let's explore a few case studies and examples:

E-commerce: Language Variations and Product Descriptions

In the realm of e-commerce, natural language processing (NLP) encounters the intricacies of processing diverse product descriptions that exhibit variations in language, style, and format. Product descriptions provided by sellers and manufacturers may differ significantly, making it challenging for NLP systems to extract relevant information and provide accurate insights to enhance the shopping experience.

One aspect of language variation in e-commerce is the presence of abbreviations and acronyms. Sellers often use shorthand or industry-specific abbreviations to describe product features or specifications. For example, a smartphone listing might include terms like "RAM," "CPU," or "MP" to denote the device's memory, processor, and camera resolution, respectively. NLP systems need to be able to recognize and interpret these abbreviations to provide accurate and comprehensive product information.

Furthermore, misspellings and typographical errors are common in user-generated product descriptions. Customers may unintentionally introduce errors while writing reviews or descriptions, making it essential for NLP systems to handle these variations and identify the intended meaning. For example, a customer might refer to a "vaccum cleaner" instead of a "vacuum cleaner." NLP models must employ techniques like spell checking, context-based correction, or error-tolerant matching to overcome these challenges.

Subjectivity and subjective language also add complexity to e-commerce NLP. Product descriptions often include subjective terms such as "stylish," "durable," or "affordable," which can be highly subjective and vary across different individuals. NLP systems must understand the subjective context in which these terms are used and extract the intended meaning accurately. This can involve sentiment analysis to determine whether a product is described positively or negatively, as well as considering the overall tone and context of the description.

Overcoming these challenges in e-commerce NLP enables improved product search and recommendation systems, enhanced customer reviews analysis, and more efficient product categorization. By accurately understanding and extracting information from product descriptions, NLP can contribute to a more personalized and satisfactory shopping experience for consumers.

Healthcare: Clinical Text Analysis and Standardization

In the domain of healthcare, NLP plays a crucial role in analyzing and extracting valuable insights from clinical text data, including electronic health records (EHRs), medical literature, and research articles. However, NLP encounters unique challenges in accurately interpreting and standardizing clinical text due to the lack of standardized terminology, the presence of domain-specific jargon, and the complexity of medical concepts.

One significant challenge is the lack of a unified and standardized vocabulary across healthcare institutions and systems. Different healthcare providers may use diverse terminologies and abbreviations to describe medical conditions, treatments, or symptoms. For example, "Myocardial Infarction" and "Heart Attack" are two terms used interchangeably to describe the same medical condition. NLP systems must possess robust semantic understanding capabilities to recognize these variations and map them to the appropriate medical concepts.

The presence of domain-specific jargon further complicates clinical text analysis. Medical professionals often employ technical terms, abbreviations, or acronyms that may be unfamiliar to NLP models. Understanding medical jargon is critical for accurate information extraction, clinical decision support, and research analysis. NLP systems must leverage specialized medical ontologies, dictionaries, and domain-specific knowledge bases to decode and interpret these terms accurately.

Additionally, clinical text often contains implicit information, context-dependent meanings, and complex sentence structures. Extracting relevant information from clinical narratives, such as patient symptoms, treatment plans, or medication side effects, requires deep linguistic understanding and contextual interpretation. NLP models need to employ advanced techniques like named entity recognition, relationship extraction, and co-reference resolution to capture these intricate details accurately.

Addressing these challenges in healthcare NLP has wide-ranging implications. It can facilitate automatic coding and structuring of EHRs, support clinical decision-making, enable efficient data analysis for research, and contribute to the development of personalized medicine. By accurately analyzing and standardizing clinical text, NLP systems can improve healthcare outcomes and enable better patient care.

Banking and Finance: Understanding Financial News and Sentiment Analysis

In the banking and finance industry, NLP plays a crucial role in understanding financial news, analyzing market trends, and sentiment analysis. However, NLP encounters challenges in comprehending the specialized language, complex sentence structures, and nuanced sentiments present in financial texts.

Financial news articles often contain domain-specific terms and jargon that are specific to the banking and finance industry. Concepts like "dividends," "equities," or "credit default swaps" require specialized knowledge to understand their meaning and implications accurately. NLP systems need to incorporate financial dictionaries, ontologies, and domain-specific knowledge to decipher these terms and accurately interpret financial texts.

Furthermore, financial texts often exhibit complex sentence structures, making it challenging to extract key information and understand the relationships between different elements in the text. For example, financial reports might contain lengthy sentences with multiple subclauses, which can hinder accurate information extraction. NLP models need to employ syntactic parsing and dependency parsing techniques to break down complex sentences and identify the relevant entities and relationships.

Sentiment analysis in the banking and finance domain poses another challenge for NLP. Financial texts can convey nuanced sentiments, subtle market trends, and financial opinions that require sophisticated language understanding. NLP systems need to accurately capture positive or negative sentiment, identify the overall sentiment of a financial article or tweet, and assess the impact of sentiment on market behavior.

Overcoming these challenges in banking and finance NLP has significant implications for investment decisions, risk management, and financial market analysis. Accurate understanding of financial news and sentiment analysis enables more informed investment strategies, supports sentiment-based trading, and enhances risk assessment and prediction. By effectively analyzing financial texts, NLP can contribute to more efficient and accurate decision-making in the banking and finance industry.

Conclusion:

Natural Language Processing (NLP) is a fascinating field that aims to enable machines to understand and interact with human language. However, exploring the challenges and complexities of NLP reveals the intricacies involved in processing and comprehending human language. Addressing language ambiguity, cultural nuances, lack of annotated data, bias concerns, and semantic understanding are ongoing research endeavors.

Overcoming these challenges requires interdisciplinary efforts, including linguistics, machine learning, and ethics. Researchers and practitioners must continuously strive for more accurate and unbiased NLP models, leveraging techniques such as data augmentation, transfer learning, and fairness-aware algorithms. By addressing these challenges, we can unlock the full potential of NLP in various domains, promoting inclusive and effective communication between humans and machines.

## Techniques and algorithms in natural language understanding

Techniques and algorithms in natural language understanding (NLU) play a pivotal role in advancing the capabilities of artificial intelligence (AI) systems to comprehend and process human language. NLU encompasses a wide range of approaches, each tailored to handle different aspects of language understanding, such as syntax, semantics, and pragmatics. In this section, we will delve into various techniques and algorithms employed in NLU and explore their applications across different fields.

Rule-Based Approaches:

Rule-based approaches involve the use of handcrafted rules and linguistic patterns to process and understand language. These rules define the syntactic and semantic structures of sentences and guide the interpretation process. For example, in information extraction tasks, rules can be designed to identify specific entities, relationships, or events from text. While rule-based approaches provide transparency and interpretability, they are often limited in their ability to handle ambiguity and the complexities of real-world language use.

Statistical Approaches:

Statistical approaches in NLU leverage machine learning algorithms to automatically learn patterns and associations from large amounts of labeled training data. These approaches, such as the popular bag-of-words model and n-gram models, use statistical techniques to estimate the probability of word sequences and make predictions based on observed patterns. Statistical approaches excel in tasks like language modeling, sentiment analysis, and machine translation. However, they may struggle with understanding the deeper semantics and context of language.

Machine Learning:

Machine learning techniques, particularly supervised and unsupervised learning, have gained prominence in NLU. Supervised learning involves training models on labeled data, enabling them to make predictions based on learned patterns. This approach is commonly used in tasks like text classification, named entity recognition, and sentiment analysis. Unsupervised learning, on the other hand, involves training models on unlabeled data to discover hidden structures and relationships. Techniques such as clustering and topic modeling fall under unsupervised learning and find applications in document organization and topic extraction.

Deep Learning:

Deep learning has revolutionized NLU by enabling the development of deep neural networks that can learn hierarchical representations of language. Recurrent Neural Networks (RNNs) and Convolutional Neural Networks (CNNs) have been instrumental in tasks like language modeling, machine translation, and text generation. Long Short-Term Memory (LSTM) networks, a variant of RNNs, excel at capturing long-range dependencies in sequential data, making them suitable for tasks involving temporal understanding, such as speech recognition and language generation.

Transformer Models:

Transformer models, such as the groundbreaking BERT (Bidirectional Encoder Representations from Transformers), have pushed the boundaries of NLU. These models utilize self-attention mechanisms to capture contextual dependencies across words in a sentence, enabling them to understand nuances and resolve ambiguity effectively. Transformer models have achieved state-of-the-art performance in a wide range of NLU tasks, including question answering, named entity recognition, and sentiment analysis.

Pretrained Language Models:

Pretrained language models, like GPT (Generative Pre-trained Transformer) and RoBERTa (Robustly Optimized BERT approach), have emerged as powerful tools for NLU. These models are trained on vast amounts of diverse text data and capture rich linguistic knowledge and contextual understanding. By fine-tuning these models on specific tasks or domains, they can be adapted to perform various NLU tasks with remarkable accuracy and efficiency.

While these techniques and algorithms have significantly advanced NLU, it is essential to acknowledge their limitations. For example, some argue that deep learning models, although highly effective, lack interpretability and suffer from data bias issues. Others contend that rule-based approaches might struggle with the complexities of language and require extensive manual effort. Striking a balance between these different approaches and combining their strengths is an ongoing research challenge.

In the field of e-commerce, NLU techniques are employed to extract product features and sentiment from customer reviews, enabling businesses to gain valuable insights into customer preferences and improve their product offerings. In healthcare, NLU is used to process clinical text, extract medical concepts, and support clinical decision-making. In banking and finance, NLU techniques aid in sentiment analysis of financial news, analyzing market trends, and supporting investment decisions.

It is important to note that NLU is a rapidly evolving field, and researchers continue to explore innovative approaches and algorithms to overcome the challenges and further enhance the understanding of human language. By combining techniques from different domains, such as linguistics, machine learning, and deep learning, we can continue to push the boundaries of NLU and unlock new possibilities for AI systems to comprehend and interact with human language in a more nuanced and sophisticated manner.

# Sentiment analysis and language generation

Sentiment analysis and language generation are two fascinating areas within natural language processing (NLP) that have significant implications across various fields, including e-commerce, healthcare, banking, and finance. In this section, we will explore the concepts, techniques, and applications of sentiment analysis and language generation, shedding light on their potential and challenges.

## Sentiment Analysis: Understanding the Emotional Tone

Sentiment analysis, also known as opinion mining, is the process of extracting and understanding the sentiment or emotional tone expressed in text. It involves automatically categorizing text as positive, negative, or neutral, and sometimes even fine-grained sentiment analysis, which captures more nuanced emotions like happiness, sadness, anger, etc. Sentiment analysis can be applied to customer reviews, social media posts, survey responses, and other forms of user-generated content.

The importance of sentiment analysis in e-commerce cannot be overstated. By analyzing customer feedback and sentiment, businesses can gain valuable insights into customer preferences, identify areas for improvement, and tailor their marketing strategies accordingly. For example, an e-commerce company can use sentiment analysis to assess customer satisfaction with their products or services and make data-driven decisions to enhance customer experiences.

In healthcare, sentiment analysis can be utilized to monitor patient sentiment in online health forums or social media platforms, providing insights into public health concerns, patient experiences, and sentiment trends related to healthcare services. This information can help healthcare providers identify areas for improvement, enhance patient engagement, and develop targeted interventions.

In the banking and finance industry, sentiment analysis is valuable for monitoring market sentiment and assessing investor sentiment towards specific stocks, commodities, or financial products. By analyzing social media feeds, news articles, and financial forums, sentiment analysis can help investors gauge market sentiment, identify emerging trends, and make informed investment decisions.

## Language Generation: Creating Coherent and Contextual Text

Language generation, also known as text generation, focuses on generating coherent and contextually relevant text based on given prompts or conditions. It involves modeling the underlying patterns and structures of human language to produce text that resembles natural human-generated text. Language generation techniques have witnessed significant advancements with the rise of deep learning and transformer-based models.

One notable application of language generation is in chatbots and virtual assistants. These AI-powered systems use natural language generation to provide human-like responses and engage in conversational interactions with users. By understanding user inputs and generating appropriate responses, chatbots can assist customers in e-commerce platforms, provide customer support, and even simulate human-like conversation.

In healthcare, language generation techniques can be employed to generate patient reports, discharge summaries, or automated responses to common inquiries. For instance, an AI system can generate personalized health recommendations or provide information about medication usage based on patient-specific data.

In the field of banking and finance, language generation finds application in generating financial reports, market analyses, or personalized investment advice. For example, an AI-powered system can generate personalized investment recommendations based on user preferences, risk profiles, and market conditions.

Despite the progress in sentiment analysis and language generation, challenges persist. Sentiment analysis struggles with understanding sarcasm, irony, or cultural nuances, as these require a deeper understanding of context and cultural references. Language generation faces difficulties in generating diverse and creative text that goes beyond simply replicating existing patterns. Striking a balance between generating coherent text and injecting creativity and originality remains an ongoing research challenge.

In conclusion, sentiment analysis and language generation are powerful NLP techniques that enable AI systems to comprehend and generate text with emotional context and human-like fluency. Their applications span a wide range of fields, from e-commerce to healthcare and banking. As the capabilities of AI continue to advance, refining sentiment analysis techniques and developing more sophisticated language generation models will contribute to the growth and impact of AI in understanding and generating human language.

## NLP applications in different industries

Natural Language Processing (NLP) has emerged as a transformative technology with wide-ranging applications across various industries. From e-commerce to healthcare, banking, and finance, NLP is revolutionizing how organizations interact with and understand human language. In this chapter, we will explore the diverse applications of NLP in different industries, highlighting their potential impact and key challenges.

E-commerce: Enhancing Customer Experiences and Personalization:

In the realm of e-commerce, NLP holds tremendous potential for enhancing customer experiences and personalization. By leveraging NLP techniques, businesses can gain valuable insights from customer feedback, reviews, and social media interactions. One of the key applications of NLP in e-commerce is sentiment analysis, which involves analyzing customer sentiment to gauge the success of marketing campaigns or product launches.

Sentiment analysis techniques enable businesses to understand customer opinions and emotions associated with their products or services. By analyzing product descriptions and customer reviews, e-commerce platforms can identify common themes, sentiment trends, and areas of improvement. For example, suppose an online retailer discovers that customers consistently mention the poor quality of a particular product in their reviews. In that case, the retailer can take action to address the issue, such as improving the product's quality or offering better alternatives.

By harnessing NLP for sentiment analysis, e-commerce businesses can tailor their offerings to align with customer preferences and expectations. This customization can significantly enhance customer experiences, leading to improved customer satisfaction and loyalty. For instance, an e-commerce platform can recommend products based on sentiment analysis, ensuring that customers are presented with options that align with their interests and preferences. By understanding customer sentiment, businesses can also identify potential issues or concerns early on, allowing them to proactively address them and build stronger customer relationships.

Furthermore, NLP powers chatbots and virtual assistants, enabling real-time customer support and personalized recommendations. These AI-powered systems leverage natural language understanding capabilities to comprehend customer queries and provide relevant and accurate responses. For example, a customer may ask a chatbot about the availability of a specific product or request recommendations based on their preferences. By utilizing NLP algorithms, the chatbot can understand the intent behind the customer's query and provide appropriate responses or suggestions.

Additionally, NLP-based chatbots can facilitate seamless transactions by assisting customers throughout the purchasing process. They can provide product information, answer queries related to pricing or shipping, and guide customers through the checkout process. This personalized and efficient support enhances the overall customer experience, making it more convenient and satisfying.

Moreover, NLP enables e-commerce platforms to personalize shopping experiences based on customer data. By analyzing customer interactions, browsing behavior, and purchase history, NLP models can identify patterns, preferences, and individual needs. This information can be leveraged to deliver personalized recommendations, tailored product suggestions, and targeted promotions. For example, a customer who frequently purchases athletic apparel may receive personalized recommendations for new sports shoes or workout gear. This level of personalization not only improves customer satisfaction but also drives higher engagement and increases sales conversion rates.

In conclusion, NLP has brought significant advancements to the e-commerce industry, enabling businesses to enhance customer experiences and personalize interactions. By leveraging sentiment analysis, chatbots, and personalized recommendations, e-commerce platforms can understand customer preferences, address concerns, and provide a tailored shopping experience. As NLP techniques continue to evolve, e-commerce businesses have an opportunity to leverage this technology to build stronger customer relationships, drive customer loyalty, and ultimately succeed in a highly competitive marketplace.

Case Studies:

Let's explore a few case studies that illustrate the transformative impact of NLP in different industries:

E-commerce: Amazon's Product Recommendations

Amazon, one of the world's largest e-commerce platforms, has revolutionized the online shopping experience through its sophisticated recommendation system. Powered by NLP techniques, Amazon analyzes vast amounts of customer data to provide personalized product recommendations. By understanding customer preferences, purchasing history, and browsing behavior, Amazon's recommendation engine suggests relevant products that align with each customer's interests and needs.

At the heart of Amazon's recommendation system is the utilization of NLP algorithms to process and analyze product descriptions. By extracting key features, attributes, and customer reviews, Amazon gains valuable insights into each product's characteristics and user experiences. This analysis allows Amazon to understand the relationships between products, identify patterns, and make informed recommendations.

For instance, when a customer searches for a specific product or browses a category, Amazon's recommendation engine considers various factors, such as the customer's past purchases, items added to the cart, and items viewed. Leveraging NLP techniques, the system identifies similar products, complementary items, and popular choices made by customers with similar preferences. These recommendations appear on the product detail pages, personalized homepages, and even in email notifications, providing customers with relevant options and increasing the likelihood of making a purchase.

Amazon's success in recommendation systems is a testament to the power of NLP in enhancing the e-commerce experience. By employing advanced algorithms and analyzing vast amounts of customer data, Amazon can deliver personalized suggestions that improve customer satisfaction, increase engagement, and drive sales.

Expanding on Healthcare: IBM Watson's Oncology Advisor:

IBM Watson's Oncology Advisor is a prime example of how NLP can revolutionize healthcare. Oncology Advisor utilizes NLP techniques to analyze vast amounts of medical literature, patient records, and clinical guidelines. The system extracts critical information from these sources, allowing oncologists to access the latest research and evidence-based treatment recommendations.

In the field of oncology, staying up-to-date with the rapidly evolving research landscape is crucial. However, the sheer volume of medical literature and clinical guidelines makes it challenging for oncologists to keep pace with the latest advancements. IBM Watson's Oncology Advisor acts as a valuable tool by processing and understanding this vast amount of textual information, presenting oncologists with relevant treatment options and supporting their decision-making process.

By leveraging NLP algorithms, the system can interpret and analyze unstructured clinical text, such as patient records, pathology reports, and medical research papers. It extracts key insights, identifies relevant treatment options, and provides oncologists with evidence-based recommendations tailored to each patient's unique condition and characteristics.

IBM Watson's Oncology Advisor assists oncologists in navigating complex treatment decisions, taking into account factors such as disease stage, genetic markers, treatment history, and patient preferences. By analyzing a comprehensive range of data sources, this NLP-powered system enables oncologists to make well-informed decisions and deliver personalized cancer care, ultimately improving patient outcomes.

Expanding on Banking and Finance: Bloomberg's News Sentiment Analysis:

In the world of finance, NLP techniques have proven to be invaluable for analyzing vast amounts of textual data and providing real-time insights. Bloomberg, a leading financial information and media company, employs NLP algorithms to analyze financial news articles, social media feeds, and market data, allowing market participants to gauge sentiment and make informed trading decisions.

Bloomberg's news sentiment analysis utilizes NLP to process and understand the language used in financial news articles. By analyzing the sentiment expressed in these articles, such as positive or negative tones, Bloomberg can provide market participants with real-time insights into market sentiment. This information can be invaluable for traders and investors in understanding market dynamics, identifying trends, and making timely investment decisions.

Additionally, Bloomberg incorporates NLP techniques to analyze social media feeds, including platforms like Twitter, for market-related discussions and sentiments. By monitoring and analyzing these feeds, Bloomberg's NLP algorithms can identify trends, sentiment shifts, and emerging topics relevant to the financial markets. This information helps traders and investors stay informed and adapt their strategies accordingly.

News sentiment analysis empowers market participants with a deeper understanding of market dynamics, investor sentiment, and potential market-moving events. By leveraging NLP in this context, Bloomberg enables traders and investors to make data-driven decisions and navigate the complexities of the financial markets more effectively.

In conclusion, NLP has found impactful applications in various industries, including e-commerce, healthcare, and banking and finance. By harnessing NLP techniques, businesses and organizations can unlock valuable insights from textual data, enhance customer experiences, improve decision-making processes, and drive innovation in their respective fields. As NLP continues to evolve, its potential for revolutionizing industries and transforming the way we interact with information and technology is truly remarkable.

Conclusion:

NLP has opened up a world of possibilities in diverse industries, ranging from e-commerce and healthcare to banking and finance. By harnessing the power of human language, organizations can gain insights, enhance customer experiences, and make data-driven decisions. However, ethical considerations, bias mitigation, and interpretability remain crucial aspects to ensure the responsible and effective application of NLP. As technology advances and NLP techniques evolve, the potential for transformative impact across industries continues to grow, promising a future where human language becomes a powerful tool for progress and understanding.

## Ethical considerations and potential biases in NLP

As we delve deeper into the field of Natural Language Processing (NLP), it is imperative to address the ethical considerations and potential biases that arise from the development and deployment of NLP systems. While NLP brings remarkable advancements and transformative capabilities, it also poses significant ethical challenges that demand our attention. In this section, we will explore the ethical dimensions of NLP and examine the potential biases that can emerge within these systems. By critically examining these concerns, we can foster responsible AI development and ensure that NLP technologies serve society in an equitable and unbiased manner.

Ethical Considerations in NLP:

Privacy and Data Protection:

NLP systems rely on vast amounts of data, including personal information, to train and operate effectively. This raises concerns about privacy and data protection. Users must have transparency and control over the collection, storage, and use of their data. Organizations should adopt robust data protection measures, including encryption and access controls, to safeguard sensitive information. Moreover, obtaining explicit consent from users for data usage is crucial to respecting their privacy rights. Striking a balance between the benefits of NLP and preserving user privacy is essential for building trust in these systems.

Fairness and Bias:

NLP systems have the potential to perpetuate biases present in the data they are trained on, leading to unfair or discriminatory outcomes. Biases can emerge in various forms, such as gender, race, ethnicity, or socioeconomic status. It is imperative to actively address these biases and strive for fairness in NLP models. Developing fairness-aware techniques, such as debiasing algorithms, can help mitigate biases and ensure equitable outcomes. Additionally, biases in training data must be carefully managed through diverse and inclusive data collection and curation processes. By avoiding the reinforcement of societal prejudices, NLP systems can contribute to a more equitable and inclusive society.

Accountability and Transparency:

Ensuring accountability and transparency in NLP models is essential for building trust and understanding system behavior. Interpretable and explainable AI techniques enable users, regulators, and stakeholders to scrutinize the underlying processes and identify potential biases or errors. In high-stakes domains like healthcare and finance, where decisions impact lives and livelihoods, transparency is particularly crucial. By providing insights into how NLP models arrive at their decisions, organizations can be held accountable for any biases or unintended consequences. Striving for transparency and promoting open dialogue around NLP models fosters responsible AI practices and strengthens public trust.

Potential Biases in NLP:

Language and Cultural Biases:

NLP models trained on biased or unrepresentative datasets may exhibit biases in their responses and interpretations. For example, if a language model is predominantly trained on English-language data, it may struggle to accurately understand or generate content in other languages. Similarly, cultural biases in training data can result in skewed interpretations or misrepresentations of specific cultures or demographics. To address these biases, it is crucial to ensure diverse training data that represents a wide range of languages and cultural perspectives. This promotes inclusivity and avoids marginalizing or misrepresenting specific communities.

Gender and Stereotypes:

NLP models can inadvertently learn and reinforce gender stereotypes due to biases present in training data. For instance, in automated resume screening, biases may emerge that favor candidates of certain genders or penalize those from underrepresented groups. Addressing gender biases requires a careful analysis of training data and the development of techniques that promote fairness and inclusivity. Organizations should aim for gender-neutral and bias-free NLP models to ensure equal opportunities for all individuals and prevent the perpetuation of discriminatory practices.

Bias Amplification:

NLP systems have the potential to amplify existing biases present in society. For example, search engine algorithms that prioritize certain types of content may inadvertently reinforce stereotypes or limit access to diverse perspectives. This can perpetuate misinformation or hinder the dissemination of alternative viewpoints. To mitigate bias amplification, continuous monitoring and evaluation of NLP systems are necessary. This involves detecting and addressing biases that may emerge during system operation and ensuring that these systems contribute to a more inclusive and informed society.

Conclusion:

Ethical considerations and potential biases in NLP demand our attention to ensure the responsible development and deployment of AI technologies. Privacy and data protection, fairness and bias mitigation, accountability and transparency are crucial pillars that guide the ethical use of NLP. By addressing biases and promoting inclusivity, NLP can serve as a tool for positive societal impact. Striving for transparency, accountability, and fairness in NLP development enables us to harness the potential of AI while upholding fundamental values and respecting the dignity and rights of individuals.

Case Study: Amazon's Gender-Biased Hiring Algorithm:

A notable example highlighting the potential biases in NLP systems is the case of Amazon's gender-biased hiring algorithm. In an effort to automate the screening of job applicants, Amazon developed an AI-powered system that analyzed resumes to identify top candidates. However, the system exhibited biases that favored male applicants and discriminated against female applicants. This bias stemmed from the training data, which consisted mostly of resumes from male candidates who historically dominated the tech industry.

The algorithm learned to associate certain terms and patterns in resumes with high qualifications, and these patterns were more prevalent in resumes from male candidates. Consequently, the system inadvertently penalized resumes that contained terms associated with female candidates or diverse backgrounds, leading to gender discrimination in the hiring process.

This case underscores the critical importance of addressing biases in NLP systems and the need for continuous evaluation and fairness testing. It serves as a cautionary tale, reminding us of the ethical responsibilities inherent in developing and deploying AI technologies.

Mitigating Biases and Ensuring Ethical NLP:

To mitigate biases and ensure ethical NLP, several measures can be taken:

Diverse and Representative Training Data: NLP systems must be trained on diverse and representative datasets that encompass a wide range of demographics, cultures, and languages. This necessitates careful curation of data to avoid the amplification of biases present in the training set.

Bias Detection and Mitigation: Developers should integrate bias detection and mitigation techniques into NLP systems. This involves continuously monitoring system outputs, identifying biases, and implementing corrective measures to ensure fair and equitable outcomes. Techniques such as debiasing algorithms and fairness-aware training can help address biases and promote fairness.

Interdisciplinary Collaboration and Ethical Guidelines: Addressing ethical concerns and potential biases in NLP requires collaboration among researchers, practitioners, ethicists, and policymakers. Interdisciplinary dialogue and the development of ethical guidelines can help guide responsible AI practices, promote transparency, and ensure the ethical use of NLP technologies.

Conclusion:

The case of Amazon's gender-biased hiring algorithm serves as a stark reminder of the potential biases that can emerge in NLP systems. It highlights the need for developers and researchers to be proactive in addressing biases and promoting fairness. By leveraging diverse and representative training data, implementing bias detection and mitigation techniques, and fostering interdisciplinary collaboration, we can work towards ensuring that NLP systems are ethically developed and deployed.

Striving for transparency, accountability, and fairness in NLP will contribute to the development of responsible AI systems that uphold fundamental values and respect the dignity and rights of all individuals. It is our collective responsibility to mitigate biases, promote inclusivity, and ensure that NLP technologies serve as tools for positive societal impact.

# Computer Vision: Interpreting and Processing Visual Information

Computer Vision, a subfield of Artificial Intelligence, focuses on enabling machines to interpret and understand visual information, mimicking the remarkable capabilities of human vision. By harnessing the power of deep learning and advanced algorithms, Computer Vision enables machines to analyze, classify, and extract meaningful insights from images and videos. In this chapter, we will explore the fascinating world of Computer Vision, its applications in various fields, and the underlying techniques that drive its success.

## Understanding Images: Pixels, Features, and Representations

At the heart of Computer Vision lies the understanding of images as collections of pixels. Each pixel represents a specific color or intensity value, and by analyzing the patterns and relationships between pixels, machines can extract valuable information. Computer Vision algorithms identify visual features, such as edges, corners, or textures, which serve as building blocks for higher-level analysis. These features are then used to construct representations that capture the essential characteristics of an image, allowing for more sophisticated processing and interpretation.

## Image Classification and Object Recognition

One of the key tasks in Computer Vision is image classification, which involves assigning a label or category to an image based on its content. This task enables machines to recognize and differentiate objects, people, or scenes within an image. Deep learning techniques, particularly Convolutional Neural Networks (CNNs), have revolutionized image classification by learning hierarchical representations and automatically extracting relevant features. State-of-the-art models, such as the ImageNet-trained networks, have achieved remarkable accuracy, surpassing human performance in certain domains.

## Object Detection and Localization

Beyond classification, Computer Vision enables the detection and localization of objects within an image. Object detection algorithms identify and locate multiple instances of objects, often bounding boxes, within an image. This task is crucial in applications such as autonomous driving, surveillance systems, and image-based search engines. Techniques like Region-based Convolutional Neural Networks (R-CNNs) and their variants have greatly advanced object detection capabilities, allowing for real-time and accurate identification of objects in complex scenes.

## Semantic Segmentation and Scene Understanding

Semantic segmentation involves assigning a semantic label to each pixel in an image, enabling a more fine-grained understanding of the scene. This technique allows machines to differentiate between different object classes and identify object boundaries accurately. Semantic segmentation plays a crucial role in applications like medical imaging, where precise delineation of anatomical structures is essential. Advanced architectures such as Fully Convolutional Networks (FCNs) have demonstrated impressive results in semantic segmentation, enabling precise pixel-level analysis.

Visual Captioning and Image Generation

Computer Vision has also made significant progress in generating human-like descriptions or captions for images, a task known as visual captioning. By combining Computer Vision with Natural Language Processing (NLP) techniques, machines can generate coherent and contextually relevant descriptions based on the content of an image. Furthermore, recent advancements in generative models, such as Generative Adversarial Networks (GANs), have enabled the synthesis of realistic images based on given textual descriptions or concepts, expanding the possibilities of creative image generation.

Applications in Various Fields:

E-commerce: Computer Vision enhances the shopping experience by enabling visual search, where users can search for products based on images rather than textual queries. This empowers users to find similar items, discover visually appealing products, and bridge the gap between inspiration and purchase. Additionally, Computer Vision aids in quality control, ensuring accurate product categorization and detecting defects in manufacturing processes.

Healthcare: Computer Vision has transformative potential in medical imaging, assisting in the diagnosis, treatment, and monitoring of various conditions. It enables the automated analysis of medical images, such as X-rays, MRIs, and pathology slides, aiding radiologists and pathologists in detecting abnormalities and making accurate assessments. Computer Vision can also contribute to telemedicine applications by facilitating remote image interpretation and triage.

Autonomous Vehicles: Computer Vision is at the core of autonomous driving systems, enabling vehicles to perceive and understand the surrounding environment. By processing data from cameras, lidar, and radar sensors, Computer Vision algorithms detect pedestrians, vehicles, and traffic signs, allowing autonomous vehicles to navigate safely and make informed decisions. The fusion of Computer Vision with other sensor modalities, such as radar and lidar, further enhances the perception capabilities of autonomous vehicles.

Security and Surveillance: Computer Vision plays a pivotal role in security and surveillance systems, enabling real-time monitoring and threat detection. By analyzing video feeds, Computer Vision algorithms can identify suspicious activities, detect unauthorized access, and track individuals of interest. Such systems enhance public safety, assist law enforcement, and safeguard critical infrastructure.

Counterarguments and Challenges:

While Computer Vision has made remarkable progress, challenges and limitations persist. Some key considerations include:

Complexity and Interpretability: Deep learning models used in Computer Vision often exhibit complex behaviors that can be challenging to interpret and understand. The "black-box" nature of these models raises concerns about their decision-making process, limiting their applicability in safety-critical domains where interpretability and explainability are crucial.

Data Bias and Ethical Concerns: Computer Vision models are susceptible to biases present in training data, which can lead to discriminatory outcomes or reinforce societal prejudices. Ensuring fair representation and addressing biases in training data are essential to mitigate these concerns and promote ethical and inclusive Computer Vision systems.

Adversarial Attacks: Computer Vision models are vulnerable to adversarial attacks, where carefully crafted input perturbations can mislead the model's predictions. Such attacks can have severe consequences, particularly in security-sensitive applications like autonomous vehicles or facial recognition systems. Robustness against adversarial attacks remains an ongoing research challenge.

Conclusion:

Computer Vision has revolutionized the way machines interpret and process visual information, enabling a wide range of applications across diverse fields. From image classification and object detection to semantic segmentation and image generation, Computer Vision techniques continue to push the boundaries of what machines can achieve in understanding visual content. However, challenges such as model interpretability, data bias, and adversarial attacks must be addressed to ensure the ethical and responsible development and deployment of Computer Vision systems. By navigating these challenges, we can unlock the full potential of Computer Vision, empowering machines to perceive and comprehend the visual world in ways that were once solely within the realm of human perception.

# The significance of computer vision in AI systems

Computer Vision, a prominent field within the realm of Artificial Intelligence (AI), holds immense significance in enabling machines to perceive, interpret, and understand visual information, thereby bridging the gap between human-like perception and machine intelligence. This chapter delves into the multifaceted significance of Computer Vision in AI systems, exploring its applications, impact across diverse fields, and the underlying principles that drive its effectiveness.

Empowering Machines with Visual Perception:

One of the fundamental aspects that sets humans apart is their ability to perceive and make sense of the visual world. Computer Vision aims to replicate this capability in machines, enabling them to interpret and extract meaningful insights from images and videos. By analyzing visual data, machines can identify objects, recognize faces, understand scenes, and even navigate complex environments. This transformative capability opens up a world of possibilities for AI systems, empowering them to interact with the visual world in a manner that was once exclusive to humans.

Enhancing Human-Machine Interaction:

Computer Vision plays a pivotal role in enhancing human-machine interaction by enabling machines to understand and respond to visual cues. In applications such as gesture recognition, facial expression analysis, and gaze tracking, Computer Vision allows machines to interpret human behavior and respond accordingly. This enables more natural and intuitive forms of human-computer interaction, enhancing user experiences and making technology more accessible and inclusive.

Revolutionizing Fields Across Industries:

E-commerce: Computer Vision has revolutionized the e-commerce industry by enabling visual search capabilities, empowering users to search for products based on images rather than text. This allows customers to find visually similar products or discover items that align with their preferences. By leveraging Computer Vision, e-commerce platforms can enhance product recommendations, improve search accuracy, and provide a more personalized shopping experience.

Healthcare: In healthcare, Computer Vision has immense potential to improve diagnostics, aid in medical imaging analysis, and assist in surgical procedures. For example, Computer Vision algorithms can detect anomalies in medical images, identify cancerous cells, and support radiologists in making accurate diagnoses.

Additionally, Computer Vision can automate the analysis of medical scans, enabling faster and more precise interpretations, ultimately leading to improved patient care.

Autonomous Systems: Computer Vision is integral to the development of autonomous systems, such as self-driving cars and drones. By processing visual data from cameras and sensors, these systems can perceive the surrounding environment, detect obstacles, and make informed decisions in real-time. Computer Vision algorithms enable autonomous systems to navigate safely, analyze traffic patterns, and respond to dynamic situations, thereby revolutionizing transportation and logistics.

Surveillance and Security: Computer Vision has transformed the field of surveillance and security by enabling advanced video analytics. By analyzing video feeds in real-time, Computer Vision algorithms can detect suspicious activities, recognize individuals, and monitor public spaces for potential threats. This technology enhances public safety, aids law enforcement agencies, and contributes to the protection of critical infrastructure.

Counterarguments and Challenges:

Despite its remarkable advancements, Computer Vision faces several challenges and limitations:

Ambiguity and Interpretability: Visual data often contains inherent ambiguity, making it challenging for machines to accurately interpret complex scenes or handle unusual scenarios. The interpretation of visual information involves context, prior knowledge, and reasoning abilities, which remain challenging for AI systems to replicate fully. Achieving a deeper level of interpretability and explainability in Computer Vision algorithms remains an active area of research.

Data Limitations and Ethical Considerations: The quality and representativeness of training data significantly impact the performance and potential biases of Computer Vision systems. Biases in training data can lead to discriminatory outcomes or reinforce societal prejudices. Ensuring diverse and unbiased datasets is crucial to building fair and ethical Computer Vision systems that treat all individuals fairly and equitably.

Robustness and Generalization: Computer Vision algorithms may struggle with handling variations in lighting conditions, viewpoints, or occlusions. Achieving robustness and generalization across different scenarios and environments remains a challenge. Adapting Computer Vision systems to handle real-world complexities and unforeseen circumstances is a critical area of research for the widespread adoption of these technologies.

Conclusion:

Computer Vision holds immense significance in AI systems, empowering machines to interpret and understand visual information akin to human perception. Its applications span across various fields, including e-commerce, healthcare, autonomous systems, and surveillance, revolutionizing industries and improving human experiences. Despite challenges such as interpretability, data limitations, and robustness, ongoing research and advancements in Computer Vision continue to drive progress, paving the way for future innovations in AI systems that seamlessly integrate with the visual world.

## Image recognition and object detection

Image recognition and object detection are integral components of computer vision that enable machines to understand and interpret visual content. This chapter explores the concepts, techniques, and applications of image recognition and object detection, shedding light on their significance across diverse fields such as e-commerce, healthcare, banking, and finance.

Understanding Image Recognition:

Image recognition refers to the process of identifying and classifying objects or patterns within an image. It involves training machine learning models to recognize specific objects or categories by analyzing the visual features present in the image. By leveraging advanced algorithms and deep learning techniques, image recognition systems can achieve remarkable accuracy in identifying and categorizing objects.

Applications of Image Recognition:

E-commerce: Image recognition has transformed the e-commerce industry, enabling visual search capabilities and facilitating a more intuitive shopping experience. By allowing users to search for products based on images, e-commerce platforms can offer personalized recommendations, enhance product discovery, and improve user engagement. For example, customers can take a picture of an item they like and find similar products instantly, facilitating more efficient and convenient shopping.

Healthcare: In healthcare, image recognition plays a vital role in medical imaging analysis, aiding in disease diagnosis and treatment planning. For instance, radiologists rely on image recognition systems to identify abnormalities in X-rays, MRIs, or CT scans, assisting in the detection of diseases like cancer or cardiovascular conditions.

By accurately interpreting medical images, image recognition technology enhances the accuracy of diagnoses and improves patient outcomes.

Banking and Finance: Image recognition finds applications in banking and finance through processes such as check recognition and authentication. By analyzing the patterns, features, and security elements present in images of checks, image recognition systems can quickly and accurately verify their authenticity. This technology streamlines financial transactions, reduces the risk of fraud, and enhances operational efficiency.

Object Detection: Beyond Image Recognition

Object detection takes image recognition a step further by not only identifying objects within an image but also localizing their positions. Object detection algorithms can detect and outline multiple objects in an image, providing valuable information about their spatial arrangement. This capability enables various applications in fields such as autonomous driving, surveillance, and augmented reality.

Autonomous Driving: Object detection is crucial for autonomous vehicles to perceive and navigate their environment safely. By detecting and tracking objects such as vehicles, pedestrians, and traffic signs, object detection systems contribute to collision avoidance and efficient route planning. These systems rely on advanced algorithms and sensor fusion techniques to interpret the visual input from cameras and lidar sensors, enabling real-time decision-making.

Surveillance and Security: Object detection is a key component of video surveillance systems, allowing for real-time detection and tracking of individuals or suspicious activities. By analyzing video feeds, object detection algorithms can trigger alerts for security personnel, enhancing public safety and enabling rapid response to potential threats. This technology finds applications in areas such as airport security, public spaces monitoring, and crowd management.

Augmented Reality: Object detection forms the foundation of augmented reality experiences by identifying and tracking objects in real-time. By overlaying virtual content onto real-world objects, object detection enables immersive and interactive augmented reality applications. For instance, smartphone apps that superimpose virtual objects onto the physical environment rely on object detection to accurately position and anchor the virtual elements.

Challenges and Future Directions:

While image recognition and object detection have made significant strides, challenges persist:

Complex Scenes and Occlusions: Recognizing and detecting objects in complex scenes or images with occlusions remains a challenge. Occlusions occur when objects are partially hidden or obscured by other objects, making their detection and localization more challenging. Advancements in deep learning architectures and the availability of large-scale annotated datasets are driving progress in addressing these challenges.

Data Quality and Bias: The quality and diversity of training data significantly impact the performance and fairness of image recognition and object detection systems. Biases present in training data can lead to skewed results or discrimination. Ensuring diverse and representative datasets, and employing fairness-aware techniques, is crucial to mitigate biases and ensure the equitable functioning of these systems.

Conclusion:

Image recognition and object detection play pivotal roles in the field of computer vision, enabling machines to interpret and understand visual information. Their applications span across industries such as e-commerce, healthcare, banking, and finance, revolutionizing processes, enhancing user experiences, and improving decision-making. Despite challenges related to complex scenes, occlusions, data quality, and biases, ongoing research and advancements continue to push the boundaries of image recognition and object detection, paving the way for new applications and transformative innovations in AI systems.

## Image segmentation and scene understanding

Image segmentation and scene understanding are critical components of computer vision that enable machines to analyze and comprehend visual scenes with fine-grained detail. In this chapter, we delve into the concepts, techniques, and applications of image segmentation and scene understanding, highlighting their significance across diverse fields such as e-commerce, healthcare, banking, and finance.

Understanding Image Segmentation:

Image segmentation involves partitioning an image into multiple meaningful and coherent regions or objects. It aims to identify distinct boundaries and assign semantic labels to different parts of an image. By segmenting images into regions of interest, machines can analyze and understand the visual content at a granular level, enabling a range of advanced applications.

Semantic Segmentation:

Semantic segmentation aims to label each pixel in an image with the corresponding object category or class. It provides a detailed understanding of the objects present in the scene, allowing machines to differentiate between various object instances and their backgrounds. For instance, in an image of a street scene, semantic segmentation can label each pixel as road, car, pedestrian, or building, providing a comprehensive understanding of the scene's composition.

Instance Segmentation:

Instance segmentation takes image segmentation a step further by not only labeling pixels with object categories but also distinguishing between individual object instances. This technique enables machines to differentiate between multiple instances of the same object class within a scene. For example, in a crowded image with several people, instance segmentation can distinguish each person separately, assigning a unique label to each individual.

Applications of Image Segmentation:

E-commerce: Image segmentation finds applications in e-commerce by enabling product localization and extraction. By segmenting product images, e-commerce platforms can precisely extract product features, such as color, shape, or texture, and provide more accurate recommendations based on users' preferences. For instance, in the fashion industry, segmenting clothing items allows for enhanced visual search capabilities, enabling users to find similar items based on specific attributes.

Healthcare: In healthcare, image segmentation is crucial for various tasks, such as tumor detection, organ segmentation, and medical image analysis. By accurately segmenting medical images, clinicians can precisely delineate tumor boundaries, measure volumes, and plan treatments. This technology enhances diagnostic accuracy, facilitates surgical planning, and supports personalized medicine.

Banking and Finance: Image segmentation plays a role in document analysis and processing within the banking and finance industry. By segmenting documents such as invoices, checks, or identification cards, machines can extract relevant information, validate signatures, and automate document classification. This technology streamlines document management processes, reduces errors, and improves operational efficiency.

Scene Understanding:

Scene understanding involves comprehending the overall context and spatial relationships within a visual scene. It goes beyond object-level analysis to capture the scene's semantics, layout, and interactions between objects. By understanding scenes, machines can extract higher-level information and make more informed decisions.

Autonomous Vehicles: Scene understanding is critical for autonomous vehicles to navigate complex traffic scenarios. By analyzing the scene's layout, identifying road boundaries, traffic signs, and pedestrian movements, autonomous vehicles can plan safe and efficient routes, avoid obstacles, and ensure passenger safety. Scene understanding is pivotal for achieving robust and reliable autonomous driving capabilities.

Augmented Reality: Scene understanding is fundamental for delivering compelling augmented reality (AR) experiences. By understanding the scene's geometry, identifying objects, and recognizing surfaces, AR systems can overlay virtual content onto the real world in a coherent and contextually relevant manner. This technology enables immersive AR applications, such as furniture placement, virtual try-on, or interactive gaming.

Challenges and Future Directions:

Despite significant advancements, image segmentation and scene understanding present challenges that require ongoing research and innovation:

Fine-grained Details: Capturing fine-grained details and accurately delineating object boundaries remain challenging in image segmentation. Ambiguous regions, occlusions, and variations in lighting conditions can impact segmentation accuracy. Researchers are continuously exploring novel algorithms and deep learning techniques to improve the precision and robustness of image segmentation models.

Scalability and Efficiency: Image segmentation and scene understanding involve processing large amounts of visual data, requiring scalable and efficient algorithms. Real-time applications, such as autonomous driving or augmented reality, demand fast and computationally efficient solutions. Striking a balance between accuracy and efficiency is a crucial area of focus for researchers.

Data Diversity and Generalization: Image segmentation models must generalize well across diverse datasets and handle variations in object appearance, backgrounds, and imaging conditions. Ensuring models can perform effectively on unseen data is essential for their real-world deployment. Incorporating data augmentation techniques, transfer learning, and domain adaptation methods can enhance model generalization.

Conclusion:

Image segmentation and scene understanding are indispensable components of computer vision that empower machines to analyze and comprehend visual information with remarkable precision. Their applications span across industries such as e-commerce, healthcare, banking, and finance, revolutionizing processes, enhancing decision-making, and improving user experiences. While challenges persist, ongoing research and advancements continue to push the boundaries of image segmentation and scene understanding, opening up new possibilities for AI systems to gain deeper insights from visual data and contribute to transformative innovations.

## Computer vision in healthcare, manufacturing, and other industries

Computer vision has revolutionized various industries by enabling machines to perceive and interpret visual information, replicating human-like vision capabilities. In this chapter, we explore the transformative applications of computer vision in healthcare, manufacturing, and other diverse industries. We delve into the unique challenges, opportunities, and case studies that highlight the significance of computer vision in these fields.

Computer Vision in Healthcare:

Computer vision plays a pivotal role in healthcare, assisting in medical imaging analysis, disease diagnosis, surgical interventions, and patient monitoring. Here are some key applications:

Medical Imaging Analysis: Computer vision algorithms can analyze medical images, such as X-rays, MRIs, and CT scans, to assist radiologists in detecting abnormalities, segmenting organs, and identifying early signs of diseases. For instance, in mammography, computer vision techniques aid in the detection and classification of breast tumors, supporting timely diagnosis and improving patient outcomes.

Surgical Assistance: Computer vision systems enable real-time guidance and assistance during surgical procedures. By overlaying visual information from medical imaging devices onto the surgical field, surgeons can navigate complex anatomical structures, accurately locate critical areas, and perform minimally invasive surgeries with enhanced precision. These technologies help reduce risks, shorten surgical times, and improve patient safety.

Patient Monitoring: Computer vision techniques can monitor patients in hospital settings or at home, enabling remote healthcare monitoring and improving patient outcomes. Vision-based systems can track vital signs, analyze patient movements, and detect signs of distress, alerting healthcare professionals to intervene promptly. This technology promotes continuous monitoring, early intervention, and personalized care.

Computer Vision in Manufacturing:

In the manufacturing industry, computer vision has transformed production processes, quality control, and automation. Here are some notable applications:

Quality Control and Inspection: Computer vision systems can analyze visual data from production lines, detecting defects, ensuring product quality, and minimizing waste. For instance, in automotive manufacturing, computer vision techniques enable the inspection of critical components, such as engine parts or circuit boards, to identify manufacturing defects and ensure compliance with quality standards.

Robotic Guidance: Computer vision facilitates the guidance and control of robots in manufacturing environments. Vision systems can perceive the surrounding environment, detect objects, and enable robots to perform tasks with precision and adaptability. For example, in assembly lines, computer vision guides robots to locate and grasp components, enhancing productivity and reducing human error.

Process Optimization: Computer vision techniques enable the monitoring and optimization of manufacturing processes. By analyzing visual data, machines can identify bottlenecks, track production efficiency, and make real-time adjustments. This technology helps streamline operations, improve resource utilization, and enhance overall productivity.

Computer Vision in Other Industries:

Beyond healthcare and manufacturing, computer vision finds applications in a wide range of industries:

Retail: Computer vision is used in retail for inventory management, shelf monitoring, and customer behavior analysis. Vision-based systems can track product availability, optimize shelf layout, and analyze customer interactions, providing valuable insights for marketing and sales strategies.

Security and Surveillance: Computer vision is crucial in security and surveillance systems for object detection, tracking, and anomaly detection. These systems enhance public safety, identify potential threats, and aid in forensic investigations.

Autonomous Vehicles: Computer vision technologies are integral to autonomous vehicles for object detection, lane detection, and obstacle avoidance. These capabilities enable self-driving cars to navigate complex road scenarios and ensure passenger safety.

Challenges and Considerations:

Despite the immense potential of computer vision, several challenges need to be addressed:

Data Privacy and Security: Computer vision systems often require access to vast amounts of visual data, raising concerns about data privacy and security. Ensuring proper data anonymization, robust encryption, and secure storage is crucial to protect sensitive information and maintain public trust.

Ethical Considerations: Computer vision technologies must adhere to ethical principles, avoiding biases, discrimination, and misuse of personal data. Striking a balance between innovation and responsible AI practices is essential for the ethical development and deployment of computer vision systems.

Conclusion:

Computer vision has revolutionized industries such as healthcare, manufacturing, retail, security, and transportation, enabling machines to perceive and understand visual information. Its applications range from medical imaging analysis and surgical assistance to quality control in manufacturing and customer behavior analysis in retail. As computer vision continues to evolve, addressing challenges and considering ethical implications is vital to harness its full potential and ensure its responsible and beneficial integration across various industries.

# Addressing challenges and ethical considerations in computer vision

As computer vision continues to advance and find applications across various fields, it is imperative to address the challenges and ethical considerations that arise with its use. In this chapter, we delve into the complexities of computer vision and explore the measures required to ensure responsible and ethical deployment. By understanding and mitigating challenges, we can harness the full potential of computer vision while upholding fundamental values and protecting the well-being of individuals and society.

Challenges in Computer Vision:

Data Bias and Representation:

Computer vision systems heavily rely on training data, and biases present in the data can result in skewed and unfair outcomes. Biases can arise from imbalanced or unrepresentative datasets, leading to inaccurate classifications or perpetuating societal biases. For example, if a facial recognition system is trained on predominantly Caucasian faces, it may exhibit lower accuracy for individuals from other racial backgrounds. Addressing data biases requires diverse and representative datasets, rigorous data curation, and thorough evaluation of system performance across different demographics.

Privacy and Security:

Computer vision systems often process vast amounts of visual data, including personal and sensitive information. Ensuring privacy and data protection is essential to maintain trust in these systems. Proper data anonymization, robust encryption, secure storage, and compliance with privacy regulations are crucial considerations. Striking a balance between the benefits of computer vision and protecting individuals' privacy rights is paramount.

Explainability and Transparency:

The interpretability and transparency of computer vision algorithms are critical for building trust and ensuring accountability. Understanding how these algorithms arrive at their decisions is vital, particularly in domains where human lives or rights are at stake, such as healthcare or criminal justice. Developing explainable AI techniques, providing transparent decision-making processes, and enabling audits of computer vision systems can enhance trust and facilitate the identification and mitigation of biases or errors.

Adversarial Attacks:

Computer vision systems can be susceptible to adversarial attacks, where malicious actors intentionally manipulate input data to deceive or mislead the system. Adversarial attacks can lead to misclassifications or incorrect interpretations, compromising the integrity and reliability of computer vision systems. Robustness against such attacks is an ongoing challenge that necessitates the development of defense mechanisms and robust training techniques to enhance system resilience.

Data Bias and Representation:

Data bias in computer vision refers to the presence of imbalances or skewed representation in the training data, leading to biased outcomes or perpetuating societal biases. If the training data is predominantly composed of certain demographic groups, the computer vision system may struggle to accurately recognize or classify individuals from underrepresented groups. This can result in unfair outcomes and exacerbate existing societal biases.

For example, consider an automated recruitment system that uses computer vision to analyze video interviews. If the training data predominantly consists of interviews from male candidates, the system may inadvertently favor male candidates over equally qualified female candidates. This bias arises because the system has learned to associate certain visual features with successful candidates based on the biased training data. Consequently, the system perpetuates gender biases in the hiring process.

Addressing data biases requires diverse and representative datasets that encompass a wide range of demographics, ethnicities, genders, and other relevant factors. Careful data curation is necessary to ensure balanced representation and avoid reinforcing societal prejudices. Additionally, thorough evaluation of system performance across different demographic groups is crucial to identify and rectify biases. Ongoing monitoring and refinement of the training process are necessary to ensure that computer vision systems provide fair and equitable outcomes.

Privacy and Security:

Computer vision systems often deal with sensitive visual data, such as images or videos that may contain personal information. Protecting privacy and ensuring data security are essential to maintain trust in these systems. Proper data anonymization techniques should be employed to remove personally identifiable information from the visual data, reducing the risk of unauthorized access or misuse. Robust encryption methods should be implemented to secure data transmission and storage, preventing unauthorized interception or data breaches.

Compliance with privacy regulations, such as the General Data Protection Regulation (GDPR) in the European Union, is crucial. Organizations must adhere to strict guidelines regarding the collection, storage, and processing of visual data. Transparency in data handling practices, including providing clear privacy policies and obtaining informed consent, is necessary to respect individuals' rights and protect their privacy.

Striking a balance between the benefits of computer vision and privacy rights is paramount. Organizations must implement strong data protection measures while ensuring that individuals' privacy is not compromised in the pursuit of technological advancements. By establishing robust privacy frameworks and adhering to ethical principles, computer vision can be deployed in a responsible and privacy-preserving manner.

Explainability and Transparency:

Explainability and transparency are crucial aspects of computer vision systems to build trust and ensure accountability. Understanding how computer vision algorithms arrive at their decisions is particularly important in domains where human lives, rights, or critical decisions are at stake, such as healthcare or criminal justice.

Black box algorithms, where the internal workings are opaque and difficult to interpret, can raise concerns about biases, errors, or unintended consequences. Therefore, developing explainable AI techniques that provide insights into the decision-making process is essential. Explainable computer vision algorithms should provide clear explanations or visualizations of how input data is processed and how decisions are made. This allows stakeholders, including end-users, regulators, and affected individuals, to understand the rationale behind the system's outputs.

Transparency in decision-making processes is crucial for accountability. Organizations should provide documentation and public reports on the development, deployment, and performance of computer vision systems. External audits or third-party assessments can further enhance transparency and help identify potential biases or errors. By fostering explainability and transparency, computer vision systems can be scrutinized, refined, and improved, ultimately leading to more reliable and trustworthy outcomes.

Adversarial Attacks:

Adversarial attacks pose a significant challenge to computer vision systems. These attacks involve intentionally manipulating input data to deceive or mislead the system, leading to incorrect classifications or interpretations. Adversarial attacks can be designed to exploit vulnerabilities in computer vision algorithms, causing the system to misclassify objects or recognize nonexistent patterns.

For example, an adversarial attack on an autonomous vehicle's computer vision system could involve adding imperceptible perturbations to a stop sign, causing the system to misclassify it as a different object, such as a speed limit sign. This could have dangerous consequences if the vehicle fails to recognize and respond appropriately to traffic signs.

Defending against adversarial attacks requires the development of robust training techniques and defense mechanisms. Adversarial training, where models are trained on both clean and adversarial examples, can improve resilience against such attacks. Other approaches include incorporating uncertainty estimation methods or employing anomaly detection techniques to identify and flag potentially manipulated input data. Robustness against adversarial attacks is an ongoing area of research, and continued efforts are needed to develop more secure and resilient computer vision systems.

Conclusion:

Addressing challenges and ethical considerations in computer vision is crucial for responsible and trustworthy deployment. By mitigating data biases, ensuring privacy and security, promoting explainability and transparency, and defending against adversarial attacks, we can harness the full potential of computer vision while upholding ethical principles. Striking a balance between technological advancements and societal well-being is paramount to foster public trust, avoid unintended consequences, and maximize the positive impact of computer vision across various industries and domains.

Ethical Considerations in Computer Vision:

Bias and Fairness:

Ensuring fairness and avoiding discriminatory outcomes is a fundamental ethical consideration in computer vision. Bias can manifest in various forms, including racial, gender, or socioeconomic biases. Addressing biases requires proactive steps such as diverse and representative training data, fairness-aware algorithms, and continuous monitoring and evaluation to detect and mitigate biases.

Accountability and Responsibility:

Developers, researchers, and organizations utilizing computer vision technologies have a responsibility to uphold ethical standards and ensure accountability. This includes being transparent about the limitations and potential biases of computer vision systems, regularly auditing system performance, and addressing concerns raised by stakeholders or affected communities. Encouraging interdisciplinary collaboration and involving diverse perspectives can contribute to more robust and ethical decision-making.

Human Oversight and Decision-making:

While computer vision systems can provide valuable insights and automate certain tasks, human oversight and decision-making remain crucial. Humans should have the final say in critical decisions, especially in domains where human judgment, empathy, and ethical considerations are essential. Computer vision should augment human capabilities rather than replace them, with human judgment and ethical reasoning guiding the ultimate outcomes.

Case Study: Facial Recognition Technology and Biases:

One case study that highlights the ethical challenges of computer vision is the use of facial recognition technology. Facial recognition has faced scrutiny due to biases in system accuracy, disproportionately affecting individuals from certain racial or gender groups. For example, studies have shown that facial recognition systems exhibit lower accuracy rates for darker-skinned individuals or women. These biases raise concerns about unfair surveillance, discriminatory practices, and potential violations of privacy and civil liberties. Addressing biases in facial recognition technology requires robust testing, diverse training data, and ongoing evaluation to ensure equitable and unbiased outcomes.

Conclusion:

Computer vision holds immense promise in revolutionizing various fields, but it also presents significant challenges and ethical considerations. By actively addressing challenges such as data bias, privacy concerns, explainability, and adversarial attacks, we can ensure responsible and ethical deployment of computer vision systems. Upholding fairness, accountability, and human oversight is paramount to avoid unintended consequences and to harness the full potential of computer vision for the benefit of individuals and society as a whole.

# Robotics and Automation: Integrating AI with Physical Systems

In the realm of technological advancements, the integration of artificial intelligence (AI) with physical systems has paved the way for transformative innovations in the field of robotics and automation. By combining AI algorithms with physical devices, we have unlocked the potential to create intelligent machines capable of perceiving, reasoning, and acting in the real world. This chapter explores the fascinating world of robotics and automation, shedding light on how AI is revolutionizing industries and shaping the future of human-machine interactions.

The Synergy of AI and Robotics:

AI and robotics are intrinsically interconnected, with each field amplifying the capabilities of the other. AI empowers robots with the ability to perceive and interpret their environment, reason about complex tasks, and make intelligent decisions based on the available information. In return, robots provide a physical embodiment to AI algorithms, allowing them to interact with the real world and carry out tasks that were once limited to human operators.

The integration of AI and robotics has led to remarkable advancements across various industries. For instance, in the field of e-commerce, robots equipped with computer vision systems can autonomously navigate warehouses, identify and pick items, and optimize logistics operations. In healthcare, surgical robots enable minimally invasive procedures with enhanced precision, while AI-powered robotic exoskeletons aid in rehabilitation therapy. The manufacturing sector has witnessed the emergence of collaborative robots (cobots) that work alongside human operators, boosting productivity and efficiency. These examples demonstrate the transformative potential of integrating AI with physical systems.

Perception and Sensing:

One of the core aspects of robotics and automation is the ability to perceive and sense the environment. AI plays a crucial role in processing sensory inputs, such as vision, audio, and touch, to extract meaningful information. Computer vision algorithms enable robots to interpret visual data and recognize objects, people, or gestures. Natural language processing allows robots to understand and respond to human commands or interact through voice interfaces. Sensor fusion techniques integrate data from multiple sensors, such as cameras, lidars, and inertial measurement units, to build a comprehensive perception of the robot's surroundings.

Case Study: Autonomous Vehicles

A compelling case study that showcases the integration of AI and physical systems is the development of autonomous vehicles. Self-driving cars leverage AI algorithms, computer vision systems, and sensor fusion to perceive the environment, detect and track objects, and make real-time decisions. Through machine learning techniques, autonomous vehicles learn from vast amounts of data to improve their perception, decision-making, and response capabilities. By integrating AI with physical systems, autonomous vehicles have the potential to revolutionize transportation, enhancing safety, efficiency, and accessibility.

Planning and Control:

Once robots perceive and understand their environment, AI algorithms enable them to plan and execute actions to achieve desired objectives. Planning algorithms use information about the current state of the robot and its environment to generate a sequence of actions that accomplish specific tasks or goals. Reinforcement learning techniques allow robots to learn optimal policies through trial and error, enabling them to adapt to dynamic environments and improve their performance over time. Control algorithms govern the execution of actions, ensuring precise and coordinated movements of robot actuators.

Collaboration and Human-Robot Interaction:

The integration of AI with physical systems has also paved the way for enhanced collaboration between humans and robots. In collaborative scenarios, robots can understand human intentions, communicate effectively, and adapt their behavior to support human operators. AI algorithms enable robots to interpret human gestures, speech, or facial expressions, facilitating seamless interaction and collaboration. This opens up new possibilities in domains such as healthcare, where robots can assist medical professionals during complex procedures or provide support to patients in rehabilitation settings.

Ethical Considerations and Societal Impact:

As robotics and automation continue to advance, ethical considerations and societal impact become crucial areas of focus. The widespread adoption of AI-powered robots raises questions regarding job displacement, economic inequalities, and the ethical implications of autonomous decision-making. Striking a balance between technological progress and societal well-being is paramount. By ensuring transparency, accountability, and responsible deployment of AI-powered robots, we can harness the transformative potential of robotics while addressing ethical concerns and mitigating negative consequences.

Conclusion:

The integration of AI with physical systems in the field of robotics and automation has unlocked unprecedented possibilities and reshaped the way we interact with machines. By perceiving, reasoning, and acting in the real world, AI-powered robots have transformed industries such as e-commerce, healthcare, and manufacturing. However, as we venture into this realm, it is essential to address ethical considerations, ensure transparency, and carefully manage the societal impact. By doing so, we can embrace the potential of AI-powered robotics and automation to improve our lives, enhance productivity, and pave the way for a future where humans and machines collaborate harmoniously.

## The synergy of AI and robotics

The synergy of AI and robotics signifies the remarkable fusion of two distinct yet interconnected fields, resulting in transformative advancements and opening new frontiers in technology and automation. The integration of artificial intelligence (AI) and robotics has paved the way for intelligent machines capable of perceiving, reasoning, and acting in the physical world. This chapter explores the profound implications of this synergy, highlighting the profound impact it has had on various industries and the future of human-machine collaboration.

The integration of AI and robotics is a prime example of the whole being greater than the sum of its parts. AI, with its ability to process and analyze vast amounts of data, empowers robots with cognitive capabilities that mimic human intelligence. This enables robots to perceive their environment, interpret complex sensory inputs, reason about tasks, and make intelligent decisions based on the available information. In turn, robotics provides a physical embodiment to AI algorithms, enabling them to interact with the world and perform tasks that were once limited to human capabilities.

The result of this synergy is a wide range of applications across diverse fields. In the realm of e-commerce, robots equipped with computer vision systems navigate warehouses, identify products, and optimize logistics operations, revolutionizing supply chain management. In healthcare, robotic surgical systems assist surgeons with high precision and dexterity, enhancing the quality of surgical procedures and patient outcomes. Manufacturing has witnessed the emergence of collaborative robots (cobots) that work side by side with human operators, improving productivity and efficiency on the factory floor. These examples demonstrate the transformative potential of integrating AI with physical systems.

Perception lies at the core of this synergy, enabling robots to understand and interact with their surroundings. AI algorithms provide robots with the ability to process sensory inputs such as vision, touch, and sound. Computer vision algorithms allow robots to interpret visual data, recognize objects, and navigate their environment. This has applications in fields as diverse as autonomous vehicles, where AI-powered vision systems enable self-driving cars to perceive the road, detect obstacles, and make real-time decisions based on the environment. Natural language processing allows robots to understand and respond to human commands or engage in dialogue, facilitating human-robot interaction.

The planning and control capabilities enabled by AI further enhance the capabilities of robotic systems. Robots can plan and execute actions to achieve desired objectives based on the information they perceive and reason about. Planning algorithms generate sequences of actions that enable robots to perform complex tasks efficiently. Reinforcement learning techniques enable robots to learn from interactions with the environment, allowing them to adapt their behavior and improve their performance over time. Control algorithms govern the execution of actions, ensuring precise and coordinated movements of robot actuators. These capabilities are essential for applications ranging from autonomous navigation to industrial automation.

Another important aspect of this synergy is collaboration between humans and robots. Advances in AI and robotics have led to the development of robots that can understand human intentions, communicate effectively, and adapt their behavior to support human operators. This opens up opportunities for collaboration in various domains, such as healthcare, where robots can assist medical professionals in complex surgeries or aid in patient rehabilitation. Human-robot collaboration also has potential in fields like disaster response, where robots can assist in search and rescue operations, minimizing risks to human rescuers.

While the synergy of AI and robotics presents immense opportunities, it also raises important ethical considerations. As robots become more autonomous and integrated into society, questions of accountability, transparency, and ethical decision-making arise. For instance, in critical domains like healthcare or autonomous vehicles, the responsibility for decisions made by AI-powered robots becomes a crucial consideration. Striking a balance between technological progress and societal well-being is essential to ensure that the benefits of this synergy are harnessed responsibly.

In conclusion, the integration of AI and robotics has given rise to a powerful synergy that has revolutionized industries and transformed the way humans interact with machines. The cognitive capabilities of AI enable robots to perceive, reason, and act, while robotics provides a physical embodiment to AI algorithms, enabling them to interact with the physical world. This synergy has led to remarkable advancements in e-commerce, healthcare, manufacturing, and many other fields. However, ethical considerations and responsible deployment of AI-powered robots must be prioritized to ensure a harmonious integration of this transformative technology into our society.

## Applications of AI-driven robotics in different sectors

The integration of artificial intelligence (AI) with robotics has brought about groundbreaking advancements across various sectors, revolutionizing industries and transforming the way tasks are performed. In this chapter, we delve into the wide-ranging applications of AI-driven robotics in sectors such as e-commerce, healthcare, manufacturing, and agriculture, showcasing the transformative potential of this synergy and its impact on efficiency, productivity, and human-machine collaboration.

### E-commerce: Enhancing Logistics and Customer Experience

In the realm of e-commerce, AI-driven robotics has had a significant impact on logistics and supply chain management. Robots equipped with computer vision and autonomous navigation capabilities are employed in warehouses to automate tasks such as inventory management, order fulfillment, and product sorting. These robots can navigate complex environments, identify objects, and optimize the movement of goods, improving operational efficiency and reducing human error.

One prominent example is Amazon's fulfillment centers, where robots work alongside human employees to streamline order processing. These robots, known as Autonomous Mobile Robots (AMRs), autonomously navigate the fulfillment centers, carrying shelves of products to human workers for packing and shipping. By automating repetitive and physically demanding tasks, AI-driven robots increase productivity, shorten order processing times, and enhance the overall customer experience.

### Healthcare: Assisting Medical Professionals and Enhancing Patient Care

AI-driven robotics has the potential to transform healthcare by augmenting the capabilities of medical professionals and improving patient care. Surgical robots equipped with advanced vision systems and precision instruments assist surgeons in performing complex procedures with enhanced precision, control, and dexterity. For instance, the da Vinci Surgical System enables minimally invasive surgeries, reducing patient trauma and improving surgical outcomes.

Robots are also utilized in rehabilitation settings to aid patients in regaining motor functions and facilitating therapeutic exercises. These robots can provide personalized assistance, monitor progress, and collect data for analysis, empowering healthcare providers to deliver targeted rehabilitation programs.

In remote or underserved areas, telepresence robots allow healthcare professionals to remotely interact with patients, providing consultations and monitoring vital signs. This enables access to quality healthcare services, regardless of geographical limitations. AI-driven robotics holds immense promise for improving efficiency, accuracy, and accessibility in healthcare, ultimately enhancing patient care and outcomes.

Manufacturing: Driving Automation and Flexibility

The manufacturing industry has embraced AI-driven robotics to enhance automation, flexibility, and efficiency in production processes. Collaborative robots, or cobots, work side by side with human workers, augmenting their capabilities and improving productivity. Cobots can perform tasks that require precision, strength, or repetitive motions, relieving human workers from physically demanding or monotonous activities.

In automotive manufacturing, AI-driven robots are used for tasks such as assembly, welding, and painting, optimizing production speed and quality. These robots can adapt to different product variations, enabling flexible manufacturing and rapid reconfiguration of production lines.

Moreover, AI-driven robotics has enabled the emergence of "lights-out" factories, where fully automated systems operate without human intervention. These lights-out factories leverage AI algorithms to optimize production schedules, monitor equipment performance, and handle maintenance tasks, resulting in increased productivity and reduced operational costs.

Agriculture: Improving Efficiency and Crop Management

In the agricultural sector, AI-driven robotics is transforming farming practices, enabling precision agriculture and improving yield optimization. Autonomous robots equipped with computer vision systems and AI algorithms can perform tasks such as planting, harvesting, and crop monitoring with high accuracy and efficiency.

For instance, robotic harvesting systems have been developed to pick delicate fruits and vegetables with minimal damage. These robots use AI algorithms to detect ripe produce, plan optimal picking paths, and execute precise movements, reducing labor costs and increasing harvesting speed.

AI-driven robots also play a crucial role in crop monitoring and management. They can analyze plant health, detect pests or diseases, and optimize irrigation and fertilizer application. By collecting and analyzing data on soil conditions, weather patterns, and crop growth, these robots enable farmers to make data-driven decisions, resulting in improved crop yields and resource utilization.

While the applications of AI-driven robotics in different sectors bring numerous benefits, challenges remain. Ethical considerations, job displacement concerns, and the need for human oversight are among the factors that need to be addressed to ensure responsible deployment and maximize the benefits of this technology.

In conclusion, AI-driven robotics has proven to be a transformative force across various sectors, offering enhanced efficiency, productivity, and human-machine collaboration. From e-commerce and healthcare to manufacturing and agriculture, the integration of AI algorithms with robotic systems has led to advancements that optimize operations, improve outcomes, and revolutionize industries. As this synergy continues to evolve, careful consideration of ethical implications and responsible implementation will be vital to harness the full potential of AI-driven robotics and ensure a future that benefits both humans and machines.

## Implications for productivity, safety, and employment

The integration of artificial intelligence (AI) and robotics has profound implications for productivity, safety, and employment across various industries. While this synergy offers significant benefits in terms of efficiency, accuracy, and cost-effectiveness, it also raises concerns regarding job displacement, workplace safety, and the future of work. In this section, we delve into these implications, presenting a balanced analysis of the potential advantages and challenges associated with AI-driven robotics.

Productivity Enhancement:

One of the primary advantages of AI-driven robotics is its potential to enhance productivity in diverse sectors. By automating repetitive and physically demanding tasks, robots can work tirelessly, leading to increased output and operational efficiency. This enables human workers to focus on higher-level tasks that require creativity, problem-solving, and decision-making abilities, thereby augmenting overall productivity.

For example, in manufacturing, AI-driven robots can perform complex assembly tasks with precision and speed, leading to improved production rates and product quality. In e-commerce, robots equipped with computer vision systems streamline warehouse operations, enabling faster order fulfillment and reducing processing times. These examples demonstrate how AI-driven robotics can contribute to significant productivity gains, benefiting businesses and the economy as a whole.

Safety Enhancement:

AI-driven robotics has the potential to enhance workplace safety by reducing human exposure to hazardous environments and minimizing the risk of accidents. Robots can perform tasks that are dangerous, strenuous, or require working in extreme conditions without compromising human well-being. This leads to a safer work environment and a reduction in occupational hazards.

In the healthcare sector, AI-driven surgical robots assist surgeons in performing minimally invasive procedures, reducing the risk of complications and patient trauma. In industries such as mining, AI-driven robots can navigate and inspect hazardous environments, mitigating risks to human workers. These applications showcase the potential of AI-driven robotics to prioritize safety and protect human lives.

Employment Dynamics:

While the benefits of AI-driven robotics are evident, concerns about job displacement and the future of work have also emerged. The automation of certain tasks may lead to a decrease in demand for low-skilled and routine jobs, potentially resulting in job losses or a shift in required skill sets. However, it is essential to recognize that AI-driven robotics also creates new employment opportunities in emerging fields such as robot programming, maintenance, and system supervision.

Historically, technological advancements have led to job reallocation rather than outright job loss. As certain tasks become automated, the workforce can transition to more complex and value-added roles. This necessitates the development of new skills and continuous education to adapt to the changing labor market.

Moreover, AI-driven robotics can augment human capabilities and facilitate human-machine collaboration. By leveraging the strengths of both humans and robots, synergistic partnerships can be formed, leading to increased productivity and innovation. For example, in healthcare, robots can assist medical professionals, enabling them to deliver more personalized and efficient care.

To address the potential impact on employment, it is crucial to foster policies and initiatives that facilitate workforce reskilling and upskilling. Governments, educational institutions, and businesses should collaborate to ensure that workers have access to the necessary training and support to adapt to the changing landscape.

Balancing Benefits and Challenges:

While AI-driven robotics offers numerous benefits in terms of productivity enhancement and safety, it is essential to address the associated challenges and ethical considerations. The potential displacement of certain job roles calls for proactive measures to support affected workers and facilitate their transition to new opportunities. Additionally, ensuring the safety and ethical use of AI-driven robotics requires robust regulations, standards, and responsible deployment practices.

By striking a balance between harnessing the productivity potential of AI-driven robotics and addressing the societal challenges, we can maximize the benefits while minimizing the potential negative impacts. Collaboration between industry stakeholders, policymakers, and researchers is crucial to ensure responsible adoption, safeguard human well-being, and create a future of work that is inclusive, equitable, and sustainable.

In conclusion, the integration of AI and robotics has far-reaching implications for productivity, safety, and employment. While AI-driven robotics has the potential to enhance productivity and workplace safety, it also poses challenges related to job displacement and workforce dynamics. By proactively addressing these challenges and fostering responsible adoption, we can navigate the transformative power of AI-driven robotics to create a future that maximizes the benefits for individuals, businesses, and society as a whole.

## Ethical considerations and safety protocols in robotics

As robotics technology advances and becomes increasingly integrated into various industries, it is imperative to address the ethical considerations and safety protocols associated with these autonomous machines. Robots have the potential to impact human lives, interact with individuals, and make critical decisions, raising ethical questions about their behavior, accountability, and the potential risks they pose. In this section, we will explore the ethical dimensions and safety protocols that should be considered when designing and deploying robotic systems.

Ethical Considerations:

Human-Robot Interaction: Ensuring that robots interact with humans in an ethical and respectful manner is paramount. Robots should respect human autonomy, privacy, and dignity, and their behavior should align with societal norms and values. For example, in healthcare settings, robots should maintain patient confidentiality and obtain informed consent when collecting personal information.

Transparency and Explainability: The ability to understand and interpret the decisions made by robots is crucial for building trust and accountability. Employing explainable AI techniques allows humans to comprehend the reasoning behind a robot's actions. This is particularly important in high-stakes domains, such as healthcare and autonomous vehicles, where the transparency of decision-making processes is essential.

Bias and Fairness: Robots should be designed and trained to mitigate biases and ensure fairness in their actions. Bias in data or algorithms can result in discriminatory outcomes, such as biased hiring or loan approval decisions. By carefully curating training data and regularly auditing algorithms, biases can be minimized, and fair outcomes can be promoted.

Impact on Employment: The widespread adoption of robots raises concerns about potential job displacement and its societal implications. While automation can lead to increased efficiency and productivity, it is important to develop strategies to address the potential impact on employment, such as reskilling programs and creating new job opportunities in emerging fields.

Safety Protocols:

Risk Assessment and Management: Assessing and managing the risks associated with robotic systems is critical. This involves identifying potential hazards, understanding their consequences, and implementing appropriate safety measures. Risk management protocols should be developed and followed throughout the entire lifecycle of a robot, including design, testing, deployment, and maintenance.

Human-Robot Collaboration: Establishing effective collaboration between humans and robots is essential for ensuring safety. Robots should be equipped with sensors and algorithms that enable them to detect and respond to human presence and potential risks. Collaborative robots, or cobots, are designed to work alongside humans safely, allowing for shared tasks and workspace without compromising human safety.

Cybersecurity: Robotic systems are vulnerable to cyberattacks that can have significant consequences. Implementing robust cybersecurity measures, such as encryption, authentication, and intrusion detection, is crucial to safeguarding robots from malicious exploitation. Secure communication protocols should be established to protect sensitive data and prevent unauthorized access to robot systems.

Emergency Stop Mechanisms: Robots should be equipped with emergency stop mechanisms that allow humans to quickly halt their operations in case of unforeseen hazards or emergencies. This ensures that human operators have the ability to intervene and take control if necessary.

Case Studies:

One example of ethical considerations and safety protocols in robotics is the use of autonomous vehicles. Companies developing self-driving cars must address ethical dilemmas, such as how the vehicle should prioritize the safety of passengers versus pedestrians in a potential accident scenario. Additionally, safety protocols in autonomous vehicles involve collision avoidance systems, real-time monitoring of road conditions, and fail-safe mechanisms to ensure the safety of passengers and other road users.

In the healthcare sector, surgical robots are becoming more prevalent. Ethical considerations include informed consent from patients, maintaining patient privacy and confidentiality, and addressing the potential for errors or malfunctions during surgeries. Safety protocols involve robust training for medical professionals, redundant systems to prevent accidents, and continuous monitoring of the robot's performance during surgical procedures.

Conclusion:

Ethical considerations and safety protocols play a crucial role in shaping the development and deployment of robotics. Addressing ethical concerns related to human-robot interaction, transparency, fairness, and employment dynamics is necessary to ensure that robots align with societal values. Simultaneously, implementing safety protocols, risk assessment and management, cybersecurity measures, and emergency stop mechanisms are vital for mitigating risks and safeguarding human well-being. By integrating ethical considerations and safety protocols into the design and operation of robotic systems, we can unlock the full potential of robotics while upholding ethical standards and ensuring human safety.

# Collaborative approaches for humans and robots

In recent years, the field of robotics has witnessed a paradigm shift from isolated and autonomous systems to collaborative approaches that involve close interaction between humans and robots. This collaborative framework holds immense potential in a wide range of industries, including manufacturing, healthcare, logistics, and more. By combining the unique capabilities of both humans and robots, collaborative approaches aim to enhance productivity, improve efficiency, and tackle complex tasks that are challenging for either party alone. In this section, we will explore the significance and benefits of collaborative approaches in human-robot interaction.

Enhancing Efficiency and Productivity:

Collaborative approaches enable humans and robots to work together synergistically, leveraging each other's strengths to achieve greater efficiency and productivity. While robots excel at repetitive, precise, and physically demanding tasks, humans possess cognitive abilities, creativity, and adaptability. By combining these capabilities, collaborative systems can streamline operations, optimize resource allocation, and accelerate production cycles. For example, in e-commerce warehouses, collaborative robots, or cobots, work alongside human workers to facilitate order fulfillment, reducing errors and increasing throughput.

Improving Safety and Ergonomics:

One of the key motivations for adopting collaborative approaches is to enhance safety in shared workspaces. Collaborative robots are designed to be inherently safe, employing sensors and algorithms that detect human presence and respond appropriately to prevent collisions or accidents. This ensures that humans and robots can work in close proximity without compromising safety. Collaborative approaches also contribute to improved ergonomics by automating physically strenuous tasks, reducing the risk of repetitive strain injuries and allowing human workers to focus on more intellectually demanding aspects of the job.

Flexibility and Adaptability:

Collaborative systems offer flexibility and adaptability, allowing for seamless task allocation and dynamic reconfiguration based on changing requirements. This is particularly valuable in industries with rapidly evolving production lines or unpredictable environments. For instance, in manufacturing, collaborative robots can easily switch between different tasks, adapt to variations in product specifications, and quickly reconfigure workflows based on real-time demand. This flexibility empowers businesses to respond effectively to market fluctuations and customer demands.

Skill Enhancement and Learning:

Collaborative approaches provide opportunities for skill enhancement and learning for both humans and robots. Humans can acquire new technical skills related to robot programming, maintenance, and supervision. By working alongside robots, human workers gain exposure to advanced technologies and automation, fostering their professional development and expanding their skill set. On the other hand, robots can learn from human workers through imitation, reinforcement learning, or shared control, enabling them to improve their performance and adapt to novel situations.

Case Studies:

One notable example of collaborative approaches is the use of surgical robots in healthcare. Surgeons and robots work collaboratively during procedures, with the robot assisting the surgeon with precise movements and providing enhanced visualization. This collaboration allows for minimally invasive surgeries, shorter recovery times, and improved patient outcomes.

In the banking industry, collaborative approaches involve the use of chatbots and virtual assistants to enhance customer service. These AI-powered systems can handle routine customer inquiries, providing quick and accurate responses, while human employees can focus on more complex and personalized interactions. This collaboration ensures efficient customer support and allows employees to devote more time to building relationships with clients.

Conclusion:

Collaborative approaches for humans and robots hold significant promise in revolutionizing various industries. By combining the unique strengths of humans and robots, these approaches can enhance efficiency, improve safety, enable flexibility, and foster skill enhancement. The successful integration of collaborative systems requires careful consideration of task allocation, safety protocols, and training programs. As technology continues to advance, collaborative approaches will play a pivotal role in reshaping work environments, driving innovation, and unlocking the full potential of human-robot collaboration.

# AI Ethics and Societal Impact

Artificial Intelligence (AI) has emerged as a transformative technology with the potential to revolutionize various aspects of our lives. From healthcare and finance to transportation and education, AI-powered systems are increasingly integrated into our society, shaping the way we live and work. However, as AI continues to advance, it brings forth a myriad of ethical considerations and societal impacts that demand careful examination and deliberation. In this section, we will explore the fundamental concepts of AI ethics and delve into the profound societal implications of this groundbreaking technology.

Understanding AI Ethics:

AI ethics refers to the study and application of moral principles and values in the development, deployment, and use of AI systems. It encompasses a wide range of ethical concerns, such as fairness, transparency, accountability, privacy, bias, autonomy, and the potential for unintended consequences. The goal of AI ethics is to ensure that AI technologies are developed and deployed in a manner that aligns with societal values, respects human rights, and promotes the well-being of individuals and communities.

Societal Impact of AI:

The rapid proliferation of AI technologies has profound implications for various aspects of society, including the economy, workforce, privacy, healthcare, and decision-making processes. Understanding and navigating these impacts is crucial to harnessing the benefits of AI while mitigating potential risks and challenges.

Economic Disruption and Workforce Transformation:

AI has the potential to reshape industries and labor markets, leading to significant economic disruption and changes in the workforce. Automation driven by AI technologies may replace certain job roles, raising concerns about unemployment and inequality. However, AI also creates new job opportunities, enhances productivity, and enables the development of innovative products and services. Careful consideration of the economic and societal implications is essential to ensure a just and inclusive transition to an AI-driven economy.

Privacy and Data Protection:

AI systems rely on vast amounts of data for training and operation, raising concerns about privacy and data protection. As AI technologies become increasingly embedded in everyday life, there is a need to establish robust frameworks for data governance, consent, and transparency. Striking a balance between the benefits of AI and safeguarding individual privacy rights is crucial to maintain public trust and ensure ethical practices.

Bias and Fairness:

AI systems can inherit and perpetuate biases present in the data they are trained on, leading to unfair or discriminatory outcomes. For instance, biased algorithms in hiring processes or loan approvals can amplify societal inequalities. Addressing bias and promoting fairness in AI systems require diverse and representative training data, rigorous evaluation of algorithms, and ongoing monitoring to detect and mitigate biases. It is essential to develop AI systems that are not only accurate but also unbiased and fair.

Ethical Decision-Making and Accountability:

AI technologies are increasingly involved in decision-making processes that have significant societal impact. From autonomous vehicles to healthcare diagnostics, these systems must make ethically sound decisions while being accountable for their actions. Building AI systems that are transparent, explainable, and subject to scrutiny is crucial for maintaining accountability and ensuring that decisions align with societal values and norms.

Case Studies:

One notable case study is the use of facial recognition technology in law enforcement. The widespread adoption of such technology raises concerns about privacy, accuracy, and potential bias. For example, the misidentification of individuals from certain racial or ethnic backgrounds can have severe consequences. These ethical concerns have sparked debates on the appropriate use and regulation of facial recognition technology.

Another case study is the application of AI in finance, where automated trading algorithms and robo-advisors have gained prominence. While these systems offer efficiency and convenience, there are concerns regarding their impact on market stability and the potential for algorithmic biases. Striking a balance between algorithmic decision-making and human oversight is crucial to ensure the integrity and fairness of financial markets.

Conclusion:

As AI technologies continue to advance and permeate various domains, understanding and addressing the ethical considerations and societal impacts is of paramount importance. Ethical principles and guidelines must guide the development and deployment of AI systems to ensure they align with societal values, respect human rights, and promote the well-being of individuals and communities. Collaboration between researchers, policymakers, industry leaders, and the public is essential to navigate the complex landscape of AI ethics and societal impact. By fostering a multidisciplinary dialogue, we can collectively shape the future of AI in a manner that maximizes benefits while safeguarding against potential risks and challenges.

# Ethical considerations in AI development and deployment

As Artificial Intelligence (AI) technologies continue to advance and permeate various domains, it is imperative to address the ethical considerations that arise in the development and deployment of AI systems. AI has the potential to greatly benefit society, but it also raises complex ethical questions that demand careful examination. In this section, we will explore the ethical dimensions of AI development and deployment, focusing on key considerations such as transparency, fairness, accountability, privacy, and the potential impact on employment.

Transparency and Explainability:

Transparency in AI refers to the openness and comprehensibility of AI systems and their decision-making processes. It is crucial to understand how AI systems arrive at their conclusions, especially in high-stakes domains like healthcare or finance. Explainability is closely tied to transparency and involves providing human-readable explanations of AI systems' reasoning. Ensuring transparency and explainability in AI can foster trust, facilitate accountability, and help identify and address potential biases or errors.

Case Study: In the healthcare domain, the use of AI algorithms for diagnosis and treatment recommendation requires transparency and explainability. When an AI system suggests a particular treatment plan for a patient, it is essential for healthcare professionals to understand the underlying factors and reasoning behind the system's recommendation to make informed decisions.

Fairness and Bias Mitigation:

Fairness in AI refers to the equitable treatment of individuals and the avoidance of biases in decision-making. AI systems trained on biased data can perpetuate existing social inequalities or discriminate against certain groups. Bias mitigation techniques, such as diverse and representative training data, fairness-aware algorithms, and continuous monitoring, are essential to ensure that AI systems are fair and do not reinforce societal prejudices.

Case Study: In e-commerce, algorithms used for product recommendations or pricing must be fair and not discriminate against certain customers based on their demographics or personal characteristics. Failure to address biases in these systems can result in unfair treatment, reduced opportunities, and perpetuation of existing economic disparities.

Accountability and Responsibility:

Accountability in AI refers to the ability to attribute responsibility for the actions and decisions made by AI systems. As AI technologies become more autonomous, it becomes crucial to establish mechanisms for holding developers, deployers, and users accountable for the outcomes of AI systems. This includes clear guidelines, standards, and regulatory frameworks that govern the development, deployment, and use of AI technologies.

Case Study: In autonomous vehicles, determining liability in the event of an accident involving a self-driving car raises complex legal and ethical questions. Establishing clear guidelines and legal frameworks that assign responsibility and accountability is necessary to navigate these challenges.

Privacy and Data Protection:

AI systems often rely on vast amounts of data, including personal and sensitive information. Respecting privacy rights and protecting data is of paramount importance. Striking a balance between the benefits of AI and protecting individuals' privacy requires robust data anonymization techniques, secure storage and transmission protocols, and compliance with privacy regulations such as the General Data Protection Regulation (GDPR).

Case Study: In banking and finance, AI algorithms analyze individuals' financial data to assess creditworthiness. Safeguarding the privacy and security of this sensitive information is critical to maintain customer trust and ensure compliance with data protection regulations.

Ethics in Employment and Social Impact:

The widespread adoption of AI technologies raises concerns about the impact on employment and societal well-being. While AI can enhance productivity and create new job opportunities, it also has the potential to automate certain tasks, leading to job displacement. Ethical considerations include ensuring a just transition for affected workers, promoting lifelong learning, and addressing potential socioeconomic inequalities that may arise from AI-driven automation.

Case Study: In manufacturing, the deployment of AI-powered robots can lead to job displacement for human workers. Ethical approaches involve upskilling and reskilling programs to ensure a smooth transition for workers, as well as considering the social impact on affected communities.

Conclusion:

Ethical considerations play a crucial role in the development and deployment of AI systems. Transparency, fairness, accountability, privacy, and the impact on employment and society are fundamental aspects to be addressed. By incorporating ethical principles into AI development, we can ensure that AI technologies align with societal values, respect human rights, and promote the well-being of individuals and communities. A multidisciplinary approach, involving collaboration among researchers, policymakers, industry leaders, and the public, is necessary to navigate the complex ethical landscape and maximize the positive impact of AI while mitigating risks and challenges.

## Fairness, transparency, and accountability in AI systems

As Artificial Intelligence (AI) continues to permeate various aspects of society, ensuring fairness, transparency, and accountability in AI systems becomes paramount. The deployment of AI systems in domains such as E-commerce, Healthcare, Banking, and Finance raises complex ethical and societal implications. In this section, we will delve into the importance of fairness, transparency, and accountability in AI systems, exploring the challenges they present and discussing potential solutions.

Fairness in AI Systems:

Fairness in AI systems refers to the equitable treatment of individuals, regardless of their personal characteristics such as race, gender, or socioeconomic status.

Fairness is crucial to ensure that AI systems do not perpetuate existing biases or discriminate against certain groups. However, achieving fairness in AI systems is challenging due to inherent biases in training data and algorithmic decision-making.

One approach to address fairness is through algorithmic fairness, which involves designing AI algorithms that minimize disparate impacts on different groups. For example, in E-commerce, algorithms used for product recommendations should not favor specific demographics, ensuring fair access to products and services for all customers.

However, defining fairness itself can be complex, as different perspectives and definitions exist. The trade-offs between different fairness criteria, such as equal opportunity and equalized odds, need careful consideration. Achieving fairness often requires striking a balance between competing goals and being mindful of unintended consequences.

Transparency in AI Systems:

Transparency in AI systems refers to the ability to understand and explain how AI systems arrive at their decisions. Transparent AI systems enhance trust, enable accountability, and allow users to assess the reliability and potential biases of the system. Lack of transparency can lead to mistrust, especially in critical domains like Healthcare or Finance, where the consequences of AI decisions can be significant.

Transparency can be achieved through techniques such as interpretable machine learning, which aims to provide human-understandable explanations for AI decisions. For example, in Healthcare, when an AI system recommends a treatment plan for a patient, it is essential for healthcare professionals to understand the underlying factors and reasoning behind the recommendation.

However, achieving transparency can be challenging, particularly for complex AI models like deep neural networks, which operate as black boxes. Balancing model complexity with interpretability is an ongoing research area. Striking the right balance is essential to ensure that AI systems are not only accurate but also transparent and understandable to human users.

Accountability in AI Systems:

Accountability in AI systems refers to the responsibility and liability for the actions and decisions made by AI systems. As AI technologies become more autonomous, it is crucial to establish mechanisms to attribute responsibility when AI systems fail or produce unintended consequences. Accountability is vital for ensuring trust, identifying and addressing errors or biases, and enabling recourse for individuals adversely affected by AI decisions.

Establishing accountability in AI systems requires a combination of legal frameworks, regulatory oversight, and ethical guidelines. These frameworks should outline the roles and responsibilities of developers, deployers, and users of AI systems. In the Banking and Finance sector, for instance, regulatory bodies may impose strict requirements to ensure that AI algorithms are audited and adhere to fairness and transparency standards.

However, determining accountability in AI systems is not straightforward. The complex interplay between human decision-making and AI algorithms raises challenging legal and ethical questions. Clear guidelines and frameworks that consider the nuances of AI decision-making are essential for establishing accountability and addressing potential risks.

Conclusion:

Fairness, transparency, and accountability are critical pillars in the development and deployment of AI systems. Striving for fairness ensures equitable treatment, transparency fosters trust and understanding, and accountability establishes responsibility for AI outcomes. Achieving these goals requires interdisciplinary collaboration, involving researchers, policymakers, industry leaders, and the public.

While challenges exist, advancements in algorithmic fairness, interpretability, and regulatory frameworks are driving progress in ensuring the ethical use of AI. By promoting fairness, transparency, and accountability, we can harness the transformative power of AI while minimizing the risks and maximizing the benefits for individuals and society as a whole.

## Addressing biases and discrimination in AI algorithms

In the realm of Artificial Intelligence (AI), algorithms play a central role in making decisions and influencing outcomes across various domains. However, the increasing reliance on AI algorithms raises concerns about biases and discrimination embedded within these systems. In this section, we will delve into the challenges associated with biases and discrimination in AI algorithms and explore strategies to address and mitigate these issues.

Understanding Biases in AI Algorithms:

Biases in AI algorithms can arise from multiple sources, including biased training data, flawed algorithm design, or unintended biases introduced during the development process. These biases can lead to discriminatory outcomes, perpetuating social inequalities or reinforcing existing prejudices. For instance, in E-commerce, biased product recommendations based on users' demographic information can reinforce stereotypes and limit choices for individuals.

One critical aspect of addressing biases in AI algorithms is recognizing that algorithms themselves do not hold biases. Instead, biases are learned from the data used to train the algorithms. Therefore, it becomes crucial to carefully curate and diversify training data to avoid skewed representations of different groups.

Mitigating Biases in AI Algorithms:

Diverse and Representative Data: Ensuring diversity and representativeness in training data is essential to mitigate biases in AI algorithms. By including a wide range of examples from various demographics and social groups, algorithms can be trained to make fair and unbiased decisions. For example, in Healthcare, training AI algorithms on data that reflects the diversity of patient populations helps ensure equal access to healthcare services.

Bias Detection and Evaluation: Implementing robust mechanisms for bias detection and evaluation is crucial in identifying and understanding biases in AI algorithms. By continuously monitoring algorithmic outputs and evaluating their impact on different groups, biases can be detected and addressed. This involves using fairness metrics and conducting thorough audits to identify discriminatory patterns.

Algorithmic Fairness Techniques: Researchers have developed algorithmic fairness techniques to mitigate biases in AI algorithms. These techniques aim to quantify and reduce disparities across different groups. For instance, in Banking and Finance, fairness-aware algorithms can be employed to ensure that loan approval decisions are not influenced by factors that unfairly disadvantage certain groups.

Human-in-the-Loop Approach: Incorporating human judgment and oversight in AI decision-making processes can help mitigate biases. Human experts can provide critical insights, evaluate algorithmic outputs, and override decisions that may be biased or discriminatory. This collaborative approach, known as the human-in-the-loop, combines the strengths of AI algorithms with human expertise to achieve fair and unbiased outcomes.

Counterarguments and Challenges:

Critics argue that achieving complete fairness in AI algorithms is a complex and elusive goal. They contend that defining fairness itself is subjective, and different fairness criteria may conflict with each other. Additionally, addressing biases in AI algorithms can be challenging when there is a lack of transparency or interpretability in complex algorithmic models.

Moreover, some argue that the responsibility for addressing biases lies not only with developers but also with society as a whole. They emphasize the need for broader societal discussions, regulations, and diverse perspectives to ensure fairness and accountability in AI algorithms.

Conclusion:

Addressing biases and discrimination in AI algorithms is a crucial endeavor that requires concerted efforts from researchers, developers, policymakers, and society. By curating diverse and representative data, employing bias detection and evaluation techniques, leveraging algorithmic fairness methods, and adopting a human-in-the-loop approach, we can work towards mitigating biases and promoting fairness in AI systems.

However, the path to achieving complete fairness is not without challenges. Striking the right balance between competing fairness criteria and ensuring transparency and interpretability remain ongoing areas of research. Nonetheless, by recognizing the importance of addressing biases and discrimination in AI algorithms, we pave the way for more inclusive, equitable, and responsible AI systems that benefit individuals and society as a whole.

## AI and the future of work: Impact on employment and skills

Artificial Intelligence (AI) has become a transformative force across various industries, revolutionizing the way we work and interact with technology. As AI continues to advance, there are concerns about its impact on employment and the skills required in the future workforce. In this section, we will explore the potential implications of AI on employment, the evolving nature of work, and the skills needed to thrive in the AI-driven future.

Automation and Job Displacement:

One of the primary concerns surrounding AI is the potential for automation to replace human workers. Automation, driven by AI technologies, can perform routine tasks more efficiently and accurately than humans. As a result, certain jobs that involve repetitive and predictable tasks may become obsolete. For instance, in the field of E-commerce, AI-powered chatbots are increasingly being used to handle customer inquiries, reducing the need for human customer service representatives.

However, it is essential to note that while automation may lead to job displacement in certain areas, it also creates new opportunities. As routine tasks become automated, individuals can focus on higher-value and more creative work. This shift in job roles necessitates the acquisition of new skills to adapt to the changing demands of the labor market.

Skill Demands in the AI Era:

The emergence of AI technology demands a new set of skills to thrive in the future workforce. While routine tasks may be automated, there is an increasing demand for skills that complement AI systems. These skills include:

Technical Literacy: A basic understanding of AI technologies, algorithms, and data analysis is crucial in the AI era. This literacy enables individuals to collaborate effectively with AI systems and leverage their capabilities. For example, in Healthcare, medical professionals with knowledge of AI can utilize AI-powered diagnostic tools to improve patient outcomes.

Critical Thinking and Problem Solving: AI can analyze vast amounts of data and provide insights, but human judgment and critical thinking remain vital. Individuals who can interpret and validate AI-generated results, identify biases, and make informed decisions will be highly valued. In Banking and Finance, professionals skilled in analyzing AI-generated predictions can enhance investment strategies and risk management.

Creativity and Innovation: AI may excel at repetitive tasks, but it still lacks human creativity and innovation. Individuals who can think outside the box, generate new ideas, and apply creative problem-solving approaches will be in demand. In E-commerce, AI can assist in identifying customer preferences, but creative minds are needed to design innovative marketing campaigns that resonate with consumers.

Adaptability and Lifelong Learning: The rapid pace of AI advancement requires individuals to embrace lifelong learning and adaptability. Continuous upskilling and reskilling will be essential to remain competitive in the job market. This adaptability allows professionals to navigate changing technologies and take advantage of emerging opportunities.

Counterarguments and Perspectives:

Critics argue that the rise of AI will lead to significant job displacement and exacerbate existing inequalities. They raise concerns about the impact on low-skilled workers who may face challenges transitioning to new roles. Additionally, the ethical considerations of AI's influence on labor practices, such as the gig economy and worker rights, are important aspects to consider.

Moreover, some experts argue that AI will create new job roles and industries that we cannot yet envision. They emphasize the need to focus on developing uniquely human skills, such as emotional intelligence, empathy, and ethical decision-making, which are less likely to be automated.

Conclusion:

The integration of AI into the workforce brings both opportunities and challenges. While automation may lead to job displacement in some areas, it also opens up new avenues for human creativity, critical thinking, and innovation. As AI becomes increasingly prevalent, it is crucial for individuals to acquire technical literacy, critical thinking skills, creativity, and adaptability to thrive in the AI-driven future.

Furthermore, it is essential for policymakers, educators, and businesses to invest in reskilling and upskilling initiatives to ensure a smooth transition for workers affected by automation. By fostering a culture of lifelong learning and supporting workers in developing the skills necessary for the AI era, we can embrace the potential of AI while ensuring a fair and inclusive future of work.

Ultimately, AI is a tool that, when combined with human skills and capabilities, can lead to greater productivity, efficiency, and innovation. By embracing the opportunities and addressing the challenges, we can shape a future where AI augments human potential, creating a more prosperous and fulfilling society.

## Balancing progress with societal well-being

As the development and integration of artificial intelligence (AI) continue to advance at a rapid pace, it becomes imperative to consider the balance between technological progress and the well-being of society. While AI offers immense potential to improve various aspects of our lives, it also raises ethical, social, and economic concerns. In this section, we will explore the challenges associated with balancing progress and societal well-being and discuss potential approaches to address them.

Ethical Considerations:

Ethics play a crucial role in ensuring that AI technologies are developed and deployed in a manner that aligns with societal values and promotes well-being. It is essential to address ethical considerations such as fairness, transparency, accountability, and privacy in AI systems.

Fairness: AI algorithms can inadvertently perpetuate biases present in training data, leading to unfair outcomes. For example, in the field of finance, if a lending algorithm is trained on biased historical data, it may result in discriminatory lending practices. It is vital to develop algorithms that are fair and unbiased, considering diverse perspectives and ensuring equal opportunities for all.

Transparency: AI systems should be transparent in their decision-making processes. The "black box" nature of some AI algorithms makes it challenging to understand how they arrive at their conclusions. To build trust and accountability, efforts must be made to develop explainable AI techniques that provide insights into the reasoning behind AI-generated decisions.

Accountability: It is important to establish accountability mechanisms to address the impact of AI systems on individuals and society. Organizations and developers should be responsible for the ethical design, development, and deployment of AI technologies. Clear guidelines, regulations, and oversight can help ensure that AI systems are held accountable for their actions.

Privacy: The collection and use of personal data in AI systems raise privacy concerns. Striking a balance between the benefits of AI and protecting individuals' privacy rights is paramount. Robust data anonymization, encryption, and adherence to privacy regulations are essential considerations to safeguard personal information.

Social Impact:

AI has the potential to reshape industries, the job market, and society as a whole. While technological advancements bring efficiency and convenience, they can also lead to job displacement and socioeconomic inequalities.

Job Displacement: Automation driven by AI technologies may replace certain job roles, leading to concerns about unemployment and job insecurity. However, historical examples of technological advancements, such as the Industrial Revolution, have shown that new job opportunities are also created. It is crucial to support workers in transitioning to new roles through retraining and upskilling programs.

Socioeconomic Inequalities: AI has the potential to exacerbate existing socioeconomic inequalities if access to AI technologies and benefits is not equitably distributed. For instance, in the healthcare sector, access to advanced AI-driven diagnostic tools may be limited to certain communities, perpetuating disparities in healthcare outcomes. Efforts should be made to ensure equal access and mitigate inequalities through inclusive policies and initiatives.

Approaches to Balancing Progress and Societal Well-being:

Ethical Frameworks: Developing and adhering to ethical frameworks can guide the design, development, and deployment of AI technologies. Organizations and researchers should adopt principles such as fairness, transparency, accountability, and privacy to ensure that AI systems are aligned with societal values and promote well-being.

Multi-stakeholder Collaboration: Balancing progress and societal well-being requires collaboration among various stakeholders, including researchers, policymakers, industry leaders, and civil society. Engaging in open dialogue, sharing best practices, and considering diverse perspectives can help shape responsible AI development and deployment.

Regulation and Oversight: Governments and regulatory bodies play a crucial role in setting guidelines and regulations to ensure the ethical use of AI. Establishing clear standards, certification processes, and oversight mechanisms can help address potential risks and ensure accountability.

Public Awareness and Education: Promoting public awareness and education about AI technologies and their implications is essential. By fostering AI literacy and providing platforms for public participation, individuals can actively engage in shaping the direction of AI development and ensuring its alignment with societal well-being.

Conclusion:

The balance between progress and societal well-being is a complex and ongoing challenge in the era of AI. Ethical considerations, such as fairness, transparency, accountability, and privacy, must guide the development and deployment of AI technologies. Additionally, addressing the social impact of AI, including job displacement and socioeconomic inequalities, requires proactive measures, such as retraining programs and inclusive policies.

By adopting ethical frameworks, fostering collaboration, implementing regulations, and promoting public awareness, we can strive for a future where AI advancements benefit society as a whole. Balancing progress with societal well-being ensures that AI remains a force for positive change, empowering individuals, and promoting a fair and inclusive society.

# Looking Ahead: Emerging Trends and Future Possibilities

As artificial intelligence (AI) continues to advance, we stand at the cusp of a transformative era that holds immense potential for various fields, including E-commerce, Healthcare, Banking and Finance, and beyond. In this section, we will explore some of the emerging trends and future possibilities that lie ahead in the realm of AI, offering a glimpse into the exciting possibilities that await us.

Advancements in Natural Language Processing (NLP):

Natural Language Processing, a branch of AI focused on enabling computers to understand and interact with human language, is experiencing rapid advancements. We can expect to see significant progress in language translation, sentiment analysis, and chatbot technologies. NLP-powered virtual assistants, capable of understanding and responding to complex human queries, will revolutionize customer support in industries like E-commerce and Banking.

For instance, imagine an E-commerce platform that utilizes advanced NLP to provide personalized product recommendations based on natural language descriptions provided by customers. This technology can enhance the customer experience by accurately understanding their preferences and needs.

Reinforcement Learning and Autonomous Systems:

Reinforcement learning, a subfield of machine learning, empowers AI systems to learn and improve through interaction with their environment. This approach has the potential to unlock the development of highly autonomous systems capable of making complex decisions and performing tasks without explicit programming.

In industries like Healthcare, autonomous robotic systems can assist in surgical procedures, offering precision and reducing human error. In the Banking sector, reinforcement learning algorithms can optimize investment strategies and portfolio management by adapting to changing market conditions. These autonomous systems have the potential to improve efficiency, productivity, and outcomes in various domains.

AI in Personalized Healthcare:

AI has the potential to revolutionize the healthcare industry by enabling personalized and precision medicine. By analyzing vast amounts of patient data, AI algorithms can assist in diagnosing diseases, predicting treatment outcomes, and recommending personalized therapies.

In the field of genomics, AI can analyze DNA sequences to identify genetic predispositions and develop tailored treatment plans. AI-driven imaging technologies can help detect early signs of diseases, such as cancer, by analyzing medical images with high accuracy and speed.

AI and Climate Change:

As concerns about climate change intensify, AI can play a crucial role in addressing environmental challenges. AI algorithms can analyze large datasets to identify patterns, optimize energy consumption, and predict environmental risks.

In the E-commerce sector, AI-powered supply chain management systems can optimize logistics, reduce waste, and minimize carbon footprints. AI can also assist in forecasting renewable energy generation and optimizing distribution, contributing to a more sustainable future.

Responsible AI and Ethical Considerations:

As AI technologies become more pervasive, the need for responsible AI development and deployment becomes increasingly important. Ethical considerations, such as fairness, transparency, accountability, and privacy, must be at the forefront of AI advancements.

Efforts are underway to develop ethical frameworks, guidelines, and regulations to ensure the responsible use of AI. Multidisciplinary collaborations involving experts from various fields, including AI researchers, policymakers, and ethicists, are shaping the development of AI technologies that align with societal values and promote well-being.

Conclusion:

The future of AI holds tremendous potential for transformative advancements across various industries. As we look ahead, emerging trends such as advancements in Natural Language Processing, reinforcement learning, personalized healthcare, AI's role in addressing climate change, and the importance of responsible AI development will shape the path forward.

However, it is crucial to balance the excitement with a cautious approach, addressing ethical considerations, and ensuring that AI technologies are developed and deployed responsibly. By harnessing the power of AI while being mindful of its potential risks and ethical implications, we can pave the way for a future that benefits society as a whole.

As we embark on this journey, continuous research, collaboration, and open dialogue will be crucial to navigate the evolving landscape of AI and shape its future in a manner that aligns with our collective values and aspirations.

## Advances in AI research and development

The field of artificial intelligence (AI) is constantly evolving, driven by ongoing research and development efforts that push the boundaries of what machines can achieve. In this section, we will delve into some of the remarkable advances in AI, exploring key breakthroughs, methodologies, and their implications across various domains.

Deep Learning and Neural Networks:

Deep learning, a subset of machine learning, has revolutionized AI research by enabling machines to learn from large amounts of data with complex structures. Deep neural networks, inspired by the human brain's neural connections, have demonstrated exceptional capabilities in tasks such as image recognition, natural language processing, and speech synthesis.

For example, in E-commerce, deep learning algorithms can analyze vast customer data to personalize product recommendations, improving customer satisfaction and sales. In Healthcare, deep learning models can aid in medical image analysis, enhancing diagnostic accuracy and speeding up treatment decisions.

Reinforcement Learning and Self-Learning Systems:

Reinforcement learning, a branch of AI, focuses on training agents to make sequential decisions in dynamic environments. Through a trial-and-error process, agents learn to maximize rewards and achieve specific goals. This approach has led to significant advancements in autonomous systems, robotics, and game playing.

An illustrative example comes from the finance sector. Reinforcement learning algorithms can optimize investment strategies by adapting to market conditions, maximizing returns while managing risks. In the gaming industry, AI agents trained through reinforcement learning have surpassed human performance in complex games like Go and chess.

Generative Models and Creative AI:

Generative models have gained attention in recent years for their ability to create new content, including images, music, and text. These models, such as generative adversarial networks (GANs) and variational autoencoders (VAEs), have shown promise in generating realistic and diverse outputs.

In the field of art and design, creative AI applications leverage generative models to produce original artworks, generate music compositions, and even assist in architectural design. In E-commerce, generative models can aid in creating personalized shopping experiences by generating product recommendations tailored to individual preferences.

Explainable AI and Interpretable Models:

As AI systems become more complex, there is a growing need for transparency and interpretability. Explainable AI aims to provide insights into how AI algorithms arrive at their decisions, ensuring accountability and building trust.

In Healthcare, explainable AI can help clinicians understand and trust the reasoning behind AI-assisted diagnoses, leading to improved patient care. In Banking and Finance, explainable AI models can provide justifications for credit scoring or fraud detection decisions, aiding in regulatory compliance and customer understanding.

AI and Cross-Domain Collaborations:

Advancements in AI research often benefit from collaborations between different fields. Interdisciplinary collaborations between AI and domains such as biology, physics, and social sciences have led to breakthroughs in diverse areas, including drug discovery, climate modeling, and social network analysis.

For example, in Healthcare, AI researchers working with biologists have developed AI-driven drug discovery approaches that accelerate the identification of potential treatments for various diseases. In the field of E-commerce, collaborations between AI and social sciences have led to the development of recommender systems that consider social influence and user behavior.

Conclusion:

The advances in AI research and development have the potential to reshape industries, enhance decision-making processes, and drive innovation across domains. Deep learning and neural networks enable machines to learn complex patterns, while reinforcement learning empowers autonomous systems. Generative models foster creative AI, and explainable AI ensures transparency and trust.

As AI continues to advance, interdisciplinary collaborations will play a crucial role in pushing the boundaries of what is possible. By combining expertise from different fields, researchers can tackle complex challenges and unlock the full potential of AI applications.

However, it is essential to remain mindful of the ethical considerations associated with these advances, such as privacy, bias, and societal impact. A responsible and ethical approach to AI research and development will ensure that the benefits of AI are harnessed while mitigating potential risks and ensuring a positive impact on society as a whole.

As the field of AI continues to evolve, it is important to stay informed and engaged, fostering a culture of continuous learning and exploration. By embracing these advances and responsibly applying them, we can harness the transformative power of AI to tackle real-world challenges, create innovative solutions, and shape a better future for humanity.

## Exploring cutting-edge technologies like reinforcement learning and generative AI

In the ever-evolving landscape of artificial intelligence (AI), researchers and practitioners are constantly exploring cutting-edge technologies to push the boundaries of what machines can achieve. Two of the most exciting areas of research and development in AI are reinforcement learning and generative AI. In this section, we will delve into these technologies, exploring their principles, applications, and potential implications across various domains.

Reinforcement Learning:

Reinforcement learning is a branch of machine learning that focuses on training intelligent agents to make sequential decisions in dynamic environments. Unlike supervised learning, where the agent is provided with labeled data, and unsupervised learning, where the agent discovers patterns in unlabeled data, reinforcement learning involves an agent interacting with an environment and learning through trial and error.

At the heart of reinforcement learning is the concept of rewards and punishments. The agent learns to take actions that maximize cumulative rewards over time, guided by a reward signal provided by the environment. Through exploration and exploitation, the agent develops strategies to achieve specific goals.

Reinforcement learning has shown remarkable success in a variety of domains. For instance, in robotics, reinforcement learning enables robots to learn complex motor skills and perform tasks that are challenging to program explicitly. In finance, reinforcement learning algorithms can optimize investment portfolios by adapting to changing market conditions and maximizing returns. In gaming, reinforcement learning has led to breakthroughs, with AI agents surpassing human performance in games like Go and poker.

Generative AI:

Generative AI involves the creation of new content, such as images, music, or text, by AI models. Unlike traditional AI models that are designed for specific tasks, generative models are trained to learn the underlying patterns and structure of the data and generate new samples that resemble the training data.

One of the most prominent generative AI techniques is generative adversarial networks (GANs). GANs consist of two neural networks: a generator network that creates new samples and a discriminator network that distinguishes between real and generated samples. The generator and discriminator networks engage in a competitive process, with the generator improving its ability to generate realistic samples and the discriminator becoming better at distinguishing them.

Generative AI has found applications in various domains. In E-commerce, generative models can create personalized product recommendations based on individual preferences, leading to improved customer satisfaction and sales. In art and design, generative AI enables the creation of unique and original artworks, music compositions, and even fashion designs. In healthcare, generative models can assist in medical imaging, synthesizing realistic medical images for training and research purposes.

Implications and Challenges:

While reinforcement learning and generative AI offer exciting possibilities, they also come with their fair share of challenges and ethical considerations.

One key challenge is the requirement for significant computational resources and large amounts of data for training these models effectively. Training reinforcement learning agents can be computationally expensive and time-consuming. Similarly, generative models often require extensive datasets to capture the underlying distribution of the data accurately.

Another challenge is the potential for biases and unintended consequences in the generated outputs. Reinforcement learning agents can learn biases present in the training data, leading to biased decision-making. Generative models may generate outputs that inadvertently perpetuate societal biases or generate misleading information.

Ethical considerations are also crucial when using these technologies. Reinforcement learning algorithms must adhere to ethical guidelines and avoid harmful actions or behaviors. Generative AI models should respect copyright and intellectual property rights, and their outputs should be used responsibly.

Conclusion:

Reinforcement learning and generative AI represent cutting-edge technologies that hold immense promise and potential in various fields. Reinforcement learning enables intelligent agents to learn from interaction with the environment, leading to impressive capabilities in robotics, finance, and gaming. Generative AI allows for the creation of new and original content, benefiting industries like E-commerce, art, and healthcare.

As these technologies continue to advance, it is important to address challenges such as computational requirements, biases, and ethical considerations. Responsible and ethical development and deployment of reinforcement learning and generative AI systems will ensure that their benefits are harnessed while minimizing potential risks.

By exploring and understanding these cutting-edge technologies, researchers, practitioners, and policymakers can actively shape their development, ensuring that they contribute positively to society and align with our values and aspirations. Embracing the potential of reinforcement learning and generative AI while navigating their challenges will pave the way for groundbreaking applications and advancements in the field of AI.

Potential applications and implications of AI in the future

As artificial intelligence (AI) continues to advance at a rapid pace, its potential applications and implications in various fields are becoming increasingly significant. From E-commerce and healthcare to banking and finance, AI holds the promise of transforming industries and revolutionizing the way we live and work. In this section, we will explore the potential applications of AI in the future and discuss their implications for society, economy, and ethics.

E-commerce:

AI has already made a significant impact on the E-commerce industry, improving customer experiences, personalization, and efficiency. In the future, AI-powered virtual assistants and chatbots will become more sophisticated, providing personalized recommendations and guiding customers throughout their shopping journey. Advanced machine learning algorithms will enable retailers to predict customer preferences accurately, optimize pricing strategies, and enhance supply chain management. For example, AI algorithms can analyze vast amounts of customer data to identify emerging trends and tailor marketing campaigns accordingly, leading to higher conversion rates and increased sales.

Implications: While AI has the potential to revolutionize E-commerce, it also raises concerns about privacy and data security. The collection and analysis of large amounts of customer data may infringe upon privacy rights if not handled responsibly. Striking a balance between personalization and data privacy will be a crucial consideration in the future.

Healthcare:

AI has the potential to revolutionize healthcare by enabling more accurate diagnoses, personalized treatments, and improved patient outcomes. Machine learning algorithms can analyze medical records, patient data, and research studies to provide physicians with valuable insights and support in making informed decisions. AI-powered systems can detect patterns and anomalies in medical images, aiding in early detection of diseases such as cancer. Additionally, AI chatbots can assist patients in triaging symptoms and provide reliable healthcare information, enhancing accessibility to medical advice.

Implications: The use of AI in healthcare raises concerns about the ethical implications of relying solely on algorithms for medical decision-making. While AI can augment the capabilities of healthcare professionals, it is essential to maintain a human-centered approach and ensure that the final decisions are made by medical experts. Moreover, ensuring the security and privacy of patient data is crucial to maintain trust in AI-driven healthcare systems.

Banking and Finance:

The banking and finance sector stands to benefit greatly from AI advancements. AI-powered chatbots and virtual assistants can provide personalized financial advice, assist in investment decisions, and automate routine tasks, such as customer inquiries and transaction processing. Machine learning algorithms can analyze vast amounts of financial data to detect fraud, predict market trends, and optimize investment portfolios. Moreover, AI can facilitate the development of decentralized finance (DeFi) systems, enabling efficient, secure, and transparent financial transactions without intermediaries.

Implications: The adoption of AI in banking and finance raises concerns about potential biases in decision-making algorithms. If not carefully monitored, AI algorithms may perpetuate existing biases in lending practices or investment decisions. Implementing robust transparency and fairness mechanisms will be crucial to mitigate such risks and ensure equitable outcomes.

Conclusion:

The potential applications of AI in the future are vast and far-reaching, with the power to transform industries and reshape our daily lives. However, it is crucial to consider the implications of AI advancements to ensure a responsible and ethical integration into society.

Addressing privacy concerns, data security, and bias mitigation will be critical to maintaining trust in AI systems. Ethical considerations should guide the development and deployment of AI technologies, ensuring that they align with societal values and respect individual rights.

As we move forward, a collaborative approach involving researchers, policymakers, and industry leaders will be essential to navigate the potential risks and harness the transformative power of AI for the benefit of humanity. By balancing innovation with ethical considerations, we can create a future where AI enhances our lives, promotes economic growth, and fosters a fair and inclusive society.

# The importance of ongoing research, collaboration, and responsible AI development

Research plays a fundamental role in advancing AI capabilities and addressing its limitations. Ongoing research allows us to push the boundaries of what is possible, refine existing algorithms, and develop new techniques and methodologies. It enables us to explore cutting-edge technologies, such as reinforcement learning and generative AI, to unlock new possibilities and applications. By continually questioning assumptions and challenging the status quo, researchers can drive innovation and make significant breakthroughs in the field of AI.

Examples of ongoing research can be found in various domains. In E-commerce, researchers are developing advanced recommendation systems that leverage deep learning and natural language processing to provide more accurate and personalized product suggestions to customers. In healthcare, ongoing research focuses on improving disease diagnosis and treatment through AI-powered algorithms that can analyze complex medical data. In banking and finance, researchers are exploring AI techniques to enhance fraud detection and risk assessment models.

Collaboration:

Collaboration is essential in the field of AI as it brings together diverse perspectives, expertise, and resources. No single entity or organization can tackle the complex challenges posed by AI on their own. Collaborative efforts between academia, industry, policymakers, and civil society can foster knowledge exchange, promote interdisciplinary research, and ensure the ethical and responsible development and deployment of AI technologies.

For example, collaborations between AI researchers and healthcare professionals can lead to the development of AI systems that assist in diagnosing rare diseases and suggest personalized treatment plans. Partnerships between financial institutions and AI experts can drive innovation in the banking sector, leading to more efficient fraud prevention and improved customer experiences. Cross-industry collaborations can also facilitate the sharing of best practices and the establishment of ethical guidelines to guide the responsible use of AI across sectors.

Responsible AI Development:

Responsible AI development is crucial to ensure that AI technologies align with societal values and respect ethical principles. It involves considering the broader implications of AI systems on individuals, communities, and society as a whole. Responsible AI development encompasses various aspects, including addressing biases and discrimination, ensuring transparency and accountability, and safeguarding privacy and security.

To address biases and discrimination, developers need to be mindful of the data used to train AI models, ensuring that it is diverse, representative, and free from inherent biases. Regular audits and evaluations of AI systems can help identify and rectify potential biases or unfair outcomes. Transparency and explainability are essential in building trust in AI technologies. By providing clear explanations of how AI systems arrive at their decisions, users can better understand and challenge their outputs when necessary. Additionally, responsible AI development requires robust privacy and security measures to protect individuals' data and mitigate potential risks.

Counterarguments and Dissenting Opinions:

It is worth noting that not all perspectives on ongoing research, collaboration, and responsible AI development are unanimous. Some argue that excessive collaboration and sharing of research findings may lead to the misuse of AI technologies or even the development of AI weapons. Others express concerns about the monopolization of AI research and the potential negative impacts on employment due to automation.

Conclusion:

In conclusion, ongoing research, collaboration, and responsible AI development are indispensable for the continued progress and ethical implementation of AI technologies. Through ongoing research, we can push the boundaries of AI capabilities and address its limitations. Collaboration facilitates knowledge sharing, interdisciplinary approaches, and the establishment of ethical guidelines. Responsible AI development ensures that AI technologies align with societal values, promote fairness, and protect privacy and security. By embracing these pillars, we can maximize the benefits of AI while mitigating potential risks, paving the way for a future where AI serves as a powerful tool for the betterment of humanity.

# Chapter 8: How AI Algorithms Work in Simple Terms

In this chapter, we will demystify the inner workings of AI algorithms and provide a clear understanding of how they operate. While AI algorithms can be complex and sophisticated, it is possible to explain them in simple terms without sacrificing accuracy. By gaining a fundamental understanding of these algorithms, we can appreciate their capabilities and limitations, enabling us to make informed decisions about their use and potential impact.

The Basics of AI Algorithms:

AI algorithms are the computational instructions that drive AI systems. They are designed to process large amounts of data, recognize patterns, and make intelligent predictions or decisions. At their core, AI algorithms are mathematical models that learn from data and improve their performance over time. The key principle behind AI algorithms is that they can extract meaningful information and patterns from data without being explicitly programmed.

Machine Learning Algorithms:

Machine learning algorithms are a subset of AI algorithms that enable computers to learn from data and improve their performance without being explicitly programmed. These algorithms are the backbone of many AI applications, ranging from image recognition and natural language processing to recommendation systems and autonomous vehicles.

One of the most common types of machine learning algorithms is the supervised learning algorithm. In supervised learning, the algorithm is trained on labeled data, where each data point is associated with a corresponding label or outcome. By learning from this labeled data, the algorithm can make predictions or classifications on new, unlabeled data.

For example, in e-commerce, a supervised learning algorithm can analyze historical customer data, including past purchases and browsing behavior, to predict which products a customer is likely to be interested in. In healthcare, these algorithms can analyze medical records and diagnostic images to assist in disease diagnosis and treatment planning.

Deep Learning Algorithms:

Deep learning algorithms are a specialized type of machine learning algorithm inspired by the structure and functioning of the human brain. These algorithms are designed to process and analyze complex, high-dimensional data by creating artificial neural networks with multiple layers of interconnected nodes, also known as artificial neurons.

Deep learning algorithms excel in tasks such as image and speech recognition, natural language processing, and generative modeling. They have achieved remarkable success in various fields, including autonomous driving, virtual assistants, and drug discovery.

For example, in finance, deep learning algorithms can analyze vast amounts of financial data and news articles to predict stock market trends and inform investment decisions. In e-commerce, these algorithms can be used to analyze customer reviews and sentiments, enabling businesses to gain insights into customer preferences and improve their products or services.

Reinforcement Learning Algorithms:

Reinforcement learning algorithms are a type of machine learning algorithm that learns through interactions with an environment. These algorithms aim to maximize a reward signal by taking specific actions in different states of the environment. They learn through trial and error, receiving feedback in the form of rewards or penalties based on their actions.

Reinforcement learning algorithms have demonstrated impressive capabilities in areas such as game playing, robotics, and optimization problems. For instance, in healthcare, reinforcement learning can be used to optimize treatment plans for patients by learning from the outcomes of different interventions.

Limitations and Ethical Considerations:

While AI algorithms have shown tremendous potential, it is important to acknowledge their limitations and address ethical considerations. AI algorithms heavily rely on the data they are trained on, and biases present in the data can lead to biased or unfair outcomes. For example, if a facial recognition algorithm is trained on a dataset that is predominantly composed of a particular racial group, it may exhibit higher error rates for other racial groups, leading to biased results.

Additionally, transparency and interpretability are crucial considerations. Some AI algorithms, especially deep learning algorithms, are often considered black boxes, making it challenging to understand how they arrive at their decisions or predictions. This lack of transparency raises concerns about accountability, especially in critical applications such as healthcare and autonomous vehicles.

Conclusion:

In conclusion, AI algorithms are the driving force behind AI systems, enabling them to process data, recognize patterns, and make intelligent predictions or decisions. By understanding the basics of machine learning, deep learning, and reinforcement learning algorithms, we can appreciate their capabilities and potential applications in fields like e-commerce, healthcare, banking, and finance.

However, it is essential to be aware of the limitations and ethical considerations associated with AI algorithms. Addressing biases, ensuring transparency, and promoting accountability are crucial steps toward responsible AI development. By fostering a deeper understanding of AI algorithms and their impact, we can harness their power while ensuring they align with societal values and serve the best interests of humanity.

**The Building Blocks of AI: Explaining the fundamental components of AI algorithms, such as input data, features, and output predictions.**

In this chapter, we will delve into the essential components that form the building blocks of AI algorithms. Understanding these fundamental elements is crucial for grasping how AI algorithms process information, make predictions, and generate output. By dissecting the inputs, features, and output predictions of AI algorithms, we can unravel the inner workings of these powerful computational tools.

Input Data: The Foundation of AI Algorithms

Expanding on the importance of input data in AI algorithms, we find that it serves as the bedrock for the learning and decision-making processes. Input data is the raw material that AI algorithms analyze to uncover valuable patterns, correlations, and insights. The nature of this data can vary across different domains and applications, encompassing a wide range of formats such as images, text, audio, or structured numerical data.

Let's explore the role of input data in two specific domains: e-commerce and healthcare. In e-commerce, a product recommendation system relies heavily on input data to provide personalized recommendations to customers. The input data for such a system comprises diverse information sources, including customer browsing history, purchase behavior, demographic details, and product attributes. By analyzing this data, AI algorithms can identify patterns that reveal customers' preferences, interests, and purchase tendencies. For example, if a customer frequently buys athletic shoes and sports apparel, the algorithm can leverage this information to recommend similar items, enhancing the customer's shopping experience. The quality, diversity, and representativeness of the input data in e-commerce significantly impact the accuracy and effectiveness of the recommendation system.

In the healthcare domain, input data plays a crucial role in disease diagnosis algorithms. These algorithms utilize a wide range of data sources, such as patient medical records, diagnostic images, lab results, and genetic information, to aid healthcare professionals in accurate and timely diagnoses. By analyzing this comprehensive set of input data, AI algorithms can identify patterns and indicators of specific diseases, potentially assisting healthcare providers in making informed decisions. For instance, a disease diagnosis algorithm may analyze patient symptoms, medical history, and genetic markers to support the identification of a rare genetic disorder. The quality and diversity of the input data, as well as its representation of various patient populations, significantly impact the algorithm's ability to make accurate diagnoses and avoid potential biases.

In both e-commerce and healthcare, the quality of the input data is of paramount importance. High-quality data, free from errors, inconsistencies, or missing values, ensures the reliability and effectiveness of AI algorithms. Moreover, the diversity and representativeness of the input data are critical for avoiding biases and generating fair and equitable outcomes. If the input data is skewed or unrepresentative of the population, AI algorithms may perpetuate existing biases or make inaccurate predictions. For example, if a disease diagnosis algorithm is primarily trained on data from a specific demographic group, it may exhibit lower accuracy when applied to individuals from different backgrounds.

To overcome these challenges, researchers and practitioners strive to gather comprehensive and diverse datasets that capture the complexity and nuances of the real-world scenarios. They also employ techniques such as data augmentation, data cleaning, and data preprocessing to enhance the quality and representation of the input data. By addressing these considerations, AI algorithms can improve their performance, increase accuracy, and minimize potential biases, leading to more reliable and trustworthy outcomes.

In conclusion, the input data forms the foundation for AI algorithms, enabling them to learn, analyze, and make decisions. In e-commerce and healthcare, input data can range from customer behavior and product attributes to patient records and diagnostic images. The quality, diversity, and representativeness of the input data significantly influence the performance and accuracy of AI algorithms. By prioritizing high-quality and diverse datasets, researchers and practitioners can mitigate biases, improve accuracy, and unlock the full potential of AI in various domains.

Features: Transforming Data into Actionable Information

Expanding on the concept of features in AI algorithms, we find that raw input data is often too complex for algorithms to effectively process and extract meaningful patterns. To address this challenge, AI algorithms rely on feature extraction, which involves transforming the raw data into a more manageable and informative format.

Features can be seen as specific representations or characteristics derived from the input data that capture relevant information. They serve as the building blocks that enable AI algorithms to focus on the most discriminative and predictive aspects of the data. By selecting and creating relevant features, algorithms can effectively learn and make accurate predictions.

The process of feature extraction encompasses various techniques, including dimensionality reduction, feature scaling, and feature engineering. Dimensionality reduction techniques aim to reduce the number of features while retaining the most important information. This is particularly useful when dealing with high-dimensional data, as it helps to mitigate the "curse of dimensionality" and improve computational efficiency. Feature scaling techniques, on the other hand, ensure that features are on a comparable scale, preventing certain features from dominating the learning process due to their larger magnitude. Feature engineering involves manually designing and creating features based on domain knowledge and understanding of the problem at hand. This step allows for the incorporation of prior knowledge and can enhance the algorithm's performance.

Let's consider some examples to illustrate the concept of features in different domains. In image recognition tasks, features could include edges, textures, or color histograms that represent distinct visual patterns. For instance, an AI algorithm tasked with classifying images of animals may extract features related to the presence of specific edges or textures characteristic of different animals. These extracted features then serve as inputs to the algorithm for classification.

In natural language processing, features can take the form of word frequencies, syntactic structures, or sentiment scores. For sentiment analysis, an AI algorithm may extract features related to the frequency of positive or negative words in a text, the presence of specific grammatical structures, or the overall sentiment expressed. These features provide important cues for the algorithm to understand and classify the sentiment conveyed in a piece of text.

The choice and design of features are crucial in shaping the performance and interpretability of AI algorithms. Well-selected and informative features can lead to better algorithm performance and enhance the understanding of the underlying patterns and relationships within the data. Conversely, inadequate or irrelevant features may hinder the algorithm's performance or introduce biases.

The process of feature selection and engineering is both an art and a science. It requires a deep understanding of the problem domain, data characteristics, and the specific requirements of the AI algorithm. Researchers and practitioners continually explore new techniques and approaches to effectively extract features that capture the most relevant information and improve the performance of AI systems.

In conclusion, features play a critical role in AI algorithms by enabling the transformation of raw input data into a more manageable and informative format. Through techniques such as dimensionality reduction, feature scaling, and feature engineering, AI algorithms can extract relevant characteristics from the data and focus on the most predictive aspects. The choice and design of features significantly impact the algorithm's performance and interpretability in various domains, such as image recognition and natural language processing. By carefully selecting and creating features, researchers and practitioners can enhance the capabilities and effectiveness of AI algorithms in solving complex problems.

Learning and Prediction: From Features to Output Predictions

Once the input data has undergone processing and feature extraction, AI algorithms rely on learning techniques to make predictions or decisions based on the transformed information. Learning is the process by which the algorithm is trained using labeled or unlabeled data, allowing it to develop a model that captures the underlying patterns and relationships within the data.

Supervised learning is a widely used approach in which the algorithm learns from labeled data, where each data point is associated with a corresponding output or class label. By analyzing the relationships between the input data and the provided labels, the algorithm can generalize and make predictions on new, unseen data. Supervised learning is particularly effective in scenarios where the desired outputs are known and can be used to guide the learning process.

For example, in the field of fraud detection in banking, supervised learning algorithms can be trained using historical transaction data that has been labeled as either fraudulent or non-fraudulent. By analyzing the patterns and features extracted from this labeled data, the algorithm can learn to identify patterns indicative of fraudulent activities. Once trained, the algorithm can then process new, incoming transactions and classify them as either fraudulent or non-fraudulent based on the learned patterns.

In contrast, unsupervised learning involves finding patterns or clusters within unlabeled data, where no explicit output labels are provided. The goal is to discover hidden structures or relationships within the data without any prior knowledge of the expected outcomes. Unsupervised learning algorithms seek to identify similarities, differences, or groupings within the data, allowing for the discovery of previously unknown patterns or insights.

In the context of e-commerce, unsupervised learning algorithms can be applied to customer segmentation. By analyzing customer browsing and purchasing behavior, these algorithms can identify distinct customer segments based on similarities in their preferences and behaviors. This information can then be used to personalize recommendations, target marketing campaigns, or improve customer satisfaction.

It is important to note that both supervised and unsupervised learning approaches have their strengths and limitations. While supervised learning is effective when labeled data is available and desired outputs are known, it requires significant effort in data labeling and may be limited by the availability of labeled examples. Unsupervised learning, on the other hand, does not require labeled data but relies on the algorithm's ability to uncover patterns solely from the input data.

Furthermore, there exist other learning paradigms beyond supervised and unsupervised learning. Reinforcement learning, for example, involves an agent learning to interact with an environment through trial and error to maximize a reward signal. This approach has shown remarkable success in areas such as game playing and robotics.

In conclusion, learning techniques form a crucial component of AI algorithms, enabling them to make predictions or decisions based on the processed input data and extracted features. Supervised learning leverages labeled data to associate inputs with corresponding outputs, while unsupervised learning seeks to uncover patterns or clusters within unlabeled data. Both approaches have a wide range of applications in various fields, such as fraud detection in banking and customer segmentation in e-commerce. By harnessing the power of learning algorithm

Iterative Improvement: Feedback Loops and Model Refinement

AI algorithms are not static entities but continuously evolve and improve through feedback loops. This iterative improvement is achieved through techniques such as model evaluation, validation, and refinement. By incorporating new data and feedback, AI algorithms adapt their models to enhance their performance and accuracy.

For example, in recommendation systems, AI algorithms track user feedback, such as ratings or clicks, and refine their recommendations accordingly. In healthcare, AI algorithms can be continuously updated with new medical research and patient outcomes to improve diagnosis accuracy and treatment recommendations.

Furthermore, AI algorithms may incorporate reinforcement learning, where they receive feedback in the form of rewards or penalties based on their actions. This enables them to learn optimal decision-making strategies through trial and error.

Conclusion:

Understanding the fundamental components of AI algorithms—input data, features, and output predictions—provides a solid foundation for comprehending the inner workings of AI systems. The quality and diversity of input data, the selection and engineering of relevant features, and the learning and prediction processes all contribute to the performance and effectiveness of AI algorithms across various fields.

As AI continues to advance, researchers and practitioners are constantly refining these building blocks, developing new techniques, and exploring innovative ways to enhance the accuracy, interpretability, and fairness of AI algorithms. By unraveling the intricacies of AI algorithms, we can harness their power responsibly and ethically, leveraging their potential for the benefit of society.

**Supervised Learning: Introducing the concept of supervised learning, where AI models are trained with labeled data to make accurate predictions.**

In the vast landscape of artificial intelligence, supervised learning stands as a cornerstone methodology that allows machines to learn from labeled data and make accurate predictions or classifications. At its essence, supervised learning operates on the principle of learning from examples. By providing the AI model with a dataset where each data point is paired with a known output or class label, the model can discern patterns, correlations, and underlying structures that enable it to make informed decisions. In this chapter, we will delve into the intricacies of supervised learning, exploring its fundamental principles, applications across various domains, and potential limitations.

The Essence of Supervised Learning

Supervised learning lies at the core of AI algorithms, enabling machines to learn from labeled data and make accurate predictions or classifications. At the heart of this methodology is the availability of a labeled dataset, which serves as a guiding beacon for the AI model. The labeled data pairs each input data point with its corresponding output label or class, providing crucial information for the learning process.

The goal of supervised learning is for the AI model to discern and understand the relationships between the input data and the provided labels. By examining these labeled examples, the model can uncover patterns, statistical dependencies, and relevant features that contribute to accurate predictions or classifications.

To illustrate the concept, consider a supervised learning scenario in the field of e-commerce. Let's say we have a dataset that includes information about customer browsing behavior, purchase history, and demographic data, along with labels indicating whether the customer made a purchase or not. The input data consists of various features such as the number of products viewed, time spent on the website, and customer age. The corresponding labels indicate whether the customer ultimately made a purchase or not.

By analyzing this labeled dataset, the supervised learning algorithm can identify patterns and dependencies between the input features and the purchase behavior. It may discover that customers who spend more time on the website and view a larger number of products are more likely to make a purchase. Through the learning process, the AI model learns to recognize these patterns and can use them to make accurate predictions on new, unseen data.

The availability of labeled data plays a pivotal role in supervised learning. It provides a clear indication of what the correct output or class should be for a given input. This labeled data serves as a benchmark for the AI model to train and fine-tune its internal parameters, optimizing its ability to generalize from the training examples to unseen data.

Furthermore, examining the labeled examples allows the AI model to extract relevant features from the input data. These features capture essential characteristics or representations of the data that contribute to accurate predictions. Feature extraction is a crucial step in supervised learning, as it helps to transform raw input data into a more meaningful and manageable format. For example, in image recognition, features could include edges, textures, or color histograms that represent distinct visual patterns. In natural language processing, features could be word frequencies, syntactic structures, or sentiment scores.

By learning from labeled data, supervised learning algorithms are capable of making accurate predictions or classifications on new, unseen data. The model generalizes from the patterns and relationships it has learned during training, allowing it to make informed decisions based on previously unseen input data.

In summary, the essence of supervised learning lies in the availability of labeled data, which guides the AI model in understanding the relationships between the input data and the corresponding output labels. By analyzing the labeled examples, the model uncovers patterns, statistical dependencies, and relevant features that contribute to accurate predictions or classifications. Supervised learning unlocks the power of labeled data and enables AI systems to make informed decisions in a wide range of applications and domains.

The Process of Supervised Learning

The process of supervised learning can be encapsulated in several key steps:

Data Collection: Acquiring a high-quality labeled dataset is essential for the success of supervised learning. This dataset should be carefully curated to ensure it represents the problem domain and covers a broad range of scenarios and variations.

Data Preprocessing: Before the AI model can be trained, the data often requires preprocessing to ensure its quality, consistency, and suitability for analysis. This preprocessing stage may involve tasks such as data cleaning, feature scaling, or handling missing values.

Model Training: With the labeled dataset in hand, the AI model embarks on its training journey. During this stage, the model learns from the labeled examples, analyzing the input features and the associated output labels to establish a functional relationship between them. Through iterations and optimization techniques, the model fine-tunes its internal parameters to minimize prediction errors and maximize its ability to generalize.

Model Evaluation: After the model has been trained, it is essential to evaluate its performance on unseen data. This evaluation helps assess the model's generalization capabilities and provides insights into its predictive accuracy. Various evaluation metrics, such as accuracy, precision, recall, and F1 score, can be employed to measure the model's effectiveness.

Model Deployment: Once the model has demonstrated satisfactory performance, it can be deployed to make predictions or classifications on new, unseen data. This deployment phase marks the practical application of the trained model in real-world scenarios.

Applications of Supervised Learning

Supervised learning has proven to be a versatile and powerful tool in a wide range of domains. Let us explore some notable applications:

E-commerce: In the realm of e-commerce, supervised learning algorithms can be employed to build personalized recommendation systems. By learning from past purchase history, browsing behavior, and customer preferences, these systems can suggest relevant products, enhancing customer satisfaction and driving sales.

Healthcare: The healthcare industry can benefit immensely from supervised learning. By training models on labeled medical records, diagnostic images, and patient data, AI algorithms can assist in disease diagnosis, treatment planning, and prognosis prediction. For example, supervised learning can aid in the early detection of diseases like cancer by analyzing medical images and identifying abnormal patterns.

Finance: Supervised learning has made significant contributions to the field of finance. It can be employed for credit scoring, fraud detection, and stock market prediction. By learning from labeled historical financial data, AI models can assess creditworthiness, detect fraudulent transactions, and identify potential market trends, enabling better decision-making and risk management.

Natural Language Processing: In the realm of natural language processing, supervised learning techniques are instrumental in tasks such as sentiment analysis, text classification, and language translation. By training on labeled text data, AI models can comprehend and interpret human language, opening doors for applications like sentiment analysis in social media monitoring or language translation in cross-cultural communication.

Limitations and Challenges

While supervised learning holds immense promise in enabling accurate predictions and classifications, it is essential to acknowledge its limitations and the challenges it presents. Understanding these limitations is crucial for deploying AI systems responsibly and effectively. Let us explore some of the key challenges associated with supervised learning:

Availability of Labeled Data: Supervised learning heavily relies on the availability of labeled data, where each data point is paired with its corresponding output label or class. However, acquiring labeled data can be a time-consuming and resource-intensive process. In some domains, such as healthcare or finance, obtaining labeled data may require expert domain knowledge or involve sensitive information. The scarcity or unavailability of labeled data can limit the application of supervised learning approaches in certain contexts.

Bias in Labeled Data: Labeled data can unintentionally inherit biases present in society or reflect cultural prejudices. These biases, if not adequately addressed, can be learned and perpetuated by AI models, leading to biased predictions or decisions. For example, in a hiring application, if historical hiring decisions were biased towards certain demographics, the trained model may perpetuate those biases when making future hiring recommendations. Mitigating bias requires proactive efforts in data curation, fairness assessment, and bias mitigation techniques to ensure that AI systems treat individuals fairly and equitably.

Generalization to Unseen Data: The performance of a supervised learning model on the training data does not guarantee its ability to generalize well to unseen data. Overfitting is a common challenge, where the model becomes overly complex and captures noise or idiosyncrasies specific to the training data, rather than learning the underlying patterns. As a result, the model may struggle to make accurate predictions on new, unseen data. Techniques such as regularization, cross-validation, and early stopping are employed to combat overfitting and improve generalization.

Need for Continuous Learning: In dynamic environments where data distributions change over time, supervised learning models may face difficulties in adapting to these shifts. This phenomenon, known as concept drift, can lead to degraded performance if the model does not adapt to the evolving patterns in the data. Continuous learning approaches, such as online learning or incremental learning, allow the model to adapt and update its knowledge as new labeled data becomes available. Regular monitoring and updating of the model are crucial to ensure its accuracy and relevance in changing circumstances.

Addressing these challenges requires a multi-faceted approach that combines data collection and curation, algorithmic advancements, and ethical considerations. Collaborative efforts between researchers, practitioners, and policymakers are necessary to develop robust techniques that mitigate biases, improve generalization, and enable continuous learning.

Conclusion

Supervised learning stands as a cornerstone technique in the field of artificial intelligence, allowing machines to learn from labeled data and make accurate predictions or classifications. By leveraging the relationships between input data and provided labels, AI models can discern patterns, correlations, and underlying structures, enabling them to make informed decisions. With applications spanning diverse domains such as e-commerce, healthcare, finance, and natural language processing, supervised learning has proven its value in solving real-world problems. However, challenges such as the availability of labeled data, biases, generalization to unseen data, and the need for continuous learning must be addressed to ensure the ethical and effective use of supervised learning in AI systems. With ongoing research and responsible development, supervised learning continues to shape the landscape of artificial intelligence and propel advancements in various industries.

**Unsupervised Learning: Describing unsupervised learning and how AI models find patterns and structures within unlabeled data.**

In the realm of AI, unsupervised learning stands as a fascinating paradigm that enables AI models to extract meaningful patterns and structures from unlabeled data. Unlike supervised learning, which relies on labeled examples to guide the learning process, unsupervised learning operates on the premise that data itself contains inherent patterns waiting to be uncovered. Let us delve into the intricacies of unsupervised learning and explore how AI models make sense of unlabeled data.

At its core, unsupervised learning aims to identify hidden patterns, similarities, and clusters within a dataset. Without explicit labels or guidance, AI models embark on a journey of exploration, seeking to understand the underlying structure and organization of the data. Through sophisticated algorithms, these models detect statistical dependencies, group similar data points together, and uncover intrinsic relationships that might otherwise remain unnoticed.

Clustering, one of the primary techniques employed in unsupervised learning, aims to group similar data points into distinct clusters based on their shared characteristics. This approach allows AI models to uncover natural groupings within the data, providing insights into different categories or segments. For instance, in e-commerce, clustering algorithms can identify distinct customer segments based on their purchasing behavior, enabling businesses to tailor their marketing strategies accordingly.

Another key technique within unsupervised learning is dimensionality reduction, which aims to capture the essential information within a dataset while reducing its complexity. By compressing the data into a lower-dimensional representation, dimensionality reduction techniques help reveal the most significant features or components that contribute to the data's variance. This can be particularly useful in fields such as image or text analysis, where high-dimensional data can be challenging to process and interpret.

An illustrative example of unsupervised learning's power lies in anomaly detection. By learning the normal patterns within a dataset, AI models can flag unusual or anomalous instances that deviate significantly from the learned norms. This technique finds applications in various domains, including fraud detection in banking, network intrusion detection in cybersecurity, and equipment failure prediction in manufacturing.

However, it is important to acknowledge the inherent challenges and limitations associated with unsupervised learning:

Lack of Ground Truth: Without labeled data to guide the learning process, evaluating the performance and quality of unsupervised learning algorithms becomes inherently more challenging. Assessing the discovered patterns and structures often relies on qualitative analysis and expert judgment.

Interpretability and Subjectivity: Unsupervised learning outputs may lack direct interpretability, as the learned patterns are not tied to explicit labels. Deciphering the meaning and relevance of identified clusters or components can be subjective and domain-dependent, requiring human intervention and domain knowledge.

Overcoming Noise and Outliers: Unsupervised learning algorithms must contend with noise and outliers present in real-world datasets. Robust techniques are necessary to mitigate the impact of noisy data points that could distort the discovered patterns or skew the clustering results.

Despite these challenges, unsupervised learning presents vast opportunities for uncovering hidden knowledge and gaining insights from unlabeled data. By leveraging the power of unsupervised learning, AI models can autonomously explore and reveal the underlying structure and organization within complex datasets, offering new perspectives and driving innovation in diverse fields such as genetics, social network analysis, and market segmentation.

In conclusion, unsupervised learning represents a remarkable approach in AI that unlocks the potential of unlabeled data. By harnessing sophisticated algorithms and techniques, AI models can identify hidden patterns, clusters, and relationships, providing valuable insights and driving decision-making processes. While challenges exist, the rewards of unsupervised learning are vast, fostering advancements in various domains and opening doors to new frontiers of knowledge.

**Reinforcement Learning: Explaining reinforcement learning, where AI models learn through trial and error and receive feedback to optimize their actions.**

In the realm of AI, there exists a captivating paradigm known as reinforcement learning, which allows AI models to learn and optimize their behavior through trial and error. This unique approach draws inspiration from the way humans and animals learn from their environment by receiving feedback on their actions. In this section, we shall explore the intricacies of reinforcement learning and uncover how AI models navigate the path of learning and optimization.

Reinforcement learning is rooted in the notion of an agent interacting with an environment. The agent seeks to achieve a specific objective or maximize a notion of cumulative reward by taking actions within the environment. These actions lead to changes in the agent's state and subsequently impact the rewards it receives. Through repeated interactions, the agent learns to identify the actions that yield the most favorable outcomes, thereby optimizing its behavior.

At the core of reinforcement learning lies the concept of a reward signal. This signal serves as a feedback mechanism, guiding the agent's learning process. When the agent takes actions that align with the desired objective, it receives positive rewards. Conversely, actions leading to suboptimal outcomes are met with negative rewards or penalties. By maximizing the cumulative reward over time, the agent learns to make better decisions and achieve its goals more effectively.

To facilitate the learning process, reinforcement learning employs the use of a policy, which is a strategy or set of rules guiding the agent's actions in different states. The policy determines the action taken by the agent in response to its current state. Reinforcement learning algorithms explore different policies and update them based on the feedback received from the environment. Through this iterative process, the agent refines its policy, gradually converging towards an optimal strategy.

One key aspect of reinforcement learning is the exploration-exploitation trade-off. In the early stages of learning, the agent explores various actions and their consequences to gain a better understanding of the environment. This exploration allows the agent to discover new strategies and potentially more rewarding paths. However, as the agent learns and accumulates knowledge, it shifts towards exploiting the actions that have demonstrated higher rewards. Balancing exploration and exploitation is crucial to achieving optimal performance in reinforcement learning.

A notable characteristic of reinforcement learning is its ability to handle dynamic and complex environments. Unlike supervised or unsupervised learning, reinforcement learning models do not rely on pre-labeled or explicit guidance. Instead, they adapt and learn through direct interaction with the environment, making it well-suited for scenarios where explicit labeling may be impractical or costly. Examples of such scenarios include robotic control, autonomous driving, and game playing.

The application of reinforcement learning spans diverse fields. In e-commerce, reinforcement learning can be used to optimize pricing strategies, where an agent adjusts prices based on customer responses to maximize profits. In healthcare, reinforcement learning can aid in optimizing treatment plans for patients with chronic conditions, adapting the therapy based on patient outcomes. In finance, reinforcement learning can be employed to develop optimal trading strategies that adapt to changing market conditions.

While reinforcement learning holds great promise, it also presents challenges and considerations:

Reward Design: Designing appropriate reward functions that capture the desired objective is crucial. Care must be taken to ensure that the rewards align with the true goals, as misaligned rewards can lead to unintended behavior and suboptimal outcomes.

Exploration Strategies: Finding effective exploration strategies that strike a balance between exploring new actions and exploiting known strategies is an ongoing area of research. Exploration techniques such as epsilon-greedy, Thompson sampling, or Monte Carlo Tree Search aim to strike this delicate balance.

Sample Efficiency: Reinforcement learning algorithms typically require a significant amount of interaction with the environment to learn effectively. Improving sample efficiency, reducing the number of interactions needed, is an active area of research to make reinforcement learning more practical and applicable to real-world scenarios.

In conclusion, reinforcement learning stands as a fascinating approach in AI, allowing models to learn through trial and error, optimize their behavior, and adapt to complex environments. By leveraging rewards and iterative feedback, agents learn to navigate the path of learning and achieve their objectives. While challenges exist, reinforcement learning continues to drive innovation across various domains, from robotics to healthcare and beyond, enabling AI models to excel in tasks that require autonomous decision-making and adaptive behavior.

**Deep Learning: Introducing deep learning, a subset of ML that uses neural networks with multiple layers to process complex data.**

Within the realm of machine learning, a captivating field emerges—a field that pushes the boundaries of complexity and unleashes the full potential of artificial intelligence. This field is none other than deep learning, a subset of machine learning that harnesses the power of neural networks with multiple layers. In this section, we embark on a journey to unravel the intricacies of deep learning and explore how it enables us to process and understand complex data like never before.

Deep learning represents a paradigm shift in the way we approach the analysis of data. Traditional machine learning algorithms rely on handcrafted features, which often require domain expertise and significant human effort to engineer. Deep learning, on the other hand, leverages the remarkable capacity of neural networks to automatically learn relevant features from the raw input data. This characteristic allows deep learning models to effectively handle intricate patterns, subtle dependencies, and high-dimensional data.

At the core of deep learning lies the neural network—a computational model inspired by the intricate network of neurons in the human brain. A neural network is composed of layers of interconnected artificial neurons, also known as nodes or units. Each neuron receives inputs, performs a computation, and produces an output signal. The network's architecture consists of an input layer, one or more hidden layers, and an output layer. These layers work in concert to process and transform the input data into meaningful outputs.

What sets deep learning apart from traditional neural networks is the depth of its architecture. Deep neural networks comprise multiple hidden layers, each consisting of numerous neurons. These additional layers allow the model to learn hierarchical representations of the data, with each layer capturing progressively more abstract and complex features. By employing multiple layers, deep learning models can effectively model intricate relationships and extract meaningful representations from the data.

Training a deep learning model involves the process of optimizing its weights and biases, enabling it to make accurate predictions or classifications. This optimization is achieved through a technique called backpropagation, which involves propagating errors backward through the network and adjusting the weights accordingly. The backpropagation algorithm efficiently computes the gradients of the model's parameters, facilitating the process of updating the weights and fine-tuning the network's performance.

The power of deep learning is perhaps best exemplified by its remarkable success in computer vision tasks. Deep convolutional neural networks (CNNs) have revolutionized image recognition, achieving superhuman performance on challenging datasets such as ImageNet. By learning hierarchical representations of visual features, deep CNNs can discern intricate details, recognize objects, and even localize specific regions within an image. This breakthrough has fueled advancements in fields like autonomous driving, medical imaging analysis, and surveillance systems.

Beyond computer vision, deep learning has found application in natural language processing (NLP) and speech recognition. Recurrent neural networks (RNNs) and transformer models have been instrumental in achieving state-of-the-art performance in tasks such as language translation, sentiment analysis, and speech synthesis. These models capture the temporal dependencies and contextual information present in sequential data, enabling them to comprehend and generate human-like language.

The potential applications of deep learning span across various domains. In e-commerce, deep learning can be used to personalize product recommendations based on user browsing behavior and purchase history. In healthcare, deep learning models can aid in medical diagnosis, leveraging electronic health records, medical imaging, and genetic data. In finance, deep learning can be applied to fraud detection, anomaly detection in financial transactions, and stock market prediction.

While deep learning has propelled the field of AI to new heights, it is essential to acknowledge its challenges and limitations:

Data Requirements: Deep learning models often require large amounts of labeled data to achieve optimal performance. Acquiring and annotating such datasets can be resource-intensive and time-consuming, particularly in specialized domains.

Computational Resources: Training deep learning models can be computationally demanding, necessitating powerful hardware and substantial computational resources. GPUs (Graphics Processing Units) and specialized hardware accelerators have become crucial for training deep neural networks efficiently.

Interpretability: The black-box nature of deep learning models poses challenges in understanding and interpreting their decisions. Interpreting the inner workings of complex neural networks is an ongoing area of research, aiming to enhance transparency and trustworthiness.

In conclusion, deep learning stands as a remarkable achievement in the realm of artificial intelligence, empowering us to tackle complex data and extract meaningful insights. By leveraging the depth and hierarchical representations of neural networks, deep learning models have revolutionized fields like computer vision and natural language processing. While challenges exist, the potential applications of deep learning are vast, offering transformative solutions across diverse domains. Through ongoing research and innovation, we continue to unlock the full potential of deep learning and witness its profound impact on the future of AI.

# Chapter 9: The Development and Rapid Growth of AI

As we embark on the journey through the chapters of this book, we have witnessed the profound impact of artificial intelligence (AI) across various fields, from e-commerce and healthcare to banking and finance. In this chapter, we delve into the development and rapid growth of AI, exploring the key factors that have propelled its advancement and the implications of this transformative technology.

The Evolution of AI

Artificial intelligence has a rich history that stretches back decades. From its early beginnings as a theoretical concept in the 1950s to its practical applications today, AI has undergone a remarkable evolution. The rapid progress in computing power, the availability of big data, and breakthroughs in algorithmic techniques have fueled the growth of AI, leading to unprecedented achievements in machine learning, natural language processing, computer vision, and robotics.

One of the pivotal moments in the development of AI was the advent of machine learning, which shifted the focus from rule-based programming to systems that could learn from data. Machine learning algorithms, such as supervised learning, unsupervised learning, and reinforcement learning, have become the bedrock of AI applications, enabling computers to recognize patterns, make predictions, and adapt their behavior based on feedback.

The Rise of Big Data

The proliferation of digital technologies and the Internet has given rise to an explosion of data. This abundance of data, often referred to as big data, has become the lifeblood of AI. From customer transaction records and social media interactions to medical records and sensor data, the sheer volume, velocity, and variety of data have opened up new possibilities for AI systems.

With big data, AI models can be trained on vast amounts of information, allowing them to extract meaningful insights and make accurate predictions. For example, in e-commerce, companies leverage customer behavior data to personalize product recommendations, optimize pricing strategies, and enhance customer experiences. In healthcare, AI algorithms analyze large-scale patient data to improve diagnostics, detect diseases at an early stage, and develop personalized treatment plans.

Advancements in Computing Power

The exponential growth of computing power has played a pivotal role in the rapid development of AI. Moore's Law, which states that the number of transistors on a microchip doubles approximately every two years, has driven the miniaturization and increased performance of computer processors. This has enabled the execution of complex AI algorithms at scale and reduced the time required for training and inference.

Moreover, specialized hardware, such as graphics processing units (GPUs) and tensor processing units (TPUs), have been specifically designed to accelerate AI computations. These hardware advancements have significantly boosted the processing capabilities of AI systems, enabling the training of deep neural networks with millions or even billions of parameters.

The Impact of AI on Various Industries

The impact of AI on industries such as e-commerce, healthcare, banking, and finance cannot be overstated. In e-commerce, AI-powered recommendation systems enhance customer engagement and drive sales by suggesting relevant products based on individual preferences and browsing history. In healthcare, AI algorithms assist in disease diagnosis, drug discovery, and personalized medicine, revolutionizing patient care and improving health outcomes.

In banking and finance, AI is transforming fraud detection, credit risk assessment, and algorithmic trading. AI-powered chatbots and virtual assistants are enhancing customer service experiences, automating routine tasks, and providing personalized recommendations. The potential applications of AI span across numerous fields, promising efficiency gains, cost savings, and transformative solutions to complex problems.

Ethical Considerations and Future Implications

As AI continues to advance, it is vital to address ethical considerations and potential implications. The increasing automation of tasks raises concerns about job displacement and the impact on the workforce. Ensuring the ethical use of AI, mitigating biases, protecting privacy, and ensuring transparency are crucial for building trust and accountability in AI systems.

Furthermore, the rapid growth of AI necessitates ongoing research and collaboration. The exploration of AI's ethical, legal, and societal implications requires interdisciplinary efforts, engaging experts from various fields. By fostering responsible

AI development, we can harness the transformative power of this technology while addressing potential risks and challenges.

In conclusion, the development and rapid growth of AI have ushered in a new era of technological innovation. The convergence of big data, advancements in computing power, and breakthroughs in algorithmic techniques has propelled AI to new heights, transforming industries and reshaping the way we live and work. However, as we venture into this era, it is crucial to navigate the ethical considerations and future implications of AI with prudence, ensuring that this powerful tool remains a force for good in our rapidly evolving world.

**The Birth of AI: Tracing the origins of AI from early concepts to the Dartmouth Workshop and the Turing Test.**

Artificial Intelligence (AI) has emerged as one of the most transformative and intriguing fields of study in the modern era. In this chapter, we delve into the early origins of AI, exploring its conceptual foundations and the key milestones that shaped its development. From the initial ideas that sparked the imagination of visionaries to the pivotal Dartmouth Workshop and the renowned Turing Test, we trace the fascinating birth of AI.

The Conceptual Foundations

The roots of AI can be traced back to the early 20th century, a time when forward-thinking visionaries began contemplating the audacious idea of creating machines that could emulate human intelligence. This revolutionary concept laid the groundwork for the birth of Artificial Intelligence, a field that would eventually reshape the technological landscape.

One of the pivotal figures in the development of AI was the brilliant mathematician and logician, Alan Turing. Turing's work, particularly his influential paper "On Computable Numbers, with an Application to the Entscheidungsproblem" published in 1936, marked a significant milestone in the conceptual underpinnings of AI. In this seminal work, Turing introduced the concept of a universal computing machine, which later became known as the "Turing machine." This theoretical construct was capable of simulating any algorithmic computation, embodying the essence of computation itself. Turing's groundbreaking ideas provided a theoretical foundation for the possibility of creating machines that could replicate human thought processes.

Another key figure in the development of AI was the collaboration between Warren McCulloch, a neurophysiologist, and Walter Pitts, a logician. In 1943, they published a seminal paper titled "A Logical Calculus of Ideas Immanent in Nervous Activity," which presented a groundbreaking concept that would shape the future of AI—neural networks. This influential work proposed a mathematical model that emulated the behavior of the human brain through interconnected nodes, or "neurons," and the transmission of electrical signals. The paper provided a theoretical framework for the idea of building machines that could learn and exhibit intelligent behavior, drawing inspiration from the complex workings of the human nervous system.

McCulloch and Pitts' paper laid the foundation for the field of neural networks, which would become a cornerstone of AI research and development. Their work demonstrated that the human brain's ability to process information and make decisions could potentially be replicated in artificial systems. The concept of neural networks served as a catalyst for further exploration into building intelligent machines that could learn from data and adapt their behavior.

These early contributions by visionaries such as Turing, McCulloch, and Pitts set the stage for the remarkable advancements in AI that followed. Their pioneering ideas and theoretical frameworks provided the necessary tools to explore and realize the potential of creating machines with human-like intelligence. The subsequent decades witnessed an exponential growth in AI research and applications, with neural networks evolving into deep learning models and the development of sophisticated algorithms capable of processing vast amounts of data.

The work of these early pioneers not only paved the way for the development of AI but also sparked a paradigm shift in our perception of what machines can achieve. Their ideas continue to resonate in the field today, as researchers push the boundaries of AI and strive to create intelligent systems that can solve complex problems, make informed decisions, and interact with humans in a meaningful way.

In summary, the roots of AI can be traced back to the visionary thinkers of the early 20th century who contemplated the idea of creating machines that could mimic human intelligence. The seminal contributions of Alan Turing and his concept of the universal Turing machine, along with the groundbreaking work of Warren McCulloch and Walter Pitts in neural networks, laid the foundation for the development of AI. These ideas sparked a revolution in computing and set the stage for the rapid growth and advancements that have propelled AI to the forefront of technological innovation.

The Dartmouth Workshop: Birth of AI as a Field of Study

The true birth of AI as a distinct field of study can be traced back to a seminal event that took place in the summer of 1956—a gathering of scientists and researchers at Dartmouth College in Hanover, New Hampshire. This historic event, known as the Dartmouth Workshop, brought together some of the most influential figures in the field, including John McCarthy, Marvin Minsky, Nathaniel Rochester, and Claude Shannon.

The Dartmouth Workshop was a catalyst for the formal establishment of AI as an interdisciplinary field of study. It marked the first concerted effort to explore the possibility of building machines that could exhibit intelligent behavior. The participants recognized the potential of AI to revolutionize computing and sought to create a platform for collaboration and knowledge sharing.

During the workshop, the attendees engaged in deep discussions and debates, delving into fundamental questions and challenges related to AI. They explored topics such as natural language processing, problem-solving, and machine learning. These early pioneers recognized the importance of interdisciplinary collaboration, bringing together experts from mathematics, psychology, computer science, and other relevant disciplines. By bridging these diverse fields of knowledge, they aimed to tackle the complex and multifaceted nature of intelligence.

While the outcomes of the Dartmouth Workshop were modest in terms of immediate breakthroughs or concrete solutions, its significance lies in the vision it ignited and the community it formed. The workshop generated a wave of enthusiasm and inspired researchers to continue exploring the frontiers of AI. It provided a platform for researchers to exchange ideas, share their progress, and establish a common language and framework for the field.

The Dartmouth Workshop laid the foundation for subsequent research and development in AI. It sparked a new era of exploration, where scientists and researchers began to experiment with novel algorithms, develop innovative approaches, and push the boundaries of what was considered possible. The workshop's impact can be seen in the numerous breakthroughs and advancements that have propelled AI to its current prominence.

Furthermore, the Dartmouth Workshop's interdisciplinary nature set the tone for AI as a field that transcends traditional disciplinary boundaries. It recognized that understanding and replicating human intelligence required insights from a variety of domains, including computer science, psychology, linguistics, and philosophy. This interdisciplinary approach remains a defining characteristic of AI today, as researchers continue to draw inspiration and insights from a wide range of disciplines.

In summary, the Dartmouth Workshop held in 1956 marked the true birth of AI as a distinct field of study. It brought together leading figures in the field and set the stage for interdisciplinary collaboration and exploration. While the workshop's immediate outcomes were modest, its significance lies in the enthusiasm it generated and the foundation it established for subsequent research and development. The Dartmouth Workshop paved the way for AI to evolve into the vibrant and rapidly advancing field it is today.

The Turing Test: A Milestone in AI

In 1950, Alan Turing proposed a groundbreaking test to assess a machine's ability to exhibit intelligent behavior. Known as the Turing Test, it involved a human judge engaging in natural language conversations with a machine and a human. If the judge could not reliably distinguish between the machine and the human, the machine would be deemed to possess human-like intelligence.

The Turing Test captured the imagination of researchers and the public alike, becoming a prominent benchmark for evaluating AI systems. While the test has sparked debates and controversies over its validity as a measure of true intelligence, it remains a significant milestone in the history of AI, highlighting the quest to create machines that can exhibit human-like cognition.

Implications and Future Directions

The birth of AI marked the beginning of a journey that continues to shape the world we live in today. From the early concepts and foundational research to the Dartmouth Workshop and the Turing Test, AI has evolved into a vibrant and multidisciplinary field with far-reaching implications.

AI has found applications across diverse domains, from e-commerce and healthcare to banking and finance. It has transformed the way we interact with technology, enabling voice assistants, recommendation systems, and autonomous vehicles. Moreover, AI has revolutionized industries, unlocking new possibilities for innovation, efficiency, and problem-solving.

However, as AI continues to advance, it raises profound ethical and societal questions. Concerns surrounding privacy, bias, job displacement, and the impact on human autonomy demand careful consideration and responsible development. As we navigate the future of AI, interdisciplinary collaborations, regulatory frameworks, and ethical guidelines will play a crucial role in harnessing the potential of AI while mitigating risks.

In conclusion, the birth of AI can be traced back to the visionary ideas of brilliant thinkers, the collaborative efforts of the Dartmouth Workshop, and the concept of the Turing Test. This chapter has provided a glimpse into the origins of AI, setting the stage for the remarkable advancements and ethical challenges that lie ahead. As we move forward, we must remain vigilant, ensuring that AI remains a force for progress and empowerment, while being guided by ethical considerations and a commitment to the well-being of humanity.

**AI Winter and Resurgence: Explaining the periods of "AI Winter" when funding and interest in AI declined, followed by resurgences due to technological advancements and breakthroughs.**

The field of artificial intelligence has experienced periods of intense excitement and enthusiasm, followed by periods of skepticism and reduced interest. These cycles, known as "AI Winters," have had a profound impact on the development and progress of AI. Understanding these cycles is crucial for appreciating the challenges and opportunities that have shaped the field.

The first AI Winter occurred in the 1970s and 1980s. During this time, initial optimism surrounding AI gave way to disappointment as the high expectations of what AI could achieve collided with the limitations of the available technology. The prevailing sentiment was that AI had overpromised and underdelivered, leading to a significant reduction in funding and interest in the field. This AI Winter was characterized by a decline in research projects, closures of AI labs, and a general skepticism towards the feasibility of AI.

One of the factors contributing to the AI Winter was the perception that early AI systems were not living up to the grand visions that had been painted. The limitations of computational power, memory, and data availability hindered progress. Additionally, the inability of AI systems to handle complex real-world problems effectively led to disillusionment.

However, the AI Winter eventually gave way to a resurgence in the late 1990s and early 2000s. This resurgence was fueled by several factors, including advancements in computing power, the availability of large datasets, and breakthroughs in machine learning algorithms. These developments reignited interest in AI and led to significant progress in areas such as computer vision, natural language processing, and robotics.

Technological breakthroughs played a crucial role in the AI resurgence. For example, the advent of deep learning, a subfield of machine learning, revolutionized the field by enabling AI models to learn hierarchical representations from vast amounts of data. This breakthrough led to remarkable advancements in image and speech recognition, language translation, and autonomous driving.

The resurgence of AI has had a profound impact on various industries. In e-commerce, AI-powered recommendation systems have transformed the way products are recommended to customers, leading to improved personalization and customer satisfaction. In healthcare, AI has shown promising applications in disease diagnosis, drug discovery, and personalized medicine. Banking and finance have also witnessed the adoption of AI for fraud detection, algorithmic trading, and risk assessment.

Despite the recent AI resurgence, it is important to approach the field with caution and realistic expectations. Some skeptics argue that we are currently in the midst of another AI Winter, as the hype around AI has led to inflated expectations and a focus on short-term gains rather than long-term research. They argue that many AI systems still lack a deep understanding of context and are prone to biases and errors. Additionally, ethical concerns surrounding AI, such as privacy and transparency, continue to be topics of debate.

In conclusion, the history of AI is marked by cycles of hope and disillusionment. The AI Winter periods have highlighted the challenges and limitations of the field, while the resurgences have showcased the remarkable progress and potential of AI. As we navigate the future of AI, it is essential to strike a balance between enthusiasm and caution, promoting responsible research, ethical considerations, and long-term sustainable development. By learning from the past and addressing the challenges that arise, we can ensure that AI continues to thrive and benefit humanity in a responsible and meaningful manner.

**Big Data and AI: Discussing the role of big data in fueling AI development and how large datasets are crucial for training sophisticated AI models.**

In the era of AI, one cannot overlook the pivotal role played by big data. The exponential growth of digital information has created a vast ocean of data that serves as the lifeblood for AI algorithms. Big data provides the fuel that powers the development and advancement of sophisticated AI models, enabling them to learn from patterns, make accurate predictions, and deliver transformative insights.

The term "big data" refers to the immense volume, variety, and velocity of data generated from various sources, including social media, sensors, online transactions, and more. The availability of such vast quantities of data has revolutionized the field of AI, unlocking new possibilities and driving unprecedented progress.

One of the fundamental requirements for training AI models is access to substantial amounts of relevant data. Large datasets are crucial for enabling AI algorithms to identify meaningful patterns and correlations, and to make accurate predictions or classifications. The abundance of data allows AI models to capture a comprehensive understanding of the underlying processes and phenomena they aim to emulate.

To illustrate the significance of big data in AI, let us consider some examples across different fields. In e-commerce, companies leverage customer data, including browsing behavior, purchase history, and demographic information, to build personalized recommendation systems. By analyzing vast amounts of data, these systems can accurately predict customer preferences and tailor recommendations accordingly, leading to improved customer satisfaction and increased sales.

In healthcare, big data plays a vital role in medical research, drug development, and patient care. Researchers can analyze large-scale genomic datasets to identify genetic variations associated with specific diseases, enabling the development of targeted treatments. Patient records, combined with real-time monitoring data, can provide valuable insights for early disease detection and personalized treatment plans.

In the banking and finance sector, big data analytics is used for fraud detection, risk assessment, and algorithmic trading. By analyzing large volumes of transaction data, AI models can identify suspicious patterns and flag potential fraudulent activities. Moreover, AI algorithms can analyze market trends and historical data to make informed investment decisions and optimize trading strategies.

While the potential benefits of big data in AI are vast, it is important to address the challenges associated with its utilization. One key challenge is the quality of the data. Ensuring data accuracy, completeness, and relevance is crucial for obtaining reliable insights and avoiding biases. Additionally, data privacy and security must be carefully addressed to protect sensitive information and maintain public trust.

Critics of big data argue that the sheer volume of data can lead to information overload and noise, making it challenging to extract meaningful insights. They also raise concerns about the potential for surveillance and privacy breaches when dealing with sensitive data.

In conclusion, big data has emerged as a critical enabler of AI advancements. The availability of vast amounts of data empowers AI algorithms to learn, adapt, and provide valuable insights across a range of industries. However, careful attention must be given to data quality, privacy, and ethical considerations. By harnessing the potential of big data responsibly, we can unlock the transformative power of AI and pave the way for a future of data-driven innovation and progress.

**The Impact of Computing Power: Explaining the significance of increased computing power in enabling AI's growth and the rise of GPUs for parallel processing.**

The remarkable progress of artificial intelligence (AI) owes much to the ever-increasing power of computing technology. The availability of powerful computing resources has played a pivotal role in enabling AI algorithms to tackle complex problems, process vast amounts of data, and achieve unprecedented levels of performance. In this section, we will explore the significance of increased computing power and the role of graphics processing units (GPUs) in AI's growth.

The relationship between computing power and AI can be best understood through the lens of computational efficiency. AI algorithms, particularly those that involve deep learning, heavily rely on computational resources to train large neural networks with numerous parameters. These networks require immense computational capabilities to process and optimize the millions or even billions of interconnected weights and biases.

Historically, the development of AI was constrained by the limited computational resources available. As the complexity of AI models increased, the need for more powerful hardware became evident. In the early days of AI, researchers worked with conventional central processing units (CPUs) to train and run their models. However, CPUs were primarily designed for sequential processing, which posed a significant bottleneck for AI algorithms that demanded parallel computation.

The advent of GPUs, originally designed for rendering complex graphics in video games, brought about a paradigm shift in AI. GPUs are highly parallel processors with thousands of cores that excel at performing numerous calculations simultaneously. This parallel architecture made GPUs well-suited for the parallelizable nature of AI computations, unlocking unprecedented levels of performance and efficiency. The use of GPUs for AI accelerated the training process, allowing researchers and practitioners to experiment with larger models and process massive datasets.

The impact of increased computing power, facilitated by GPUs, can be observed across a multitude of domains. In e-commerce, AI-powered recommendation systems leverage powerful computing resources to process vast amounts of customer data in real-time, delivering personalized product suggestions. Healthcare applications, such as medical imaging and diagnostics, benefit from increased computing power to process and analyze complex medical data, aiding in the detection of diseases and the development of treatment plans. In the financial sector, high-performance computing enables AI algorithms to analyze large-scale financial datasets and make accurate predictions for investment strategies and risk management.

The exponential growth of computing power continues to shape the trajectory of AI. As computational resources become more accessible and affordable, AI applications can be deployed on a wider scale. Cloud computing platforms and specialized hardware accelerators, such as field-programmable gate arrays (FPGAs) and application-specific integrated circuits (ASICs), further contribute to the expansion of AI capabilities.

However, it is crucial to address some of the challenges that accompany the quest for increased computing power. The energy consumption of high-performance computing systems is a growing concern, as it raises environmental and sustainability issues. Efforts are being made to develop energy-efficient hardware architectures and optimize algorithms to strike a balance between computational power and energy consumption.

In conclusion, the impact of increased computing power on AI cannot be overstated. The availability of powerful hardware, particularly GPUs, has revolutionized the field by enabling the training of complex models and the processing of massive datasets. As computing power continues to advance, AI will further push the boundaries of what is possible, unlocking new frontiers of innovation and transforming industries across the globe. It is essential to continue investing in sustainable and energy-efficient computing solutions to ensure the responsible and ethical growth of AI in the years to come.

**Accelerating Innovation: Highlighting key breakthroughs and milestones that have propelled AI to new heights in recent years.**

In recent years, the field of artificial intelligence (AI) has experienced an unprecedented surge of innovation, driven by groundbreaking breakthroughs and remarkable milestones. These advancements have catapulted AI into the forefront of technological progress, revolutionizing industries and reshaping the way we live and work. In this section, we will delve into the key breakthroughs and milestones that have accelerated the growth of AI and propelled it to new heights.

One of the most significant drivers of AI innovation is the rapid development of deep learning algorithms. Deep learning, a subset of machine learning, employs neural networks with multiple layers to process complex data and extract high-level abstractions. The resurgence of deep learning can be attributed, in part, to the availability of vast amounts of labeled data and the substantial increase in computing power. This convergence of data and computational resources has empowered researchers and practitioners to train deep neural networks on massive datasets, achieving remarkable performance across various domains.

One notable breakthrough in deep learning is the successful application of convolutional neural networks (CNNs) to image recognition tasks. CNNs, inspired by the biological structure of the visual cortex, excel at analyzing visual data and extracting meaningful features. The groundbreaking success of CNNs in image classification competitions, such as the ImageNet Challenge, demonstrated the potential of deep learning in computer vision. These advancements have paved the way for applications such as autonomous vehicles, medical imaging analysis, and facial recognition systems.

Another significant milestone in AI innovation is the development of generative adversarial networks (GANs). GANs are a type of neural network architecture that involves two components: a generator and a discriminator. The generator learns to generate realistic samples, such as images or text, while the discriminator learns to distinguish between real and generated samples. GANs have shown remarkable capabilities in generating realistic synthetic data, enhancing creativity in fields like art, music, and design. They have also been used for data augmentation, improving training data diversity and generalization.

The advent of transfer learning has also played a crucial role in accelerating AI innovation. Transfer learning leverages the knowledge gained from pretraining a neural network on a large dataset to solve new, related tasks with limited labeled data. This approach has significantly reduced the need for extensive labeled datasets, allowing AI models to learn more efficiently and effectively in resource-constrained scenarios. Transfer learning has found practical applications in various domains, including natural language processing, computer vision, and healthcare, where labeled data may be scarce or expensive to obtain.

Moreover, the emergence of AI-focused hardware and specialized processing units has contributed to the acceleration of innovation. Graphics processing units (GPUs), originally designed for graphics rendering, have become a staple in AI computation due to their ability to perform parallel computations. The parallel processing power of GPUs has fueled advancements in deep learning, enabling the training and deployment of complex neural networks at scale. Additionally, the development of specialized hardware, such as tensor processing units (TPUs) and neuromorphic chips, holds the promise of even greater AI acceleration and energy efficiency.

The impact of these breakthroughs and milestones can be observed across a wide range of industries. In e-commerce, AI-powered recommendation systems leverage these advancements to deliver personalized product suggestions, enhancing the shopping experience for customers. In healthcare, AI algorithms equipped with deep learning techniques assist in medical diagnosis, drug discovery, and personalized treatment planning. The banking and finance sector relies on AI for fraud detection, algorithmic trading, and risk assessment.

While these advancements bring tremendous opportunities, it is essential to acknowledge and address the challenges they entail. Ethical considerations, such as bias in AI algorithms and the impact on employment, must be carefully examined. Transparency, fairness, and accountability should be at the forefront of AI development and deployment.

In conclusion, the accelerated pace of innovation in AI has propelled the field to new heights. Breakthroughs in deep learning, the rise of GANs, the adoption of transfer learning, and advancements in specialized hardware have transformed AI into a powerful tool with broad applications. As AI continues to evolve, interdisciplinary collaborations, ethical frameworks, and ongoing research and development will be crucial to harnessing its potential for the betterment of society. By nurturing innovation and responsible AI practices, we can unlock even greater possibilities and shape a future that benefits all of humanity.

# Chapter 10: The Role of Data in AI

In the realm of artificial intelligence (AI), data serves as the lifeblood that fuels innovation and progress. It is the foundation upon which AI algorithms are built and the fuel that powers their learning and decision-making capabilities. Without quality data, AI systems would be rendered ineffective, unable to discern patterns, make accurate predictions, or derive meaningful insights.

The Variety and Volume of Data in Today's Digital Age

In today's digital age, we are immersed in a sea of data. Every interaction, transaction, and communication generates vast amounts of information, creating an unprecedented opportunity for AI development. From social media posts and online shopping behavior to electronic health records and financial transactions, the variety and volume of data available for analysis are staggering.

The proliferation of internet-connected devices, the advent of social media platforms, and the digitization of various industries have led to an exponential growth in data generation. This wealth of data represents a treasure trove of information that can be harnessed to enhance AI capabilities and drive transformative advancements across a multitude of domains.

The Foundation: Data as the Lifeblood of AI

At the core of AI lies the reliance on data for training, learning, and decision-making. AI algorithms are designed to process and analyze data to extract patterns, correlations, and insights that are otherwise beyond human perception. The availability of high-quality, diverse, and representative data is paramount to the effectiveness and accuracy of AI systems.

Consider, for instance, an e-commerce recommendation system. Such a system relies on user browsing history, purchase behavior, demographic information, and product attributes to make personalized recommendations. The accuracy and relevance of these recommendations depend on the quality and breadth of the underlying data.

Data-Driven Decision-Making and Prediction

Data enables AI systems to make informed decisions and predictions based on historical patterns and statistical analysis. By training AI models on labeled data, they learn to recognize and generalize from patterns, allowing them to infer relationships and make predictions on new, unseen data.

In the realm of healthcare, AI algorithms can analyze vast amounts of patient medical records, diagnostic images, and genetic data to assist in disease diagnosis, treatment planning, and drug discovery. By leveraging the power of data, AI systems can provide clinicians with valuable insights and support evidence-based decision-making.

Examples of Data-Driven AI Applications in Various Fields

The impact of data in AI extends across diverse domains. In e-commerce, data-driven AI powers personalized product recommendations, dynamic pricing strategies, and targeted marketing campaigns. In healthcare, it enables early disease detection, precision medicine, and health monitoring. In banking and finance, data-driven AI facilitates fraud detection, risk assessment, and algorithmic trading.

For instance, financial institutions employ AI algorithms to analyze vast amounts of financial data, market trends, and consumer behavior to identify patterns and predict market movements. These insights help traders and investors make informed decisions and manage risks effectively.

Data Collection and Curation

Data collection for AI involves gathering relevant information from various sources, such as sensors, databases, and online platforms. This process requires careful planning, data acquisition, and data labeling to ensure the availability of high-quality data for training and testing AI models.

Challenges and Considerations in Data Collection

Data collection for AI is not without challenges. Ensuring data quality, avoiding biases, and addressing privacy concerns are crucial considerations. Biases in data can lead to biased outcomes and perpetuate existing societal inequalities. Additionally, acquiring labeled data may be time-consuming and expensive, particularly in domains that require expert domain knowledge or involve sensitive information.

Ensuring Data Quality, Integrity, and Representativeness

To derive accurate and meaningful insights, it is essential to ensure the quality, integrity, and representativeness of the data used for AI development. Data should be collected from diverse sources to avoid skewed perspectives and to capture a comprehensive view of the problem at hand. Careful attention should also be given to data preprocessing, cleaning, and transformation to remove noise, handle missing values, and normalize data distributions.

## Big Data: Fueling AI Development

The emergence of big data has significantly influenced the trajectory of AI development. Big data refers to datasets that are too large and complex to be effectively managed and analyzed using traditional data processing methods. The availability of big data has opened new avenues for AI research and application, enabling the training of sophisticated AI models with unprecedented scale and accuracy.

### The Three V's of Big Data: Volume, Velocity, and Variety

The three V's of big data—volume, velocity, and variety—highlight the defining characteristics of this data paradigm. Volume refers to the immense size of the datasets, with petabytes or even exabytes of data being generated daily. Velocity pertains to the speed at which data is generated and needs to be processed in real-time or near real-time. Variety denotes the diverse types and formats of data, including structured, unstructured, and semi-structured data.

### How Big Data Enables the Training of Sophisticated AI Models

Big data plays a crucial role in training complex AI models, such as deep neural networks, which require massive amounts of labeled data to learn intricate patterns and relationships. By leveraging big data, AI models can achieve higher accuracy, improved generalization, and enhanced predictive capabilities.

For example, in image recognition tasks, deep learning models trained on massive image datasets, such as ImageNet, have surpassed human performance in identifying objects and scenes. Similarly, in natural language processing, large-scale language models trained on extensive text corpora have achieved remarkable progress in language understanding and generation.

### Case Studies Showcasing the Impact of Big Data on AI Advancement

The impact of big data on AI advancement is exemplified by the success stories across various fields. In e-commerce, companies like Amazon and Alibaba utilize big data analytics to understand customer preferences, optimize supply chain operations, and enhance user experiences. In healthcare, organizations leverage big data to analyze patient records, clinical trials, and genetic information to unlock insights for personalized medicine and population health management. In finance, big data is harnessed to detect fraudulent transactions, manage investment portfolios, and optimize risk assessment.

By capitalizing on the vast amounts of data available, these organizations have transformed their industries and gained a competitive edge through data-driven decision-making and AI-powered innovations.

Conclusion

Data lies at the heart of AI, serving as the crucial ingredient that powers its development, training, and decision-making capabilities. The availability of diverse and high-quality data, combined with advancements in computing power and AI algorithms, has propelled AI to new heights and unlocked unprecedented possibilities across various domains.

However, the responsible use of data in AI development is of paramount importance. Ethical considerations, privacy protection, and bias mitigation must be prioritized to ensure that the potential of AI is harnessed for the benefit of all. By striking a balance between technological advancements and ethical considerations, we can continue to harness the transformative power of data and AI, paving the way for a future where intelligent systems work hand in hand with human ingenuity to shape a better world.

# The Foundation: Data as the Lifeblood of AI

In the realm of artificial intelligence (AI), data plays a pivotal role as the lifeblood that fuels innovation and progress. It serves as the foundation upon which AI algorithms are built and the catalyst that empowers their learning and decision-making capabilities. Without access to quality data, AI systems would be rendered ineffective, incapable of discerning patterns, making accurate predictions, or deriving meaningful insights.

The significance of data in AI development cannot be overstated. It is the raw material that enables AI algorithms to learn, adapt, and improve over time. Just as human intelligence relies on information gathered from our senses and experiences, AI algorithms rely on data to emulate and enhance human-like cognitive processes.

As we delve into the realm of AI, it becomes evident that the availability, quality, and diversity of data have a profound impact on the effectiveness and accuracy of AI systems. From structured databases to unstructured text, images, and sensor data, the variety and complexity of data sources provide the fuel necessary for AI algorithms to analyze, understand, and extract valuable insights.

Moreover, the exponential growth of digital technologies and the interconnectedness of our modern world have led to an unprecedented explosion of data generation. Every click, search, transaction, and interaction generates data that can be harnessed to enhance AI capabilities. Consider the e-commerce industry, where customer preferences, browsing behavior, and purchase history are meticulously collected and analyzed to provide personalized recommendations and improve customer satisfaction. In healthcare, patient records, medical imaging, and genetic information are utilized to support diagnostics, treatment planning, and drug discovery. Even in the financial sector, data-driven AI algorithms analyze market trends, customer behavior, and risk indicators to optimize investment strategies and detect fraudulent activities.

The availability of vast amounts of data, combined with advancements in computing power and AI algorithms, has fueled the rapid progress of AI in recent years. AI has transitioned from simple rule-based systems to sophisticated models that can learn from large-scale datasets, adapt to new information, and make complex decisions. This data-driven approach enables AI to tackle complex problems and achieve results that surpass human capabilities in areas such as image recognition, natural language processing, and predictive analytics.

However, the reliance on data in AI development also brings forth important considerations and challenges. The quality, accuracy, and representativeness of data must be carefully evaluated to avoid biased outcomes or erroneous conclusions. Furthermore, privacy concerns and ethical considerations surrounding data usage must be addressed to ensure that AI respects individual rights and societal values.

In this chapter, we will explore the foundational role of data in AI, delving into the fundamental importance of data in AI algorithms, the process of data collection and curation, the challenges and considerations associated with data usage, and the transformative power of big data in driving AI development. By understanding the critical role of data, we can appreciate its profound impact on AI's ability to transform industries, advance scientific research, and improve the human experience.

## The fundamental importance of data in AI algorithms

In the realm of artificial intelligence (AI), data is the cornerstone upon which the capabilities and effectiveness of AI algorithms are built. It serves as the raw material that fuels the learning and decision-making processes of these algorithms.

Understanding the fundamental importance of data in AI algorithms is essential for appreciating the transformative potential of AI and its impact across various domains.

The Learning Process of AI Algorithms:
At the heart of AI algorithms lies the process of learning from data. Just as humans acquire knowledge through observation and experience, AI algorithms learn from data to identify patterns, make predictions, and generate insights. The process of learning from data can be broadly categorized into supervised learning, unsupervised learning, and reinforcement learning.

a. Supervised Learning: In supervised learning, AI algorithms are trained using labeled data, where each data point is paired with a corresponding label or output. This labeled data serves as a guiding signal for the algorithm to learn the underlying patterns and relationships between the input and output variables. For example, in e-commerce, a supervised learning algorithm can be trained to predict customer preferences based on past purchasing behavior.

b. Unsupervised Learning: Unsupervised learning, on the other hand, involves training AI algorithms on unlabeled data, where the algorithm must uncover hidden patterns and structures within the data without any explicit guidance. This type of learning is particularly useful when the underlying patterns are unknown or when discovering novel insights. In healthcare, unsupervised learning algorithms can analyze patient data to identify clusters or subgroups with similar characteristics, enabling personalized treatment approaches.

c. Reinforcement Learning: Reinforcement learning involves training AI algorithms through a trial-and-error process, where the algorithm interacts with an environment and receives feedback in the form of rewards or penalties. The algorithm learns to optimize its actions to maximize the cumulative rewards over time. This type of learning is well-suited for training AI systems to make autonomous decisions in dynamic environments. For instance, in autonomous vehicles, reinforcement learning algorithms can learn to navigate and make driving decisions based on real-time sensor data.

Data Quality and Quantity:
The quality and quantity of data have a profound impact on the performance and capabilities of AI algorithms. High-quality data ensures the accuracy and reliability of AI models, while a large volume of data provides a more comprehensive representation of the problem space, enabling the algorithms to make more robust and accurate predictions.

a. Data Quality: Ensuring the quality of data is crucial to avoid biases, errors, or misleading conclusions. Inaccurate or biased data can lead to skewed outcomes and unethical decision-making. For example, in banking and finance, if AI algorithms are trained on biased historical loan data, they may perpetuate discriminatory lending practices. Data curation processes, including data cleaning, preprocessing, and validation, are essential to enhance the quality and integrity of data.

b. Data Quantity: The quantity of data available for AI algorithms to learn from is often a determining factor in their performance. More data enables algorithms to capture a broader range of patterns, increasing their predictive power. In e-commerce, larger datasets allow AI algorithms to extract more nuanced insights about customer behavior and preferences, enabling businesses to tailor their offerings accordingly.

The Role of Diverse and Representative Data:
To ensure the effectiveness and fairness of AI algorithms, it is crucial to have diverse and representative data. Diverse data captures a wide range of perspectives, enabling AI algorithms to account for various demographic, cultural, and contextual factors. Representative data ensures that the outcomes of AI algorithms are not biased towards specific groups or contexts. In healthcare, for instance, representative and diverse datasets are essential for developing AI algorithms that provide equitable and accurate diagnostics and treatments for different populations.

Data Privacy and Ethical Considerations:
The utilization of data in AI algorithms raises important ethical considerations regarding privacy, consent, and data ownership. It is imperative to establish robust data protection measures and frameworks to ensure that data is collected, stored, and used in a responsible and ethical manner. Respecting individuals' privacy rights, obtaining informed consent, and implementing secure data management practices are vital to maintain trust and safeguard against potential misuse or unauthorized access to sensitive data.

Conclusion:

Data lies at the core of AI algorithms, enabling them to learn, adapt, and make informed decisions. The fundamental importance of data in AI cannot be overstated, as it serves as the foundation for the development of innovative AI applications across various domains. By recognizing the significance of data quality, quantity, diversity, and ethical considerations, we can harness the power of data to drive responsible and impactful AI advancements that benefit society as a whole.

Data-driven decision-making and prediction

At the heart of AI algorithms lies the process of learning from data. Just as humans acquire knowledge through observation and experience, AI algorithms learn from data to identify patterns, make predictions, and generate insights. The process of learning from data can be broadly categorized into supervised learning, unsupervised learning, and reinforcement learning.

a. Supervised Learning: In supervised learning, AI algorithms are trained using labeled data, where each data point is paired with a corresponding label or output. This labeled data serves as a guiding signal for the algorithm to learn the underlying patterns and relationships between the input and output variables. For example, in e-commerce, a supervised learning algorithm can be trained to predict customer preferences based on past purchasing behavior.

b. Unsupervised Learning: Unsupervised learning, on the other hand, involves training AI algorithms on unlabeled data, where the algorithm must uncover hidden patterns and structures within the data without any explicit guidance. This type of learning is particularly useful when the underlying patterns are unknown or when discovering novel insights. In healthcare, unsupervised learning algorithms can analyze patient data to identify clusters or subgroups with similar characteristics, enabling personalized treatment approaches.

c. Reinforcement Learning: Reinforcement learning involves training AI algorithms through a trial-and-error process, where the algorithm interacts with an environment and receives feedback in the form of rewards or penalties. The algorithm learns to optimize its actions to maximize the cumulative rewards over time. This type of learning is well-suited for training AI systems to make autonomous decisions in dynamic environments. For instance, in autonomous vehicles, reinforcement learning algorithms can learn to navigate and make driving decisions based on real-time sensor data.

Data Quality and Quantity:

The quality and quantity of data have a profound impact on the performance and capabilities of AI algorithms. High-quality data ensures the accuracy and reliability of AI models, while a large volume of data provides a more comprehensive representation of the problem space, enabling the algorithms to make more robust and accurate predictions.

a. Data Quality: Ensuring the quality of data is crucial to avoid biases, errors, or misleading conclusions. Inaccurate or biased data can lead to skewed outcomes and unethical decision-making. For example, in banking and finance, if AI algorithms are trained on biased historical loan data, they may perpetuate discriminatory lending practices. Data curation processes, including data cleaning, preprocessing, and validation, are essential to enhance the quality and integrity of data.

b. Data Quantity: The quantity of data available for AI algorithms to learn from is often a determining factor in their performance. More data enables algorithms to capture a broader range of patterns, increasing their predictive power. In e-commerce, larger datasets allow AI algorithms to extract more nuanced insights about customer behavior and preferences, enabling businesses to tailor their offerings accordingly.

The Role of Diverse and Representative Data:
To ensure the effectiveness and fairness of AI algorithms, it is crucial to have diverse and representative data. Diverse data captures a wide range of perspectives, enabling AI algorithms to account for various demographic, cultural, and contextual factors. Representative data ensures that the outcomes of AI algorithms are not biased towards specific groups or contexts. In healthcare, for instance, representative and diverse datasets are essential for developing AI algorithms that provide equitable and accurate diagnostics and treatments for different populations.

Data Privacy and Ethical Considerations:
The utilization of data in AI algorithms raises important ethical considerations regarding privacy, consent, and data ownership. It is imperative to establish robust data protection measures and frameworks to ensure that data is collected, stored, and used in a responsible and ethical manner. Respecting individuals' privacy rights, obtaining informed consent, and implementing secure data management practices are vital to maintain trust and safeguard against potential misuse or unauthorized access to sensitive data.

Conclusion:

Data lies at the core of AI algorithms, enabling them to learn, adapt, and make informed decisions. The fundamental importance of data in AI cannot be overstated, as it serves as the foundation for the development of innovative AI applications across various domains. By recognizing the significance of data quality, quantity, diversity, and ethical considerations, we can harness the power of data to drive responsible and impactful AI advancements that benefit society as a whole.

# Examples of data-driven AI applications in various fields

The integration of artificial intelligence (AI) and data has revolutionized numerous fields, enabling transformative advancements and unlocking new possibilities. This section explores diverse examples of data-driven AI applications across various domains, including e-commerce, healthcare, banking, and finance. These examples illustrate the profound impact of AI in solving complex problems, improving efficiency, and enhancing decision-making processes.

E-commerce:

In the realm of e-commerce, AI algorithms utilize vast amounts of customer data to personalize user experiences, enhance recommendations, and optimize pricing strategies. For instance, online retail giants leverage data on customers' past purchases, browsing behavior, and demographic information to generate tailored product recommendations. These personalized recommendations have become a cornerstone of e-commerce platforms, enhancing customer satisfaction and driving sales. Additionally, AI-powered chatbots leverage natural language processing techniques to provide instant customer support, improving response times and enhancing the overall shopping experience.

Healthcare:

AI has made significant strides in healthcare, transforming the diagnosis, treatment, and management of diseases. By analyzing patient data, including medical records, genomic information, and imaging data, AI algorithms can provide accurate and timely diagnoses, identify treatment options, and predict patient outcomes. For example, in radiology, AI algorithms assist radiologists in detecting abnormalities and improving diagnostic accuracy. AI-powered healthcare systems can also monitor patient vitals in real-time, enabling early detection of deteriorating conditions and prompt intervention.

Banking and Finance:

The banking and finance industry has witnessed the integration of AI to improve fraud detection, risk assessment, and customer service. AI algorithms analyze vast amounts of financial data, including transaction records, credit scores, and market trends, to detect fraudulent activities and mitigate risks. Moreover, AI-powered chatbots and virtual assistants provide personalized customer support, facilitating quick and efficient resolution of queries and enhancing customer satisfaction. AI algorithms can also analyze market data and optimize trading strategies, leading to more informed investment decisions.

Transportation and Logistics:

AI has significantly impacted the transportation and logistics sector, optimizing routes, reducing delivery times, and improving supply chain management. By analyzing historical transportation data, weather conditions, and real-time traffic information, AI algorithms can optimize delivery routes, reducing fuel consumption and improving efficiency. Furthermore, AI-powered predictive maintenance systems can monitor the health of vehicles and predict maintenance requirements, minimizing downtime and optimizing fleet management.

Social Media and Marketing:

AI plays a pivotal role in social media platforms and digital marketing. By analyzing user data, including preferences, behavior patterns, and social interactions, AI algorithms can deliver targeted advertisements, optimize ad campaigns, and improve user engagement. Social media platforms leverage AI to detect and moderate inappropriate content, ensuring a safer and more inclusive environment for users. AI-powered sentiment analysis also helps businesses gauge public opinions and sentiments regarding their products or services, enabling them to make informed marketing decisions.

Conclusion:

The examples provided in this section illustrate the wide-ranging applications of data-driven AI across various fields. From e-commerce and healthcare to banking and finance, AI algorithms harness the power of data to drive innovation, enhance decision-making processes, and deliver personalized experiences. The integration of AI and data has the potential to revolutionize industries, improve efficiency, and solve complex problems. However, it is essential to navigate the ethical considerations surrounding data privacy, bias, and responsible use of AI to ensure its positive impact on society. As AI continues to evolve, we can anticipate even more groundbreaking applications that leverage the potential of data to transform industries and shape our future.

# Data Collection and Curation

In the realm of artificial intelligence (AI), data is the lifeblood that fuels innovation and drives the development of intelligent systems. The collection and curation of data play a crucial role in shaping the effectiveness and accuracy of AI algorithms. This section explores the fundamental importance of data collection and curation in AI, highlighting the challenges, strategies, and ethical considerations involved in this process.

I. The Significance of Data in AI Development:

AI algorithms, at their core, are designed to learn from data. They rely on vast amounts of data to extract meaningful patterns, make accurate predictions, and generate valuable insights. Without access to high-quality, diverse, and representative data, AI systems would struggle to understand and interpret the complex and ever-evolving world around them.

Data serves as the foundation upon which AI models are built. It provides the raw material from which algorithms can extract valuable information and learn from it. By analyzing data, AI models can uncover hidden patterns, correlations, and dependencies that may not be apparent to human observers. This ability to detect patterns and relationships within the data is what enables AI systems to make accurate predictions and generate insights.

High-quality data is essential for the training of AI models. It needs to be representative of the real-world scenarios the AI system will encounter. This means that the data should cover a wide range of variations, including different contexts, demographics, and situations. By exposing AI models to diverse data, they can learn to generalize their knowledge and make accurate predictions in novel situations.

Moreover, the quality of the data greatly impacts the performance of AI systems. Data that is incomplete, noisy, or biased can lead to inaccurate and unreliable outcomes. AI models trained on biased data can perpetuate and amplify existing biases, leading to unfair or discriminatory results. Therefore, it is crucial to curate and preprocess the data to ensure its quality and fairness.

Data empowers AI models to acquire knowledge and generalize from examples. Through exposure to a diverse and representative dataset, AI algorithms can learn the underlying patterns and relationships that drive the data. This learning process allows AI systems to make predictions and generate insights even in the absence of explicit instructions. By learning from data, AI models can adapt and improve their performance over time, enabling them to tackle complex tasks and provide valuable solutions.

In summary, data is the lifeblood of AI algorithms. It provides the necessary information for AI systems to learn, make predictions, and generate insights. High-quality, diverse, and representative data enables AI models to understand and interpret the world around them, acquiring knowledge and generalizing from examples. Without access to such data, AI systems would be severely limited in their capabilities, underscoring the critical role that data plays in the development and success of AI.

The Challenges of Data Collection:

A. Volume and Variety:

In the digital age, we are witnessing an unprecedented explosion of data generation. Every interaction, transaction, and activity in the online world leaves behind a digital footprint, contributing to the vast landscape of data. From e-commerce transactions to healthcare records, social media interactions, and the data generated by the Internet of Things (IoT) devices, the sources of data are diverse and abundant.

E-commerce transactions, for instance, generate a tremendous amount of data as consumers browse products, make purchases, and leave reviews. This data can provide valuable insights into consumer behavior, preferences, and market trends. Companies can analyze this data to personalize recommendations, optimize pricing strategies, and enhance the overall customer experience.

Healthcare records contain a wealth of information about patient diagnoses, treatments, and outcomes. Analyzing this data can lead to significant advancements in disease prevention, diagnosis, and treatment. For example, machine learning algorithms can be trained on large-scale healthcare datasets to identify patterns and risk factors associated with specific diseases, enabling early detection and intervention.

Social media platforms generate enormous volumes of data in the form of posts, comments, likes, and shares. This data reflects people's opinions, sentiments, and social connections. By analyzing social media data, AI algorithms can gain insights into public sentiment, track trends, and even predict certain behaviors. This information is invaluable for businesses, governments, and organizations to make informed decisions, develop targeted marketing campaigns, and detect emerging social issues.

The Internet of Things (IoT) devices, such as sensors embedded in smart homes, vehicles, and industrial machinery, generate real-time data streams. This data provides valuable information about environmental conditions, energy usage, equipment performance, and more. By analyzing IoT data, AI algorithms can optimize energy consumption, detect anomalies, and improve predictive maintenance strategies.

However, the challenge lies in the collection and integration of this vast volume of data from disparate sources. Data is often scattered across multiple systems, stored in different formats, and governed by various regulations and privacy concerns. Collecting and consolidating data from these sources requires robust data management strategies, including data cleansing, transformation, and integration.

Ensuring the quality, relevance, and reliability of the data is another critical aspect. Data may contain errors, inconsistencies, or biases that can negatively impact the performance and accuracy of AI models. Data quality assurance processes, such as data validation and verification, are necessary to identify and address these issues.

Furthermore, ensuring the relevance of the data is crucial. AI models are only as good as the data they are trained on. Therefore, selecting relevant and representative data is essential to avoid skewed or biased results. Careful consideration must be given to factors such as data source diversity, sample size, and demographic representation to ensure fair and unbiased AI systems.

In conclusion, the digital age has brought about an exponential growth in data generation. E-commerce transactions, healthcare records, social media interactions, and IoT sensor data are just a few examples of the diverse sources of data. The challenge lies in collecting, integrating, and ensuring the quality, relevance, and reliability of this vast volume of data. Addressing these challenges is crucial for unlocking the full potential of AI and harnessing the power of data to drive innovation and insights.

B. Data Quality and Bias:

The quality of data plays a paramount role in the development of AI systems. AI algorithms learn from data, and the accuracy and reliability of that data directly impact the performance and outcomes of the models. Inaccurate or biased data can lead to flawed models and biased outcomes, perpetuating existing biases and reinforcing societal prejudices.

One key aspect of ensuring data quality is addressing issues such as data incompleteness, noise, and errors. Incomplete data, where important attributes or variables are missing, can introduce biases or hinder the performance of AI models. Noise and errors in the data, whether due to measurement errors or data collection processes, can introduce inaccuracies and negatively impact the model's ability to learn meaningful patterns.

Furthermore, data may inadvertently reflect societal biases and prejudices present in the real world. These biases can be present in various forms, such as gender, race, or socioeconomic biases. If not carefully addressed, AI models trained on such biased data can perpetuate and amplify these biases, leading to biased predictions or decisions. For example, a biased facial recognition system may exhibit higher error rates for certain demographic groups due to the underrepresentation or misrepresentation of those groups in the training data.

To mitigate bias and ensure fairness, rigorous data curation and fairness assessment processes are essential. Data curation involves carefully selecting and preprocessing the data to minimize biases and ensure representativeness. This process may involve techniques such as stratified sampling, data augmentation, or oversampling of underrepresented groups. Fairness assessment techniques, such as measuring disparate impact or evaluating the model's performance across different demographic groups, can help identify and address potential biases in AI systems.

Moreover, transparency and accountability in data collection and algorithmic decision-making are crucial. It is important to document the data collection processes, including any biases or limitations associated with the data sources. Additionally, monitoring and evaluating the performance of AI systems in real-world applications is necessary to detect and rectify any biases that may emerge during deployment.

To illustrate the importance of addressing bias in data, let's consider an example in the field of healthcare. Suppose an AI system is trained to assist doctors in diagnosing diseases based on medical images. If the training data predominantly consists of images from a specific demographic group, the AI system may not generalize well to other groups, leading to disparities in accurate diagnosis. By ensuring diverse and representative data in the training set, healthcare AI systems can provide more equitable and unbiased assistance to doctors, improving patient outcomes.

In conclusion, the quality of data is of paramount importance in AI development. Inaccurate or biased data can undermine the performance and fairness of AI models, perpetuating biases and leading to unjust outcomes. Addressing issues of data incompleteness, noise, and errors, as well as mitigating biases through rigorous data curation and fairness assessment processes, are crucial steps towards developing AI systems that are accurate, reliable, and fair.

Strategies for Effective Data Collection:

A. Data Gathering:

Collecting data is a complex process that involves identifying relevant sources, designing data collection protocols, and extracting or generating data from various channels. Depending on the domain and specific objectives, data collection methods may vary, but the goal remains the same: to gather meaningful and informative data that can be used to train AI models and gain insights.

In the e-commerce industry, for instance, data collection is crucial for understanding customer behavior, preferences, and purchase patterns. Relevant data can be collected through various sources, such as customer transactions, website interactions, and feedback mechanisms. Every time a customer makes a purchase or browses through products, valuable data points are generated, including item details, pricing information, customer demographics, and browsing history. These data points can be used to analyze customer preferences, recommend products, and personalize the shopping experience. Moreover, feedback mechanisms such as reviews and ratings provide additional insights into customer satisfaction and sentiment analysis.

In the healthcare sector, data collection plays a vital role in improving patient care, treatment outcomes, and medical research. Patient data can be obtained from multiple sources, including electronic health records (EHRs), wearable devices, and clinical trials. EHRs contain a wealth of information, including patient demographics, medical history, diagnoses, treatments, and laboratory results. This comprehensive data can be leveraged to develop predictive models for disease diagnosis, identify risk factors, and optimize treatment plans. Additionally, wearable devices, such as fitness trackers or smartwatches, can collect real-time physiological data, including heart rate, sleep patterns, and activity levels, providing valuable insights into patients' health and well-being. Clinical trials also generate important data by collecting information on treatment efficacy, adverse events, and patient outcomes.

The data collection process involves careful consideration of ethical and legal considerations, particularly in domains like healthcare where patient privacy and data protection are paramount. Collecting and storing data in compliance with privacy regulations, obtaining informed consent, and de-identifying sensitive information are essential steps to protect individuals' privacy and ensure data confidentiality.

Furthermore, data collection protocols need to be designed to capture relevant and representative data. Bias in data collection can lead to skewed insights and biased AI models. For example, in healthcare, if patient data is predominantly collected from a specific population or healthcare facility, the resulting AI models may not generalize well to other populations or healthcare settings. Therefore, it is crucial to ensure diversity and representativeness in the data collected to avoid biases and promote fairness.

In conclusion, collecting data involves identifying relevant sources, designing protocols, and extracting or generating data from various channels. Whether in e-commerce or healthcare, data collection plays a pivotal role in understanding customer behavior, improving patient care, and driving innovation. However, ethical considerations, privacy protection, and addressing biases are essential elements in the data collection process to ensure the quality, representativeness, and fairness of the collected data for AI applications.

B. Data Labeling and Annotation:

To enable supervised learning, data needs to be labeled or annotated with appropriate labels or tags that indicate the desired output or target variable. This process, known as data labeling, is crucial as it provides the necessary information for AI models to learn and make accurate predictions or classifications.

Data labeling can be performed through manual annotation by human experts or through automated techniques, depending on the nature of the data and the available resources. In some cases, human experts are required to review and label the data, especially when subjective judgment or domain expertise is needed. For example, in the field of healthcare, medical professionals may need to review and annotate medical records or diagnostic images with appropriate diagnoses or treatment recommendations.

In other cases, automated techniques can be employed to label the data. Natural Language Processing (NLP) algorithms can be used to automatically extract information from text and assign relevant labels or categories. For example, in sentiment analysis, NLP algorithms can automatically analyze and categorize customer reviews as positive, negative, or neutral based on the sentiment expressed.

Similarly, in computer vision tasks such as image recognition, automated techniques can be used to label images with corresponding object classes. Computer vision algorithms can analyze the visual content of images and identify objects, people, or specific features within the images. This automated image labeling process is crucial for training AI models to recognize and classify objects in real-world scenarios.

The choice between manual and automated data labeling depends on several factors, including the complexity of the task, the size of the dataset, and the availability of resources. While manual labeling ensures high accuracy and precision, it can be time-consuming and costly, especially for large datasets. On the other hand, automated labeling techniques provide scalability and efficiency but may be less accurate or require validation by human experts.

It is worth noting that data labeling is not a one-time process. As new data becomes available or as AI models evolve, additional labeling may be required to improve the model's performance or adapt to new tasks. Continuous monitoring and updating of labeled data are essential for maintaining the accuracy and relevance of AI models.

In conclusion, data labeling is a crucial step in enabling supervised learning. It involves assigning appropriate labels or tags to the data, either through manual annotation by human experts or through automated techniques. Data labeling ensures that AI models can learn from labeled examples and make accurate predictions or classifications. Whether performed manually or automatically, data labeling is an iterative process that plays a vital role in training and refining AI models.

C. Data Augmentation:

Data augmentation techniques play a crucial role in expanding the size and diversity of the available data for training AI models. By applying various transformations or perturbations to existing data, data augmentation creates new samples that effectively enhance the model's learning capabilities and improve its robustness and generalization.

In the field of computer vision, data augmentation techniques are particularly valuable. They enable AI models to learn from a wider range of variations and scenarios, making them more adept at recognizing and classifying objects in real-world settings. Some common data augmentation techniques in computer vision include rotation, scaling, flipping, cropping, and adding noise or distortions.

Rotation augmentation involves rotating images at different angles, such as 90 degrees or 180 degrees. This allows the model to learn object orientations from different perspectives, making it more invariant to rotation invariances. For example, if an AI model is trained to recognize cars, rotating the images in the dataset helps the model learn to identify cars regardless of their orientation.

Scaling augmentation involves resizing images to different scales or aspect ratios. This variation in size helps the model handle objects at different distances or with varying levels of detail. By including scaled versions of the same image in the dataset, the model can learn to recognize objects at different scales and adapt to their appearances under different viewing conditions.

Flipping augmentation involves horizontally or vertically flipping images. This technique is particularly useful when the orientation of an object is not critical for classification. For instance, in image classification tasks where the presence of an object is more important than its specific orientation, flipping the images can effectively double the dataset size and provide additional variations for the model to learn from.

Cropping augmentation involves randomly cropping portions of an image or applying spatial translations. This simulates variations in object placement within the image and helps the model become more robust to different object positions and sizes. By including cropped versions of the same image in the dataset, the model can learn to recognize objects even when they are partially visible or occluded.

In addition to these basic transformations, more advanced data augmentation techniques can be applied. These may involve introducing noise, distortions, or simulating realistic variations specific to the domain. For example, in medical imaging, augmentations can include simulated tumor growth or changes in lighting conditions to capture realistic variations in patient scans.

Data augmentation techniques not only increase the size of the dataset but also introduce diversity and variability, making the AI model more robust and less prone to overfitting. By training on a larger and more diverse dataset, the model becomes better equipped to handle variations and generalize its learning to unseen data.

However, it is important to strike a balance when applying data augmentation. While increasing the diversity of the dataset is beneficial, too much augmentation can introduce unrealistic variations or distortions that may hinder the model's ability to learn meaningful patterns. Careful consideration should be given to the domain-specific characteristics and the desired invariances or variations that need to be captured.

In conclusion, data augmentation techniques are powerful tools for expanding the size and diversity of the available data for training AI models. In computer vision, techniques such as rotation, scaling, flipping, and cropping help improve model robustness and generalization by introducing variations in object orientation, size, and position. By augmenting the dataset with transformed versions of existing data, AI models can learn to handle a wider range of scenarios and improve their performance in real-world applications.

Ethical Considerations in Data Collection and Curation:

A. Privacy and Consent:

Respecting privacy rights and obtaining informed consent are crucial ethical considerations in the collection of data for AI applications. In an era of extensive data collection and utilization, striking the right balance between data utility and privacy protection is a challenge that necessitates careful thought and adherence to data protection regulations.

Respecting privacy rights entails recognizing and upholding individuals' rights to privacy, autonomy, and control over their personal information. It involves treating data subjects as active participants who have the right to decide how their data is collected, used, and shared. Respecting privacy rights also acknowledges the potential risks associated with the misuse or mishandling of personal data, including unauthorized access, identity theft, and discriminatory practices.

Obtaining informed consent is a key aspect of respecting privacy rights. Informed consent requires individuals to have a clear understanding of the purpose and scope of data collection, the intended use of their data, the potential risks and benefits, and their rights regarding their data. This understanding enables individuals to make an informed decision about whether to provide their consent for data collection and use.

Obtaining informed consent is not merely a legal obligation; it is an ethical imperative that fosters transparency, trust, and respect for individuals' autonomy. It empowers individuals to make choices about their data and encourages organizations to be accountable and responsible for their data practices.

Striking a balance between data utility and privacy protection is a multifaceted challenge. Data utility refers to the usefulness and value that data brings in enabling AI systems to learn, make predictions, and generate insights. On the other hand, privacy protection aims to safeguard individuals' personal information and prevent its unauthorized or unintended use.

To strike this balance, organizations must navigate the complex landscape of data protection regulations and industry standards. These regulations, such as the European Union's General Data Protection Regulation (GDPR), provide a framework for handling personal data, emphasizing the principles of fairness, transparency, and accountability.

Adherence to data protection regulations involves implementing technical and organizational measures to safeguard data, ensuring data accuracy and integrity, and adopting privacy-by-design principles. Privacy-by-design emphasizes incorporating privacy considerations into the design and development of AI systems, rather than treating privacy as an afterthought.

Respecting privacy rights and finding the right balance between data utility and privacy protection is particularly challenging in the context of AI, where large volumes of data are often necessary for training sophisticated models. However, organizations must prioritize the implementation of robust privacy safeguards, including data anonymization or pseudonymization techniques, encryption, secure storage, and access controls.

Furthermore, organizations should conduct privacy impact assessments to identify and mitigate potential privacy risks associated with data collection and use. Regular audits and compliance monitoring can ensure ongoing adherence to privacy standards and identify areas for improvement.

In summary, respecting privacy rights and obtaining informed consent are essential ethical principles in data collection for AI. Striking a balance between data utility and privacy protection requires careful consideration, robust privacy safeguards, and adherence to data protection regulations. By embracing privacy-by-design principles, organizations can foster trust, transparency, and accountability in their data practices, ensuring that individuals' privacy rights are respected while benefiting from the utility of data-driven AI applications.

B. Data Security:

Data collection and storage are critical stages in the data lifecycle, and implementing robust security measures is paramount to protect sensitive information from unauthorized access or breaches. In today's digital landscape, where data breaches and cyber threats are prevalent, organizations must prioritize data security to safeguard the confidentiality, integrity, and availability of data.

One of the fundamental security measures is encryption, which involves converting data into an unreadable format using cryptographic algorithms. Encryption ensures that even if unauthorized individuals gain access to the data, they cannot decipher its contents without the corresponding decryption key. By employing encryption techniques, organizations can protect sensitive data both during transit and at rest, mitigating the risk of data exposure.

Access controls play a crucial role in data security by regulating and restricting access to data based on the principle of least privilege. Implementing strong authentication mechanisms, such as multi-factor authentication, helps ensure that only authorized individuals can access sensitive data. Role-based access controls (RBAC) and fine-grained access controls enable organizations to define and enforce access permissions at various levels, preventing unauthorized users from accessing or modifying sensitive data.

Regular security audits are essential to evaluate the effectiveness of data security practices and identify potential vulnerabilities or weaknesses in the system. These audits involve conducting comprehensive assessments of the infrastructure, applications, and security controls to detect and address any vulnerabilities promptly. By conducting regular security audits, organizations can proactively identify and mitigate security risks, minimizing the likelihood of data breaches or unauthorized access.

Data storage itself requires careful consideration to ensure security. Organizations should employ secure storage methods and technologies that protect data from physical damage, theft, or loss. This may include implementing redundant storage systems, backups, and disaster recovery plans to ensure the availability and integrity of data even in the event of hardware failures or natural disasters.

Additionally, organizations should adhere to industry best practices for data security, such as following the ISO 27001 standard or other relevant frameworks. These standards provide guidelines and controls for establishing a robust information security management system, covering areas such as data classification, risk assessment, incident response, and ongoing security monitoring.

Furthermore, organizations should stay up to date with the evolving threat landscape and implement timely security patches and updates for their data storage and management systems. Regular security awareness training for employees is also crucial to promote a culture of security and ensure that individuals understand their roles and responsibilities in protecting sensitive data.

In summary, data collection and storage must adhere to robust security measures to protect sensitive information from unauthorized access or breaches. Encryption, access controls, regular security audits, and secure storage methods are vital components of data security practices. By implementing these measures and staying vigilant against emerging threats, organizations can enhance data protection and maintain the trust of their customers and stakeholders.

C. Transparency and Explainability:

As AI systems play an increasingly influential role in decision-making processes, it becomes imperative to ensure transparency and explainability in data collection and curation processes. Transparency refers to the openness and clarity in communicating to individuals how their data is collected, used, and managed, while explainability pertains to the ability to provide understandable justifications and reasoning for the outcomes or decisions made by AI systems.

Transparency is crucial to building trust between organizations and individuals whose data is being collected. By clearly communicating data collection practices, organizations can provide individuals with a comprehensive understanding of what data is being collected, how it is being collected, and the purposes for which it will be used. Transparency also encompasses informing individuals about their rights and options, such as the ability to opt out of certain data collection activities or to request the deletion of their data.

Transparency in data collection helps individuals make informed choices about sharing their personal information. For example, in e-commerce, organizations can provide clear and easily accessible privacy policies that outline the types of data collected during online transactions and how that data is used for marketing or personalization purposes. By providing this information upfront, organizations enable individuals to make informed decisions about whether to proceed with a transaction and share their data.

Explainability is particularly important in AI systems that make automated decisions, such as loan approvals or job candidate screening. When individuals are affected by these decisions, they have the right to understand how the decision was made and the factors considered. Organizations should strive to provide explanations that are clear, understandable, and free from technical jargon. This can be achieved by designing AI systems that provide reasoning for their decisions, highlighting the key features or data points that influenced the outcome.

In addition to transparency and explainability, organizations should also consider ethical considerations when collecting and curating data. This includes ensuring that data collection practices adhere to ethical guidelines and regulations, such as obtaining informed consent, anonymizing or de-identifying data when necessary, and implementing strict security measures to protect the privacy and confidentiality of individuals.

Transparency and explainability can be further enhanced by involving individuals in the data collection process through participatory approaches. For instance, organizations can seek input from individuals on data collection practices, allow them to control certain aspects of their data sharing, or provide opportunities for feedback and redress if concerns or issues arise.

By prioritizing transparency and explainability in data collection and curation, organizations can foster trust, accountability, and responsible AI practices. Open communication, clear policies, and understandable explanations empower individuals to make informed choices about their data and hold organizations accountable for their data handling practices. Ultimately, this contributes to a more transparent, ethical, and trustworthy AI ecosystem.

Conclusion:

Data collection and curation form the bedrock of AI development, providing the raw material that enables intelligent systems to learn, reason, and make informed decisions. The challenges of data volume, variety, quality, and bias necessitate careful planning, robust strategies, and ethical considerations in the data collection and curation processes. By addressing these challenges and upholding ethical principles, we can harness the power of data to drive AI innovation and realize its full potential for the betterment of society.

# Big Data: Fueling AI Development

In the realm of artificial intelligence (AI), data plays a pivotal role in shaping the development and capabilities of AI systems. The advent of big data has revolutionized the AI landscape, enabling the creation of sophisticated models capable of processing and analyzing vast amounts of information. This section delves into the significance of big data in fueling AI development, highlighting its impact across various industries and exploring the challenges and opportunities it presents.

**Defining big data and its significance in AI**

In the era of technological advancement, the term "big data" has gained immense prominence, revolutionizing the way we approach data-driven processes and decision-making. This section aims to provide a comprehensive understanding of big data, its defining characteristics, and its profound significance in the field of artificial intelligence (AI). By delving into its multifaceted nature, we can appreciate the transformative potential it holds across various domains.

I. Defining Big Data:

A. Characteristics of Big Data:

Volume: The vast scale of data generated and collected in the digital age has reached unprecedented levels. With the proliferation of digital devices, social media platforms, online transactions, and IoT devices, data is being produced at an exponential rate. The sheer volume of data is staggering, often measured in terabytes, petabytes, or even exabytes. This abundance of data presents both opportunities and challenges for organizations, as harnessing and extracting value from such a massive amount of information requires advanced technologies and efficient processing mechanisms.

Velocity: In today's interconnected world, data is generated and processed at an astonishing speed. The velocity of data refers to the rate at which it is produced, captured, and shared in real-time. Streaming data from sources like social media feeds, sensor networks, financial transactions, and web logs flows continuously and requires rapid processing to derive meaningful insights. Real-time analysis is crucial in many domains, such as monitoring stock market trends, detecting anomalies in network traffic, or providing instant recommendations to online shoppers. The ability to handle the velocity of data is vital for organizations to stay competitive and responsive in a fast-paced environment.

Variety: Data comes in various forms and formats, ranging from structured to unstructured and semi-structured data. Structured data refers to information that is organized and fits neatly into predefined categories, such as databases and spreadsheets. Unstructured data, on the other hand, lacks a predefined structure and includes text documents, images, videos, social media posts, emails, and more. Semi-structured data lies in between, combining structured and unstructured elements, often represented in XML or JSON formats. The variety of data sources poses a challenge for organizations as they need to integrate and process diverse data types to gain a holistic understanding of their operations. Handling the variety of data requires sophisticated data management techniques and tools that can extract insights from different data formats effectively.

Veracity: The veracity of data refers to its accuracy, reliability, and trustworthiness. With the increasing volume and variety of data, ensuring data quality becomes a critical concern. Data may contain errors, inconsistencies, or biases that can lead to unreliable results and flawed decision-making. The veracity of data is especially crucial in AI, as algorithms heavily rely on accurate and trustworthy information to generate meaningful insights and make accurate predictions. Organizations need to implement robust data validation processes, data cleansing techniques, and quality control mechanisms to ensure the veracity of the data they collect and analyze.

The Three Vs of Big Data:

Volume: The exponential growth of data has outpaced the processing capabilities of traditional systems. Traditional databases and data management tools struggle to handle the sheer volume of data being generated. The advent of big data technologies such as distributed computing frameworks, parallel processing, and cloud storage has enabled organizations to scale their infrastructure and process large volumes of data efficiently. By leveraging these technologies, organizations can store, process, and analyze massive datasets that were previously unmanageable.

Velocity: The high-speed generation and continuous flow of data require real-time analysis capabilities. Traditional batch processing methods are not sufficient for handling the velocity of data. Real-time streaming and processing frameworks, such as Apache Kafka and Apache Storm, enable organizations to ingest and analyze data in motion, allowing them to respond quickly to changing conditions and make timely decisions. The velocity of data necessitates agile and responsive systems that can process and analyze data in near real-time.

Variety: The heterogeneity of data sources poses a significant challenge for organizations. Data comes from a wide range of sources, including social media platforms, websites, mobile devices, sensors, and more. Each source may produce data in different formats and structures. Handling the variety of data requires flexible processing techniques, such as schema-less databases, data integration tools, and advanced analytics algorithms. Organizations need to adopt technologies and strategies that can handle diverse data sources and effectively integrate and analyze the data to derive valuable insights.

In summary, the characteristics and three Vs of big data highlight the immense scale, velocity, variety, and veracity of data in today's digital age. Understanding and effectively managing these aspects are crucial for organizations seeking to unlock the potential of big data and harness its transformative power in AI applications across various domains. By leveraging advanced technologies, data management techniques, and analytical tools, organizations can extract meaningful insights from big data, make data-driven decisions, and gain a competitive edge in today's data-driven landscape.

II. The Significance of Big Data in AI:
A. Enabling Data-Driven AI:

The field of AI relies heavily on data to drive intelligent decision-making. Big data serves as the fuel that powers AI algorithms and models, enabling them to learn, adapt, and make accurate predictions. The abundance of data in today's digital landscape provides an unprecedented opportunity to leverage its potential for AI applications across various domains.

Enhanced Model Performance:
One of the key advantages of big data in AI is its ability to improve model performance. By utilizing large-scale datasets, AI models can be trained with a greater amount of diverse and representative data, leading to enhanced accuracy and

reliability. The sheer volume of data allows AI algorithms to uncover hidden patterns, correlations, and insights that may not be apparent in smaller datasets. This enables organizations to make more informed decisions and gain a deeper understanding of complex phenomena.

Uncovering Hidden Patterns and Insights:
Big data encompasses diverse data sources, including structured, unstructured, and semi-structured data. By combining and analyzing data from multiple sources, AI algorithms can uncover hidden patterns and insights that may not be apparent when analyzing individual datasets in isolation. For example, in the field of healthcare, integrating data from electronic health records, genomic data, and wearable devices can provide valuable insights into disease prevention, personalized treatment plans, and population health management.

Deep Learning and Neural Networks:
Big data plays a pivotal role in training deep learning models, which are a subset of AI algorithms that simulate the neural networks of the human brain. Deep learning models require a massive amount of data to train their multiple layers of interconnected nodes. The availability of big data has enabled significant breakthroughs in complex tasks such as image recognition, natural language processing, and speech recognition. For instance, deep learning models have revolutionized image recognition technology, enabling accurate identification of objects, facial recognition, and autonomous driving systems.

Personalization and Customization:
The abundance of data allows AI systems to personalize and customize their interactions based on individual preferences and needs. By analyzing extensive user data, such as browsing history, purchase behavior, and demographic information, AI algorithms can tailor recommendations, content, and services to each user. This level of personalization enhances customer experiences, particularly in e-commerce, entertainment, and recommendation systems. For example, e-commerce platforms utilize big data to create personalized product recommendations, improving customer satisfaction and driving sales.

In summary, big data serves as the fuel that powers AI algorithms and models. It enhances model performance by providing large-scale, diverse, and representative datasets. Big data enables AI algorithms to uncover hidden patterns and insights from multiple data sources, leading to more informed decision-making. It plays a crucial role in training deep learning models, facilitating breakthroughs in complex tasks. Additionally, big data enables personalization and customization, enhancing customer experiences in various domains. By harnessing the power of big data, organizations can unlock the full potential of AI and drive innovation in a wide range of applications.

III. Case Studies: Big Data in Action:
A. E-commerce:

In the realm of e-commerce and retail, big data plays a crucial role in delivering personalized product recommendations and implementing targeted marketing strategies. By analyzing vast amounts of customer data, including purchase history, browsing behavior, and demographic information, businesses can gain valuable insights into individual preferences and tailor their offerings accordingly.

With the help of big data analytics, companies can identify patterns and correlations in customer behavior, enabling them to make accurate predictions about future purchasing decisions. By leveraging this information, AI algorithms can generate personalized recommendations, suggesting products or services that are likely to resonate with individual customers. This approach not only enhances the customer experience by offering relevant and engaging suggestions but also increases the likelihood of conversion and customer satisfaction.

Analyzing Customer Behavior to Optimize Pricing Strategies and Improve Customer Satisfaction

Big data analytics enables businesses to analyze customer behavior and make informed decisions regarding pricing strategies. By examining data on purchasing patterns, price sensitivity, and competitor pricing, companies can optimize their pricing models to maximize revenue and customer satisfaction.

Through big data analysis, businesses can identify price thresholds, understanding at what price point customers are most likely to make a purchase. This knowledge allows companies to implement dynamic pricing strategies that adjust prices in real-time based on market conditions, customer demand, and other relevant factors. By optimizing prices, businesses can strike the right balance between profitability and customer value, ensuring competitive pricing while maintaining customer loyalty.

Healthcare: Mining Electronic Health Records and Sensor Data

The healthcare industry generates an immense amount of data through electronic health records (EHRs), wearable devices, and other sensor data. By harnessing big data analytics, healthcare providers can extract valuable insights from these vast datasets to improve disease diagnosis, treatment, and patient care.

Big data analytics allows healthcare professionals to analyze EHRs and detect patterns that can aid in disease diagnosis and personalized treatment plans. By aggregating and analyzing patient data, AI algorithms can identify correlations between symptoms, genetic markers, and treatment outcomes, leading to more accurate diagnoses and tailored treatment options.

Moreover, big data analytics facilitates predictive analytics in healthcare, enabling early disease detection and proactive intervention. By analyzing patient data over time, AI algorithms can identify patterns that may indicate the early stages of a disease or potential health risks. This early detection allows for timely interventions, improving patient outcomes and overall public health management.

Banking and Finance: Fraud Detection and Risk Assessment

In the banking and finance industry, big data plays a critical role in fraud detection and risk assessment. The extensive volume of financial transaction data, coupled with advanced analytics, allows organizations to detect fraudulent activities in real-time and prevent financial losses.

Big data analytics can identify suspicious patterns and anomalies in financial transactions, such as unusual spending patterns, unauthorized access, or fraudulent activities. By analyzing historical transaction data, AI algorithms can learn and detect patterns associated with fraudulent behavior, enabling automated fraud detection systems.

Additionally, big data analytics supports risk assessment in the banking and finance sector. By analyzing vast amounts of financial and market data, AI algorithms can identify potential risks and predict market trends. This enables financial institutions to make data-driven decisions, devise investment strategies, and mitigate risks associated with market volatility.

In conclusion, big data analytics offers numerous benefits across various industries. In e-commerce, it enables personalized product recommendations and targeted marketing, enhancing customer experiences. In healthcare, big data facilitates disease diagnosis, personalized treatment, and predictive analytics for early disease detection. In the banking and finance sector, big data analytics supports fraud detection, risk assessment, and real-time market analysis. By leveraging big data insights, organizations can make informed decisions, drive innovation, and improve overall performance in their respective fields.

The three V's of big data: volume, velocity, and variety
How big data enables the training of sophisticated AI models

The field of artificial intelligence (AI) has witnessed remarkable advancements in recent years, thanks in large part to the availability of vast amounts of data. Big data, characterized by its volume, velocity, variety, and veracity, has become the fuel that drives the training of sophisticated AI models. In this section, we will delve into the ways in which big data enables the training of AI models, unlocking their potential to make accurate predictions, gain deep insights, and exhibit intelligent behavior across various domains.

I. The Power of Big Data in AI Training:

A. Enhanced Model Performance:

One of the primary benefits of big data in AI training is the ability to improve model performance. Traditional machine learning algorithms rely on relatively smaller datasets, limiting their capacity to capture the complexity and variability of real-world scenarios. However, with big data, AI models can learn from a diverse range of examples, enabling them to generalize better and make more accurate predictions.

Consider the field of image recognition in computer vision. Big data allows AI models to train on vast image datasets, containing millions of images from different categories. As the models process and learn from this data, they become more proficient at recognizing objects, distinguishing between different classes, and even understanding context. The availability of big data enhances the model's ability to detect subtle patterns, improving its overall accuracy and robustness.

B. Uncovering Hidden Patterns and Insights:

Another key advantage of big data in AI training is the ability to uncover hidden patterns and insights that may not be apparent in smaller datasets. By training on diverse and comprehensive datasets, AI models can reveal underlying trends and correlations, leading to valuable discoveries and actionable insights.

In the field of e-commerce, for example, big data enables AI models to analyze customer behavior, purchase histories, and browsing patterns from millions of users. By mining this data, AI algorithms can identify patterns such as preferences, trends, and purchasing habits. This knowledge can then be leveraged to optimize product recommendations, personalize marketing strategies, and improve customer satisfaction.

C. Deep Learning and Neural Networks:

Big data plays a critical role in the training of deep learning models, which are characterized by their multiple layers of interconnected neurons. Deep learning models have demonstrated exceptional performance in complex tasks such as natural language processing, speech recognition, and image classification.

The effectiveness of deep learning relies heavily on the availability of large-scale datasets for training. By exposing deep neural networks to massive amounts of data, they can learn intricate representations and hierarchical structures, allowing them to capture and understand complex relationships within the data. Big data facilitates the training of deep learning models, enabling them to achieve breakthroughs in various domains.

For instance, in natural language processing, big data allows deep learning models to train on extensive text corpora, such as books, articles, and social media posts. This enables the models to develop a nuanced understanding of language, including syntax, semantics, and context. The result is the ability to generate coherent and contextually appropriate responses in chatbots or analyze sentiment in customer reviews.

D. Personalization and Customization:

Big data enables AI models to deliver personalized and customized experiences by leveraging extensive user data. In fields such as e-commerce, social media, and streaming platforms, user preferences, browsing history, and interaction patterns are collected and analyzed to tailor recommendations and content.

By training AI models on large-scale user datasets, personalized recommendations can be generated with high precision. For example, a streaming platform like Netflix utilizes big data to analyze the viewing habits of millions of users. Based on this data, AI algorithms can suggest personalized content, taking into account individual preferences, viewing history, and even contextual factors such as time of day or mood.

Conclusion:

Big data plays a transformative role in AI training, enabling the development of sophisticated models capable of accurate predictions, deep insights, and personalized experiences. By harnessing the power of big data, AI models can enhance their performance, uncover hidden patterns, leverage deep learning techniques, and deliver tailored solutions across various fields. As the volume of data continues to grow exponentially, the potential for training even more advanced AI models becomes increasingly promising, propelling the boundaries of what is possible in the field of AI.

## Case studies showcasing the impact of big data on AI advancement

The marriage of big data and artificial intelligence (AI) has revolutionized various industries, enabling significant advancements and transformative outcomes. In this section, we will explore several case studies that exemplify the profound impact of big data on AI advancement. These examples span diverse fields such as e-commerce, healthcare, banking, and finance, illustrating how the utilization of big data has driven remarkable achievements and opened new frontiers in AI research and application.

I. E-commerce: Amazon's Personalized Shopping Experience

One notable case study in the realm of e-commerce is Amazon, the global retail giant. Amazon harnesses the power of big data to offer personalized shopping experiences to its customers. By analyzing vast amounts of customer data, including browsing history, purchase patterns, and product ratings, Amazon's AI algorithms generate tailored product recommendations, enabling users to discover relevant items and make informed purchasing decisions.

The integration of big data in Amazon's AI algorithms allows them to identify subtle correlations and preferences, leading to more accurate and relevant recommendations. This personalized approach enhances the customer experience, increases engagement, and contributes to Amazon's success as a leading e-commerce platform.

II. Healthcare: IBM Watson's Medical Insights

IBM Watson's AI system has made significant strides in the field of healthcare by leveraging big data. With access to extensive medical records, clinical research, and scientific literature, IBM Watson uses big data to train its AI models to analyze patient data and generate valuable insights for healthcare professionals.

In one case study, IBM Watson collaborated with Memorial Sloan Kettering Cancer Center to assist oncologists in developing personalized treatment plans for cancer patients. By analyzing a vast amount of patient data, including medical histories, genomic information, and treatment outcomes, IBM Watson's AI algorithms provide evidence-based recommendations to support clinicians in making informed decisions regarding treatment options and strategies.

The utilization of big data in healthcare AI systems like IBM Watson holds the promise of enhancing diagnostic accuracy, improving treatment outcomes, and advancing personalized medicine.

III. Banking and Finance: Fraud Detection and Risk Assessment

The banking and finance industry heavily relies on AI-powered systems to combat fraud and assess risk. Big data plays a pivotal role in these endeavors by providing vast amounts of transactional data, historical patterns, and market trends.

For instance, PayPal, a leading digital payments company, employs big data analytics to detect fraudulent activities in real-time. By analyzing a vast volume of transactional data, including customer behavior, spending patterns, and device information, PayPal's AI algorithms can identify suspicious transactions and flag potential fraud attempts, protecting both customers and the company's financial interests.

Additionally, big data is instrumental in risk assessment and investment strategies. Financial institutions leverage AI models trained on extensive historical and real-time market data to make informed decisions regarding portfolio management, asset allocation, and investment recommendations. The integration of big data in AI algorithms allows for sophisticated risk analysis, enabling financial institutions to optimize returns and mitigate potential losses.

Conclusion:

The case studies presented in this section demonstrate the transformative power of big data on AI advancement in various fields. From personalized shopping experiences in e-commerce to improved healthcare decision-making and enhanced fraud detection in banking and finance, big data has emerged as the cornerstone of AI development.

By leveraging vast and diverse datasets, AI algorithms can extract valuable insights, generate accurate predictions, and optimize decision-making processes. The integration of big data in AI not only improves performance but also enables personalized experiences, efficient resource allocation, and the development of innovative solutions.

As the volume of data continues to grow exponentially, the potential for AI advancement fueled by big data becomes even more promising. However, it is crucial to address challenges related to data privacy, security, and ethical considerations to ensure responsible and beneficial utilization of big data in AI applications.

# Data Preprocessing and Feature Engineering

Data preprocessing and feature engineering play a crucial role in the success of AI algorithms. Before feeding data into AI models, it is essential to prepare and transform the data to enhance its quality, extract meaningful information, and reduce noise. This section delves into the techniques and considerations involved in data preprocessing and feature engineering, highlighting their significance in optimizing AI algorithms.

I. Data Preprocessing: Cleaning and Transforming Data

Data preprocessing involves cleaning and transforming raw data to improve its quality and usability for AI algorithms. Several techniques are employed during this process:

A. Data Cleaning: Data cleaning aims to address data quality issues such as missing values, outliers, and inconsistencies. Missing values can be imputed or eliminated based on the context and available information. Outliers, which can distort the training process, can be detected and handled through statistical methods or domain knowledge. Inconsistencies in the data can be resolved by applying logical checks or data fusion techniques.

B. Data Transformation: Data transformation involves converting data into a suitable format for analysis. Common techniques include normalization, which scales data to a common range, and standardization, which centers data around its mean and scales it by its standard deviation. Other transformations, such as logarithmic or exponential transformations, can be applied to handle data distributions that are skewed or exhibit non-linear relationships.

II. Feature Engineering: Extracting Meaningful Information from Data

Feature engineering focuses on extracting relevant and informative features from the available data. The goal is to represent the data in a way that captures the underlying patterns and relationships. Various techniques are employed for feature engineering:

A. Feature Extraction: Feature extraction involves deriving new features from existing ones. This can be achieved through mathematical operations, such as calculating ratios or differences between variables, or by aggregating data using statistical measures like mean, median, or variance. Feature extraction helps reduce dimensionality and captures essential information.

B. Feature Creation: Feature creation involves generating new features based on domain knowledge or heuristics. For example, in e-commerce, features such as "total purchase amount" or "average rating of purchased items" can be created to capture customer behavior and preferences. These engineered features provide additional insights and enhance the performance of AI models.

III. Feature Selection and Dimensionality Reduction

Feature selection and dimensionality reduction techniques aim to identify the most relevant and informative features while reducing the dimensionality of the data. This is important to mitigate the "curse of dimensionality" and improve model performance:

A. Feature Selection: Feature selection involves identifying a subset of features that have the most predictive power. It helps eliminate redundant or irrelevant features, reducing computational complexity and potential overfitting. Techniques such as correlation analysis, forward/backward selection, or regularization methods like L1 and L2 regularization are commonly used for feature selection.

B. Dimensionality Reduction: Dimensionality reduction techniques aim to reduce the number of features while preserving the most important information. Principal Component Analysis (PCA) is a popular method that transforms high-dimensional data into a lower-dimensional space while retaining the maximum variance. Other techniques, such as t-SNE (t-Distributed Stochastic Neighbor Embedding) or LDA (Linear Discriminant Analysis), can also be employed for dimensionality reduction.

Conclusion:

Data preprocessing and feature engineering are critical steps in preparing data for AI algorithms. By cleaning and transforming raw data, we enhance its quality and usability. Feature engineering extracts meaningful information and captures relevant patterns, improving the performance of AI models. Additionally, feature selection and dimensionality reduction techniques optimize computational efficiency and prevent overfitting.

The examples and techniques explored in this section demonstrate the significance of data preprocessing and feature engineering in maximizing the potential of AI algorithms. By carefully preparing and engineering data, we can uncover hidden insights, enhance predictive accuracy, and unlock the full potential of

AI systems across various domains, including e-commerce, healthcare, banking, and finance.

## The Power of Labeled Data: Supervised Learning

Supervised learning, a key approach in the field of AI, relies heavily on labeled data to train models and make accurate predictions. In this section, we will explore the significance of labeled data, its role in training AI models, and provide examples of its applications in various fields. Supervised learning enables AI systems to learn from labeled examples and generalize their knowledge to make predictions on unseen data.

A. The Importance of Labeled Data in Supervised Learning:

Labeled data is essential in supervised learning as it provides the ground truth or target values that AI models aim to predict. By associating input data with corresponding labels, we create a training dataset that serves as a basis for model learning. The availability of high-quality labeled data significantly impacts the performance and effectiveness of supervised learning algorithms.

Labeled data enables the learning process by establishing the connection between input features and their corresponding output labels. This allows the model to understand the underlying patterns and relationships in the data, facilitating accurate predictions on unseen examples. Without labeled data, the learning process would lack the necessary guidance and benchmarks to measure the model's performance.

B. Training AI Models with Labeled Data for Accurate Predictions:

Supervised learning leverages labeled data to train AI models and optimize their predictive capabilities. The training process involves presenting the model with labeled examples, which it uses to learn the mapping between input features and output labels. The model adjusts its internal parameters through an iterative process, aiming to minimize the difference between its predicted outputs and the ground truth labels.

The availability of a diverse and representative labeled dataset is crucial for training AI models. A well-annotated dataset covers a wide range of input scenarios and enables the model to generalize its learning to unseen examples. The more accurately labeled examples the model is exposed to, the better it can capture the underlying patterns and make accurate predictions.

### C. Examples of Supervised Learning Applications in Various Fields:

Supervised learning finds extensive applications across diverse fields, enabling AI systems to perform a wide range of tasks. Here are some examples of supervised learning applications:

**E-commerce:**
In e-commerce, supervised learning is utilized for personalized product recommendations. By training models on labeled data that captures customer preferences and behavior, e-commerce platforms can suggest relevant products to individual users, enhancing their shopping experience and driving sales.

**Healthcare:**
Supervised learning plays a critical role in healthcare applications, such as disease diagnosis and personalized treatment. Models trained on labeled medical data, including electronic health records and medical imaging, can assist in accurate diagnosis, predict disease progression, and recommend personalized treatment plans.

**Banking and Finance:**
In the banking and finance sector, supervised learning is employed for fraud detection and risk assessment. By training models on labeled data that represents fraudulent and non-fraudulent transactions, financial institutions can identify suspicious activities and mitigate risks.

**Natural Language Processing (NLP):**
Supervised learning is widely used in NLP tasks, such as sentiment analysis and text classification. Models trained on labeled text data can understand the sentiment behind customer reviews or classify documents into predefined categories, enabling applications like automated customer support and content filtering.

### Conclusion:

Labeled data is a crucial asset in supervised learning, enabling AI models to learn from examples and make accurate predictions. The availability of high-quality labeled data significantly impacts the performance of AI algorithms and their ability to generalize from training to unseen data. Supervised learning finds applications in various fields, including e-commerce, healthcare, banking, and natural language processing, revolutionizing industries and enhancing user experiences.

By understanding the power of labeled data and harnessing its potential, we can unlock the full capabilities of supervised learning algorithms, paving the way for advancements in AI across multiple domains.

# Unleashing Unlabeled Data: Unsupervised Learning

While labeled data plays a crucial role in training AI models through supervised learning, there exists an abundance of unlabeled data that holds untapped potential. Unsupervised learning, a branch of AI, aims to uncover patterns, structures, and hidden insights from unlabeled data. In this section, we will explore the significance of unsupervised learning, its ability to find meaningful information in unlabeled data, and provide examples of its applications across various domains.

A. Exploring the Potential of Unsupervised Learning:

Unsupervised learning is a paradigm that enables AI systems to learn from unlabeled data without explicit guidance from predefined labels. Unlike supervised learning, where the focus is on mapping input features to labeled output labels, unsupervised learning seeks to uncover intrinsic structures and relationships within the data itself. This allows for the discovery of hidden patterns and insights that may not be apparent in the labeled data.

Unsupervised learning offers great potential in tackling real-world challenges where labeled data is scarce, expensive to obtain, or simply unavailable. By relying solely on the inherent characteristics of the data, unsupervised learning can uncover valuable information and drive advancements in AI.

B. Finding Patterns and Structures in Unlabeled Data:

Unsupervised learning algorithms excel at discovering patterns, structures, and clusters in unlabeled data. These algorithms employ various techniques to analyze the data and identify meaningful representations. Clustering algorithms, for example, group similar data points together based on their intrinsic similarities. Dimensionality reduction techniques aim to capture the essential features of the data by reducing its complexity.

Through unsupervised learning, AI models can uncover underlying structures in the data that can then be leveraged for a variety of purposes. These structures can reveal insights about customer segmentation, market trends, or even genetic patterns in healthcare, among other applications.

### C. Applications of Unsupervised Learning in Different Domains:

Unsupervised learning finds applications across a wide range of domains, harnessing the power of unlabeled data to drive advancements in AI. Here are some examples:

E-commerce:
Unsupervised learning techniques can be applied to analyze customer behavior and preferences based on their browsing history and purchase patterns. By grouping customers into distinct segments, businesses can tailor their marketing strategies, recommend products, and enhance the overall customer experience.

Healthcare:
In healthcare, unsupervised learning algorithms can analyze large volumes of patient data to identify subgroups with similar characteristics or disease patterns. This can aid in disease subtyping, treatment personalization, and the discovery of novel biomarkers.

Anomaly Detection:
Unsupervised learning is particularly useful in anomaly detection, where the goal is to identify unusual patterns or outliers in data. This can be applied in various domains, such as fraud detection in banking, network intrusion detection in cybersecurity, or equipment failure prediction in manufacturing.

Natural Language Processing (NLP):
Unsupervised learning techniques are employed in NLP to discover latent structures in text data, such as topic modeling and word embeddings. These approaches enable applications like document clustering, sentiment analysis, and text summarization.

Conclusion:

Unsupervised learning provides a powerful framework for extracting valuable insights from unlabeled data. By leveraging the inherent structure and patterns within the data, unsupervised learning algorithms can reveal hidden information and drive advancements across various domains. From e-commerce and healthcare to anomaly detection and NLP, unsupervised learning plays a crucial role in unleashing the potential of unlabeled data and expanding the frontiers of AI.

As AI continues to evolve, the utilization of both labeled and unlabeled data becomes essential for maximizing the capabilities of intelligent systems. Unsupervised learning opens up new avenues for exploration, enabling AI to uncover hidden knowledge and make discoveries that may have otherwise remained obscured. By embracing unsupervised learning, we can unlock the full potential of unlabeled data and pave the way for transformative advancements in AI across diverse industries and applications.

# Ethical Considerations in Data Usage

As artificial intelligence (AI) continues to advance, the ethical implications of data collection and usage become increasingly significant. The responsible and ethical utilization of data is crucial to ensure that AI systems uphold principles such as fairness, transparency, privacy, and accountability. In this section, we will explore the ethical considerations associated with data usage in AI, including privacy concerns, bias and fairness, and the need for transparency and accountability.

A. The Ethical Implications of Data Collection and Usage in AI:

Data collection and usage in AI raise ethical concerns due to the potential impact on individuals and society. The wide-ranging availability of data and the immense power of AI algorithms necessitate thoughtful consideration of ethical principles. It is essential to respect individual privacy, prevent unauthorized data access, and consider the potential consequences of data misuse.

B. Privacy Concerns and Data Protection:

Privacy is a fundamental right that must be safeguarded in the era of AI-driven data analytics. The extensive collection of personal data raises concerns about surveillance, data breaches, and unauthorized profiling. Protecting individual privacy requires robust data protection measures, including data encryption, access controls, and adherence to privacy regulations such as the General Data Protection Regulation (GDPR) in the European Union.

Furthermore, organizations must obtain informed consent from individuals before collecting and using their data. Transparency in data collection practices and clear communication about the purpose and scope of data usage are essential to build trust and empower individuals to make informed choices about their data.

C. Addressing Bias and Fairness in AI Algorithms:

AI algorithms are only as unbiased and fair as the data they are trained on. Biased data can perpetuate societal inequalities, reinforce stereotypes, and lead to discriminatory outcomes. Addressing bias and ensuring fairness in AI algorithms require meticulous data curation, diverse representation in training data, and ongoing evaluation of algorithmic outputs.

Organizations must be vigilant in identifying and mitigating bias, both explicit and implicit, throughout the AI development process. Techniques such as bias detection, fairness assessment, and regular audits are necessary to promote fairness and mitigate the potential harm caused by biased algorithms.

D. Ensuring Transparency and Accountability in Data-Driven AI Systems:

Transparency and accountability are vital in data-driven AI systems to promote trust and enable meaningful human oversight. AI algorithms should be explainable, providing clear rationales for their decisions and actions. This is particularly important in critical domains such as healthcare and finance, where the impact of AI decisions can have profound consequences on individuals' lives.

Organizations must implement mechanisms to ensure accountability for AI systems, including clear lines of responsibility, robust governance frameworks, and avenues for redress in case of algorithmic errors or unfair outcomes. Transparent documentation of data collection and usage practices, along with regular audits, can enhance accountability and instill confidence in AI systems.

Conclusion:

The ethical considerations surrounding data usage in AI are of paramount importance. Respecting privacy rights, addressing bias and fairness, and ensuring transparency and accountability are imperative for responsible AI development and deployment. By adhering to ethical principles, organizations can harness the power of data-driven AI while upholding societal values and safeguarding individual rights.

In the ever-evolving landscape of AI, ongoing discussions, collaborations, and interdisciplinary approaches are necessary to navigate the complex ethical challenges associated with data usage. By considering the broader societal implications and engaging in ethical decision-making, we can create a future where AI technologies benefit humanity while upholding the values we hold dear.

# The Future of Data in AI

As artificial intelligence (AI) continues to evolve and shape various industries, the future of data in AI holds immense promise and potential. In this section, we will explore emerging trends, challenges, and the role of data in shaping the future of AI. Additionally, we will discuss the importance of data sharing and collaboration for AI innovation and the need for regulatory frameworks to govern data usage in AI.

### Emerging Trends and Challenges in Data-driven AI:

The future of data in AI is marked by several emerging trends and challenges. One significant trend is the increasing volume and variety of data generated by diverse sources, including Internet of Things (IoT) devices, social media platforms, and sensor networks. This abundance of data presents both opportunities and challenges in terms of storage, processing, and analysis.

Furthermore, as AI models become more complex and require larger datasets, the challenge of data scalability and efficient utilization becomes critical. Data labeling and annotation also pose challenges, as labeling large amounts of data manually can be time-consuming and costly. Hence, automated techniques for labeling and leveraging semi-supervised learning methods become essential for harnessing the full potential of data in AI.

Another emerging trend is the integration of data from multiple modalities, such as text, images, and videos. The fusion of different data types enables AI systems to learn from rich and diverse sources, leading to more comprehensive understanding and improved decision-making.

However, these trends also bring challenges related to data quality, privacy, and security. Ensuring the accuracy, reliability, and representativeness of data remain critical concerns. Protecting individual privacy while harnessing the power of data is an ongoing challenge that requires careful balancing of privacy rights and AI advancements.

### Data Sharing and Collaboration for AI Innovation:

Data sharing and collaboration are crucial for advancing AI innovation. In an interconnected world, where data is generated by multiple stakeholders, sharing data across organizations and domains can unlock new possibilities for AI applications. Collaborative efforts enable the pooling of resources, expertise, and data, leading to more robust and accurate AI models.

For example, in the field of healthcare, sharing anonymized patient data across research institutions can lead to advancements in disease diagnosis, treatment, and public health management. Similarly, in finance, collaborative data sharing allows for the identification of patterns and trends that can enhance fraud detection and risk assessment.

However, data sharing also raises concerns about intellectual property, privacy, and data ownership. Organizations must navigate these challenges by establishing clear protocols, ensuring proper anonymization and consent mechanisms, and adhering to data governance frameworks that protect the rights and interests of all stakeholders involved.

### Regulatory Frameworks and Policies to Govern Data Usage in AI:

The future of data in AI necessitates robust regulatory frameworks and policies to govern data usage. As AI systems become more pervasive and influential, it is imperative to establish guidelines that promote ethical data practices, fairness, and accountability.

Regulations such as the General Data Protection Regulation (GDPR) in the European Union and the California Consumer Privacy Act (CCPA) in the United States set standards for data protection, privacy, and consent. These regulations require organizations to obtain informed consent, provide transparency in data collection and usage, and ensure individuals' rights to access, rectify, and erase their data.

Moreover, regulatory frameworks should address the potential biases embedded in AI systems, ensuring fairness and non-discrimination. They should also provide mechanisms for auditing and accountability, enabling individuals to seek redress in case of algorithmic errors or biased outcomes.

In addition to regulatory frameworks, international collaboration and standards development are essential for harmonizing data practices across borders. By fostering global cooperation, we can establish a cohesive and responsible approach to data usage in AI.

### Conclusion:

The future of data in AI holds immense potential for advancements in various fields. Emerging trends and challenges call for scalable data processing, efficient data labeling techniques, and robust data governance practices. Collaboration and data sharing foster innovation and enable organizations to harness the collective intelligence of diverse datasets.

However, the responsible and ethical use of data remains a critical consideration. Privacy, fairness, and accountability must be prioritized to ensure that AI systems benefit individuals and society as a whole. Regulatory frameworks and policies play a vital role in setting guidelines and promoting ethical practices in data-driven AI.

As we navigate the future of data in AI, continuous dialogue, interdisciplinary collaboration, and public engagement are key to shaping a future where AI technologies unlock human potential while upholding ethical principles and societal values. By embracing these principles, we can create a future where data-driven AI empowers individuals, drives innovation, and addresses the complex challenges of our time.

## Conclusion

In conclusion, data plays an integral role in the development of artificial intelligence (AI). It serves as the lifeblood that fuels AI algorithms and models, enabling them to learn, adapt, and make intelligent decisions. Without access to high-quality, diverse, and representative data, AI systems would be limited in their ability to understand and interpret the world around them. Data serves as the foundation upon which AI models are built, allowing them to acquire knowledge and generalize from examples.

Throughout this chapter, we have explored the fundamental importance of data in AI algorithms. We have discussed how AI algorithms rely heavily on data to learn patterns, make predictions, and generate insights. We have examined the characteristics of big data, including volume, velocity, variety, and veracity, and how they impact AI development. We have also delved into the process of data collection, preprocessing, and feature engineering, highlighting their significance in training sophisticated AI models.

**Balancing Technological Advancements with Ethical Considerations:**

As we celebrate the advancements and potential of AI driven by data, it is crucial to strike a balance between technological progress and ethical considerations. AI has the power to revolutionize industries, enhance decision-making processes, and improve the quality of human life. However, it also brings forth ethical dilemmas and challenges that must be addressed.

Throughout this chapter, we have emphasized the importance of ethical considerations in data usage. We have highlighted the ethical implications of data collection and usage in AI, privacy concerns, and the need to address bias and fairness in AI algorithms. Respecting privacy rights, obtaining informed consent, and ensuring transparency and accountability are essential to building trustworthy and responsible AI systems.

**Harnessing the Power of Data to Unlock the Full Potential of AI:**

The future of AI lies in harnessing the power of data to unlock its full potential. As AI technologies continue to evolve, data will continue to be a critical driver of innovation and progress. By leveraging big data, AI models can be trained to achieve higher accuracy, uncover hidden patterns, and deliver personalized experiences across various domains.

Throughout this chapter, we have explored examples of data-driven AI applications in diverse fields such as e-commerce, healthcare, banking, and finance. We have seen how personalized product recommendations, targeted marketing, disease diagnosis, fraud detection, and real-time market analysis are empowered by the utilization of big data.

In conclusion, the journey of AI is intrinsically linked to the abundance and quality of data available. By embracing responsible data practices, addressing ethical considerations, and fostering collaboration, we can create a future where data-driven AI systems empower individuals and contribute to the betterment of society.

As we conclude this chapter, let us reflect on the immense potential of AI and the critical role of data in shaping its future. It is through a deep understanding of data and its ethical utilization that we can navigate the challenges and opportunities that lie ahead. By embracing the power of data, we can unlock the full potential of AI and create a world where technology and humanity coexist harmoniously, driving innovation and positive impact.